Medical Virology 10

A Continuation Order Plan is available for this series. A continuation order will bring delivery of each new volume immediately upon publication. Volumes are billed only upon actual shipment. For further information please contact the publisher. Plenum began publication of volumes in this series effective with Volume 8.

Medical Virology 10

Edited by
Luis M. de la Maza
and
Ellena M. Peterson
University of California, Irvine Medical Center
Orange, California

SPRINGER SCIENCE+BUSINESS MEDIA, LLC

Library of Congress Catalog Card Number 89-657524

ISBN 978-1-4613-6664-5 ISBN 978-1-4615-3738-0 (eBook)
DOI 10.1007/978-1-4615-3738-0
ISSN 1043-1837

Proceedings of the 1990 International Symposium on Medical Virology,
held October 4–6, 1990, in Newport Beach, California

Originally published by Plenum Press, New York in 1991
Softcover reprint of the hardcover 1st edition 1991

FOREWORD

This year marks the tenth anniversary of the International Symposium on Medical Virology. In the Foreword to the book of the 1980 Symposium, we stated, "However, the challenges still lying ahead are more numerous than our past accomplishments". Little did we know at the time, that within a few years the spread of human immunodeficiency virus type I was going to occur. This worldwide epidemic has, like no other disease in recent history, awakened the scientific community and the public at large. It is a reminder to all of us that regardless of our vast technical advances, Nature provides such great opportunity for biological diversity, that it will always be one step ahead of our scientific knowledge. Although our understanding of virology, molecular biology and immunology have increased by leaps and bounds over the last decade, we are still at the point of being unable to effectively control the spread of this viral infection. We hope that our Symposium this year has helped researchers to come together and exchange ideas, so that our growing knowledge of viral infections will help produce better approaches to control them.

Luis M. de la Maza
Ellena M. Peterson

Irvine, California
March, 1991

ACKNOWLEDGEMENTS

It would be impossible to single out all those individuals who helped us make this Symposium a reality, however, we would like to take this opportunity to express our appreciation for their efforts. Special recognition should go to the speakers for their excellent lectures and chapters that contributed to this book. We also want to recognize Drs. Thomas C. Cesario and Edwin E. Lennette for chairing the sessions. The participation of the attendees in the discussions, poster sessions, and informal conversations provided an exciting, intellectual and scientific experience.

Special recognition should go to Marie Pezzlo and Sandra Aarnaes who, throughout the year, provided continued support to the organization of this meeting. We are very grateful to Penny Welter for her secretarial support and typing skills in preparing the camera-ready manuscripts for this book. The Plenum Publishing Company helped us with editorial support, and particular mention should be made of Melanie Yelity, Mary Safford and Gregory Safford for their help.

Finally, we want to express our appreciation to the following organizations who provided financial support, and helped to make the Symposium a reality:

Beckman Instruments, Inc.
Bion Enterprises, LTD
Bio-Rad Laboratories, Inc.
Cambridge BioScience Corp.
Digene Diagnostics, Inc.
E.I. du Pont de Nemours & Co.
The Institute for Biological Research & Development, Inc.
Merck Sharp and Dohme
Microbix Biosystems, Inc.
Pharmacia LKB Biotechnology, Inc.
Sandoz Research Institute
Smith Kline Beecham Pharmaceuticals
Syva Co.
The Upjohn Co.
ViroMed Laboratories, Inc.
Whittaker Bioproducts, Inc.
Wyeth-Ayerst Research

CONTENTS

CHANGING TRENDS OF DIAGNOSTIC VIROLOGY IN A TERTIARY CARE MEDICAL CENTER

Thomas F. Smith and Arlo D. Wold

Mayo Clinic and Foundation
Rochester, Minnesota, USA

INTRODUCTION

Mayo Clinic (MC) has 290,000 new patient registrations each year in a multidiscipline tertiary care practice located in a small community with a population of 75,000. Many of the patients are immunosuppressed such as those undergoing treatment for neoplastic or rheumatologic diseases. Others are admitted with acquired immunodeficiency syndrome (AIDS) or receive organ transplants (cornea, kidney, bone marrow, liver, pancreas, cardiothoracic) in a program that has involved 2,362 procedures. The increasing numbers of immunosuppressed patients, compared to 10-15 years ago, has had a profound effect on both the frequency and types of viruses recovered in the clinical laboratory. This communication compares the detection of viruses during a period from 1974-1982 with our experience in 1988.

METHODS

Cell Culture

Specimens collected with a Culturette™ (Becton-Dickinson, Cockeysville, MD) swab (respiratory, dermal, genital, ocular, gastrointestinal) were extracted into 2 ml of serum-free medium and inoculated into conventional tube and/or shell vials seeded with MRC-5 cells and into primary rhesus monkey kidney cell cultures. Tissue specimens were homogenized in a Stomacher Lab-Blender (Tekmar Co., Cincinnati, OH), centrifuged and the supernatant fraction inoculated into cell culture. Leukocytes from blood were separated by Ficoll/Paque-Macrodex (Pharmacia, Piscataway, NJ) or by Sepracell (Sepratech Corp., Oklahoma City, OK) prior to inoculation into cell cultures. Body fluids such as urine and cerebrospinal fluid (CSF) were inoculated directly into the two culture systems. Viral isolates were initially detected in tube cell cultures by cytopathic effects or by hemadsorption and identified by specific antibodies in immunofluorescence tests. Similarly, monoclonal antibodies to early viral antigens were used to rapidly detect viruses in shell vials (Gleaves et al. 1984; Smith, 1985).

Medical Virology 10, Edited by L.M. de la Maza and
E.M. Peterson, Plenum Press, New York, 1991

Serology

Serum specimens were assayed for IgG class antibodies using anticomplement (herpesviruses) or indirect (measles, mumps, RSV, influenza virus types A and B) immunofluorescence methods. An aliquot of the serum specimen was reacted with goat anti-human IgG (Whittaker MA Bioproducts, Walkersville, MD), incubated at room temperature for 30 min, and then centrifuged at 700 x g for 10 min. The supernatant fraction was tested for the presence of anti-CMV IgM (Smith & Shelley, 1986).

RESULTS

Specimens

Between 1974 (10,000) and 1990 (August, 66,312; projected 100,000), our laboratory has had a 10-fold increase in the numbers of specimens submitted (Figure 1). A substantial decrease in specimen counts occurred in 1984 (13%) associated with governmental reimbursement for medical services based on Diagnostic-Related Groups (DRG). Specimens referred by Mayo Medical Laboratories (MML) represented 65% (50,002) of our total workload; however, only 21% (11,122) of these samples had requests for viral detection.

Of almost 77,000 specimens submitted during 1988, 70% were requests for viral diagnosis (Table 1). Specimens received by our laboratory for the diagno-

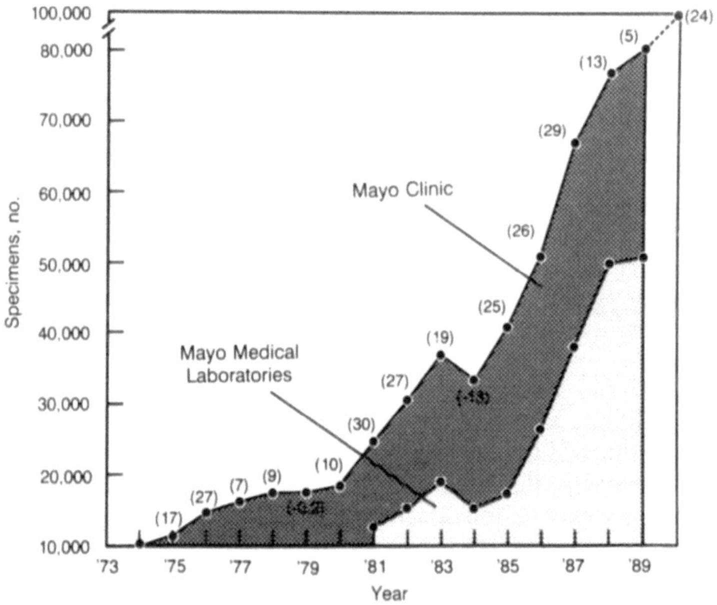

Figure 1. Specimens submitted to the laboratory for the diagnosis of viral, chlamydial and mycoplasmal infections. Total virology specimens, 1974-1990.

TABLE 1. Requests for Microbiology Testing Sent to the Virus Laboratory, Mayo Clinic, 1988

Microbiology Request	Culture	Serology	Total	% of Total Specimens
Virus	18,742	35,427[a]	54,169	70.3
C. trachomatis	2,149			
	6,443 (Micro Trak)	1,110	9,702	12.6
Mycoplasma/	81	3,711	7,776	10.1
ureaplasma	3,984			
C. difficile	4,973		4,973	6.5
P. carinii	354 (Stain)		354	0.459
Total	**36,726**	**40,248**	**76,974**	**100**

[a] Includes screening tests for viral antibodies (VZV, EBV, rubella) and rickettsial requests (n = 13,802).

sis of *Chlamydia trachomatis* (13%), *Mycoplasma/Ureaplasma* (10%), *Clostridium difficile* toxin (6%), and *Pneumocystis carinii* (1%) collectively comprised 30% of the total. Serologic tests represented 51% (40,238) of the total assays (76,974) in 1988. Eighty-eight percent (35,427) of these requests were for viral agents, 9% for *M. pneumoniae* and 3% for *C. trachomatis.*

Comparison of Viruses Detected During Years 1974-1982 with 1988

Of 4,181 viruses detected during 1974-1982, 57% were herpesviruses, herpes simplex virus (HSV) 44%; cytomegalovirus (CMV) 7%; varicella-zoster virus (VZV) 7% (Figure 2). Enteroviruses (16%), recovered in the summertime, and influenza viruses (10%), isolated exclusively in the winter months, were next in frequency to HSV during this interim. In contrast in 1988, CMV was the most prevalent isolate (Figure 3). Altogether, the herpesviruses (CMV, 43%; HSV, 37%; and VZV, 3%, accounted for 83% of the isolates during that single year. Rotavirus antigen assay (Kallestad, Austin, TX), available as a routine test in 1985 in our laboratory, provided a diagnosis in 119 instances generally associated with pediatric gastroenteritis. Similarly, another rapid enzyme immunoassay (EIA) for antigen (Abbott Laboratories, Abbott Park, IL) was instituted two years later and provided a laboratory diagnosis of 82 cases of respiratory syncytial virus (RSV) infection. Therefore, utilizing rapid EIA (rotavirus, RSV) and the shell vial assay (CMV, HSV), our laboratory provided a diagnosis within 16 h postinoculation of the specimen for 88% of the viruses detected during 1988. Two other viruses, (VZV, 3%; adenovirus, 2%) were detected 48 h after inoculation into shell vials. Thus, rapid diagnostic techniques were in place for 93% of the 2,416 viruses detected that year.

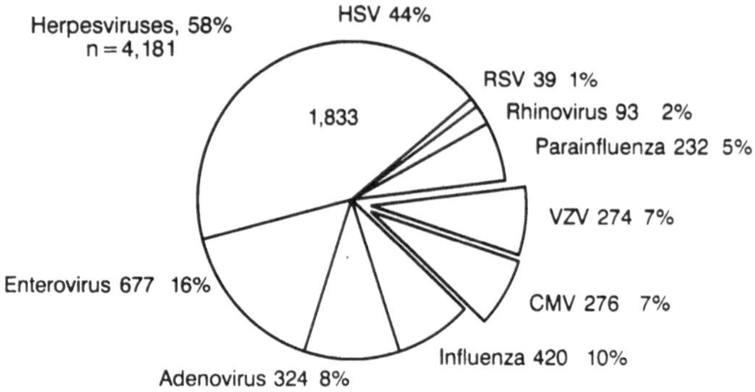

Figure 2. Detection of viruses, as percent of total. Mayo Clinic, 1974-1982.

Viruses Detected During 1988

MML accounted for 59% of the specimens (11,122/18,742) and 53% of the viruses (1,291) yielding a detection rate of 11.6% (Table 2). A higher rate of viral detection (14.8%) was obtained from specimens from local MC patients compared to the referral MML samples.

Respiratory and other (CSF and body fluids other than urine, tissues other than lung, eye, rectal) sources represented 48% of the specimens, 33% of the total viruses recovered, with an isolation rate of 10%. Conversely, dermal sites accounted for only 5% of the specimens, 11% of the viruses, but the highest yield of 30% compared to any other source (Figure 4).

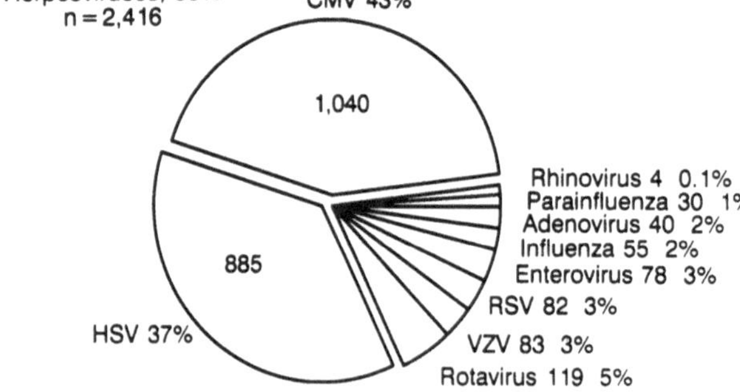

Figure 3. Detection of viruses, as percent of total. Mayo Clinic, 1988.

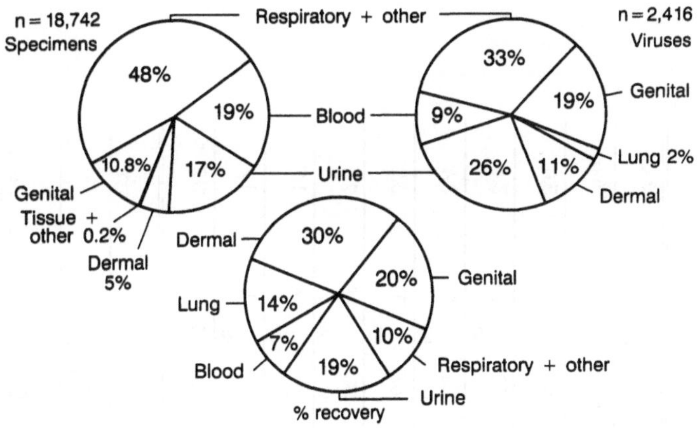

Figure 4. Specimens submitted, viruses detected, and percent recovery. Mayo Clinic, 1988.

Cytomegalovirus

Urine was the most productive source for the detection of CMV (608 isolates, 59% of the total detected) (Table 3). Importantly, among all viruses, CMV was the most predominant agent recovered from blood (219/225, 97%), bronchoalveolar lavage (BAL) (71/80, 89%), and tissue specimens (74/96, 77%) (Table 4).

Herpes Simplex Virus

As expected, genital sources provided over one-half of the total HSV isolates and more than 30% of these detected from this site were type 1 (Tables 5 and 6). HSV produced systemic disease occasionally as indicated by detection of the virus in brain, lung, and CSF. The single isolate from CSF was type 1 from a 3-year old with meningitis (Table 5). Interestingly, 113/190 (60%) of HSV isolates from dermal (nongenital) sources were type 2 (Table 6).

Rotavirus

Pediatric nursery and other outbreaks of gastroenteritis generally prompted clinicians to submit specimens for the diagnosis of rotavirus infection. Rotaviruses (119 detected) were the third most prevalent agent detected during 1988, indicating the need and importance of this test for routine use (Table 2).

Varicella Zoster Virus

VZV was recovered exclusively from dermal sites with the exception of one isolate from blood submitted through MML from a 33-year-old man (Table 7).

TABLE 2. Detection of Viruses form Specimens Submitted to the Mayo Clinic Virus Laboratory, 1988

Virus	Specimen: Genital MML[b]	Genital MC[c]	Urine MML	Urine MC	Dermal MML	Dermal MC	Lung MML	Lung MC	Blood MML	Blood MC	Other[a] MML	Other[a] MC	Total MML	Total MC	% Total
CMV	0	0	166	442	0	0	30	10	39	180	104	69	339	701	43.0
													1040		
HSV	380	68	1	4	140	50	1	0	0	0	142	99	664	221	36.6
													885		
Rotavirus	0	0	0	0	0	0	0	0	0	0	93	26	93	26	5.0
													119		
VZV	0	0	0	0	30	51	0	0	0	1	0	1	30	53	3.4
													83		
RSV	0	0	0	0	0	0	0	0	0	0	30	52	30	52	3.4
													82		
Enteroviruses	0	0	0	0	0	0	2	0	1	0	64	11	67	11	3.2
													78		
Influenza virus	0	0	0	0	0	0	1	0	0	0	16	38	17	38	2.3
													55		
Adenovirus	0	0	0	0	0	0	2	0	4	0	30	4	36	4	1.7
													40		
Parainfluenza virus	0	0	0	0	0	0	0	0	0	0	12	18	12	18	1.2
													30		
Rhinovirus	0	0	0	0	0	0	0	0	0	0	3	1	3	1	0.2
													4		
Total Viruses	380	68	167	446	170	101	36	10	44	181	494	319			
	448		613		271		46		225		813		2,416		100
Specimens	1,575	570	1,067	2,161	409	498	205	93	2,162	1,215	5,705	3,083	11,122	7,620	
	2,145		3,228		907		298		3,377		8,787		18,742		
% Virus Recovery	24.1	11.9	15.7	20.6	41.6	20.3	17.6	10.7	2.0	14.9	8.7	10.3			
	20.9		19.0		29.9		15.4		6.7		9.3		11.6%	14.8%	12.9
% Total Viruses	18.5		25.4		11.2		1.9		9.3		33.7				100

a Other: Respiratory (throat, sputum, bronchial wash and other secretions), CSF, and other body fluids (not urine), tissues other than lung, eye, and rectal.
b MML, Mayo Medical Laboratory
c MC, Mayo Clinic

TABLE 3. Detection of Cytomegalovirus, Mayo Clinic, 1988

Source of Specimen	No. Detected (%) MML	MC	Total	% From Source
Urine	166 (27)	444 (73)	610	59
Blood	39 (18)	180 (82)	219	21
BAL	28 (39)	43 (61)	71	6
Lung	30 (75)	10 (25)	40	4
Bronchial wash	38 (100)		38	4
Liver	4 (17)	19 (83)	23	2
Sputum	13 (100)		13	1
Throat	10 (100)		10	1
Colon	3 (50)	3 (50)	6	0.5
Nasal swab	4 (100)		4	0.5
Esophagus	1 (33)	2 (67)	3	0.3
Stomach	1 (50)	1 (50)	2	0.3
Eye	1 (100)		1	0.1
TOTAL	**339 (33)**	**701 (67)**	**1,040**	**100**

Ortho- and Paramyxoviruses

Respiratory syncytial virus. RSV, all from the respiratory tract (90% nasopharyngeal or nasal sources) comprised only 3.4% of the total viruses detected in 1988 (Tables 2 and 8). The low proportion of RSV cases (3.4%) likely reflects the predominant tertiary care nature of MC and the low population base of the immediate community for primary care medical services. Over 91% of the RSV cases occurred in children ≤2 years of age.

Influenza and Parainfluenza Viruses. The detection of influenza virus by specific monoclonal antibodies allowed rapid identification of strains according to serotype A or B (Table 8). The average age of patients from whom influenza virus was isolated from BAL or lung tissue was 62. These viruses accounted for 7/9 (78%) of the isolates from BAL that were not CMV (71/80, 89%) (Table 4).

Picornaviruses

Enteroviruses. Enteroviruses were identified on the basis of CPE only in MRC-5 and primary rhesus monkey kidney cells. Enterovirus was the most common viral cause of central nervous system (CNS) disease in children. Of 78

TABLE 4. Number of Viruses Recovered from BAL, Blood, CSF, Eye and Tissue Specimens, Mayo Clinic, 1988

Specimen	CMV	HSV	VZV	Entero	Influenza	Para-Influenza	Adeno	TOTAL
Blood	219		1	1			4	225
BAL	71			2	3	4		80
Eye	1	21					8	30
CSF		1		16			2	19
Tissue	74							
Lung	40	1		2	1		5	49
Liver	23						5	28
Colon	6	4						10
Esophagus	3							3
Stomach	2							2
Brain		1		1				2
Pericardium				1				1
Spleen				1				1
Total	**365**	**28**	**1**	**24**	**4**	**4**	**24**	**450**

TABLE 5. Detection of Herpes Simplex Virus, Mayo Clinic, 1988

Source of Specimen	No. Detected (%) MML	MC	Total	% From Source
Genital	380 (85)	68 (15)	448	51
Throat, sputum, mouth	116 (59)	80 (41)	196	22
Dermal	140 (74)	50 (26)	190	22
Eye	10 (48)	11 (52)	21	2
Rectal	7 (78)	2 (22)	9	1
Nose	5 (56)	4 (44)	9	1
Urine	1 (20)	4 (80)	5	0.4
Esophagus	2 (50)	2 (50)	4	0.3
CSF	1 (100)		1	0.1
Brain	1 (100)		1	0.1
Lung	1 (100)		1	0.1
TOTAL %	**664 (75)**	**221 (25)**	**885**	**100**

TABLE 6. Detection of Herpes Simplex Virus from Genital and Dermal Sites, Mayo Clinic, 1988

Genital

Source of Specimen	Total (%)	MML HSV-1 (%)	MML HSV-2 (%)	MC HSV-1 (%)	MC HSV-2 (%)
		N = 380		N = 68	
Genital	448 (70)	119 (31)	261 (69)	22 (32)	46 (68)
		N = 140		N = 50	
Dermal	190 (30)	49 (35)	91 (65)	28 (56)	22 (44)
TOTAL (%)	**638**	**520 (82)**		**118 (18)**	

TABLE 7. Detection of Varicella-Zoster Virus, Mayo Clinic, 1988

Source of Specimen	No. Detected (%) MML	MC	Total	% From Source
Dermal	30 (3.7)	52 (63)	82	99
Blood	1 (100)	0	1	1
TOTAL	31	52	83	100

TABLE 8. Detection of Ortho- and Paramyxovirus, Mayo Clinic, 1988

Source RSV	Source of Specimen	No. Detected (%) MML	MC	Total	% From Source
RSV	Respiratory tract[a]	30 (37)	52 (63)	82	100
Influenza	Respiratory tract	15, type A (30) 1, type B	35, type A (70)	51	91
	BAL		3	3	6
	Lung	1		1	3
				55	100
Para-influenza	Respiratory tract	12 (46)	14 (54)	26	87
	BAL		4	4	13
				30	

[a] Throat, nasopharyngeal, broncheal secretions

TABLE 9. Detection of Picornaviruses, Mayo Clinic, 1988

Virus	Specimen Source	No. Detected (%) MML	No. Detected (%) MC	Total	Source %
Enteroviruses	Rectal/stool	21	5	26	33
	Sputum, saliva, throat	19/(79)	5 (21)	24	31
	Nasopharyngeal				
	CSF	15/ (94)	1 (6)	16	21
	Chest tube drainage	2		2	3
	Lung	2		2	3
	BAL	2		2	3
	Pleural fluid	1		1	1
	Pericardial fluid	1		1	1
	Pericardium	1		1	1
	Blood	1		1	1
	Spleen	1		1	1
	Brain	1		1	1
		67 (86)	11 (14)	78	100
Rhinovirus		3	1	4	

TABLE 10. Detection of Adenovirus, Mayo Clinic, 1988

Source of Specimen	No. Detected (%) MML	MC	Total	% from Source
Throat, sputum, nasopharynx, etc.	17 (94)	1 (6)	18	45
Eye	6 (75)	2 (25)	8	20
Rectal/stool	4 (80)	1 (20)	5	13
Blood	4		4	10
Liver	2		2	5
Lung	2		2	5
CSF	1		1	2
Total	**36 (90)**	**4 (10)**	**40**	**100**

enterovirus isolates, 16 (21%) came from CSF specimens; this source accounted for 16/19 (84%) of the total viruses detected during 1988 from this source (Table 9).

Rhinoviruses. The scant number of rhinovirus isolates (4) likely reflects the association of these agents with self-limiting infections in nonhospitalized patients who are not specifically evaluated by diagnostic procedures for viruses (Table 9).

Adenovirus

Adenovirus has been commonly associated with upper respiratory tract and gastrointestinal infections; however, 24 of 40 (60%) isolates came from eye (8), lung (5), liver (5), blood (4), and CSF (2) (Tables 4 and 10).

Serologic Requests and Results During 1988

Eighty-three percent (17,948) of the serum specimens for nonscreening virology assays were submitted for herpesvirus determinations (Figure 5). Ninety-five percent (936) of the total diagnostic results represented the detection of IgM in acute phase serum specimens rather than increases in antibody levels in paired (acute and convalescent phase) samples (Table 11).

CONCLUSION

Almost twenty years ago, the Section of Clinical Microbiology at the MC developed a philosophy that procedures involving organisms or toxins which required cell cultures for detection be performed in the Virus Laboratory. Interestingly, these tests in 1988 were almost 20% (*C. trachomatis*, 12.6%, *C. difficile* toxin, 6.5%) of the total test volume and contributed to a mix of specimens that have allowed our laboratory to operate successfully from a financial

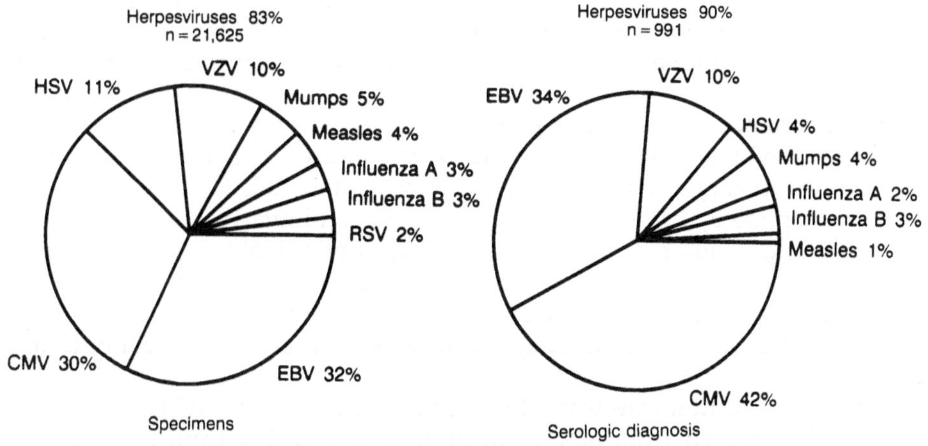

Figure 5. Diagnosis of viral infections by serology during period, 1988.

TABLE 11. Diagnosis of Viral Infections by Serology, Mayo Clinic, 1988

Virus	No. Specimens	IgM/ Seroconversion	No. of Specimens Indicating Acute Infection (%)
EBV	7,002	343/0	4.9
CMV	6,532	402/22	6.5
HSV	2,466	27/8	1.4
VZV	2,186	88/14	4.7
Mumps	1,057	34/5	3.7
Measles	864	3/2	0.6
Influenza A	678	15/4	2.8
Influenza B	585	24/0	4.1
RSV	255	0/0	0
TOTAL	**21,625**	**936/55**	**4.6**

basis for the last 15 years rather than as an institutional liability in the 1960's and early 1970's.

During the period 1974-1982, the pathogenic role of many viruses, herpesviruses, enteroviruses, myxo- and paramyxoviruses was being formulated and specimens were sent to the Virus Laboratory, mainly from patients who were immunocompetent and often, not ill enough to require hospitalization. Further, like most virology laboratories, HSV, recovered from almost every anatomical area (51% from genital sites), was our most common isolate. Viruses that generally caused upper respiratory tract infections such as enterovirus, adenovirus, influenza- and paramyxovirus, rhinovirus, and RSV comprised 42% of the total viruses during this interim. In contrast, these same agents represented only 11% of the viruses detected in 1988, CMV (and herpesviruses) became the most prevalent virus in our tertiary medical care setting consisting of much higher proportions of immunocompromised patients than 15 years ago.

The experience of our Virus Laboratory in 1988 was reflective of others which have provided services for immunocompromised patients, but different from facilities in large community or children's hospitals in which patients with acute respiratory tract infections predominate. Similarly, our approach to diagnostic virology differs from the role of State Public Health Laboratories that provide services for epidemiologic studies, but usually not rapid identification of viruses as a service component of a medical practice. Again, the viral experience of these laboratories is unique and largely unrelated, especially to our 1988 results.

Our viral detection rate was 12.9% (11.6% MML, 14.8% MC) from 18,742 specimens that ranged from 30% from dermal sites to 7% from blood specimens. A variety of specimen types, such as CSF, sputum, bronchial wash, body fluids other than urine, and several tissues, were lumped by computer and retrieved into a "respiratory and other" category without specific sorting. Collectively, these specimens represented almost one-half our workload, yet produced only a 10% yield, no doubt reflective of the difficulty of recovering viruses from CSF and tissue specimens and compatible with other reports of viral isolation from unique anatomical sites (Dagan and Menegus 1986; Darougar et al. 1984; Lipson et al. 1990; Walpita et al. 1985). In addition, the detection of viruses from specimens obtained from the respiratory tract of tertiary care patients may be lower than the rate of similar samples from children with acute infections. Importantly, high viral isolation rates typically reflect types of specimens submitted to a laboratory and the careful selection of patients with likely viral infection. For example, genital sites are productive for HSV, an agent easily propagated and recognized in cell cultures typically yielding rates of 20% to 30% (Morgan and Smith, 1984; Peterson et al. 1988). Similarly, in studies at the Mayo Clinic which included children for purposes of evaluating viral transport systems or viral flora, isolation rates were 40% to 51% (Huntoon et al. 1981; Kepfer et al. 1974). Surprisingly, 65% of the total viral specimen workload during 1988 (21,648, 61% from MML) were submitted for serologic tests. Consistent with viral culture data, both the number of requests and diagnostic results were predominantly herpesviruses. Our approach for these serologic assays has been the immediate determination of both IgG and IgM class antibodies which allowed timely detection of infections rather than the retrospective diagnosis obtained by comparing antibody levels in the conventional "acute and convalescent" phase samples.

The uses of viral serology in clinical medicine are varied and complex. For example, the determination of antibody status to CMV represents a major indication as a risk factor for subsequent CMV disease in organ transplant patients, whereas longitudinal monitoring of viral titers in these patients for evidence of acute infection is very inefficient compared to the rapid shell vial assay (Paya et al. 1989). Similarly, the presence or absence of antibodies to VZV and rubella virus has important implications regarding medical management and infection control implications. Generally, viral serologic tests provide an added dimension for the diagnosis of those infections that cannot be easily detected in the usual cell cultures such as EBV, togaviruses, and measles virus. Other extremely important serologic tests (human immunodeficiency virus, hepatitis viruses) are of obvious importance as blood-borne pathogens. Screening and diagnostic tests for these tests are sometimes performed in association with transfusion services.

Shell vial and enzyme immunoassays have enabled our laboratory to report almost 90% of our viral culture results within 1 day after inoculation. Importantly, the next level of viral diagnosis is routine application of the expanding uses of polymerase chain reaction technology for detection of viral sequences, particularly in CSF, blood, and tissue specimens for those agents that are not cultivatable in routine cell cultures.

REFERENCES

Dagan R, Menegus MA (1986) A combination of four cell types for rapid detection of enteroviruses in clinical specimens. J Med Virol 19:219-228.

Darougar S, Walpita P, Thaker U, Viswalingam N, Wishart MS (1984) Rapid culture test for adenovirus isolation. Br J Ophthalmol 68:405-408.

Gleaves CA, Smith TF, Shuster EA, Pearson GR (1984) Rapid detection of cytomegalovirus in MRC-5 cells inoculated with urine specimens by using low-speed centrifugation and monoclonal antibody to an early antigen. J Clin Microbiol 19:917-919.
Huntoon CJ, House RF Jr, Smith TF (1981) Recovery of viruses from three transport media incorporated into culturettes. Arch Pathol Lab Med 105:436437.
Kepfer PD, Hable KA, Smith TF (1974) Viral isolation rates during summer from children with acute upper respiratory tract disease and healthy children. Am J Clin Pathol 61:1-5.
Lipson SM, Costello P, Forlenza S, Agins B, Szabo K (1990) Enhanced detection of cytomegalovirus in shell vial cell culture monolayers by preinoculation treatment of urine with low-speed centrifugation. Curr Microbiol 20:39-42.
Morgan MA, Smith TF (1984) Evaluation of an enzyme-linked immunosorbent assay for the detection of herpes simplex virus antigen. J Clin Microbiol 19:730-732.
Paya CV, Smith TF, Ludwig J, Hermans PE (1989) Rapid shell vial culture and tissue histology compared with serology for the rapid diagnosis of cytomegalovirus infection in liver transplantation. Mayo Clin Proc 64:670-675.
Peterson EM, Hughes BL, Aarnaes SL, de la Maza LM (1988) Comparison of primary rabbit kidney and MRC-5 cells and two stain procedures for herpes simplex virus detection by a shell vial centrifugation method. J Clin Microbiol 26:222-224.
Smith TF (1985) Viruses: In: Washington (ed) Laboratory Procedures in Clinical Microbiology. New York, Springer-Verlag, p. 537-624.
Smith TF, Shelley CD (1988) Detection of IgM antibody to cytomegalovirus and rapid diagnosis of this virus infection by the shell vial assay. J Virol Meth 21:87-96.
Walpita P, Darougar S, Thaker U (1985) A rapid and sensitive culture test for detecting herpes simplex virus from the eye. Br J Ophthalmol 69:637-639.

DISCUSSION

Siegel C (ViroMed Laboratories, Minnetonka, MN):

Tom, do you see an increasing role for anti-viral susceptibility testing?

Smith T:

There is no doubt that it will be another dimension to our laboratories. We are in need of some rapid techniques in that regard. We certainly need something other than plaque reduction. One can use probes in order to detect the level of virus after mixture with an antiviral agent.

Amsterdam D (SUNY, Buffalo, NY):

Tom, in this rapidly changing arena of reduced available finances in health care institutions, would you comment on the application of monoclonal antibodies for verifying rapidly cytopathogenic viruses like herpes, and also on the clinical need for typing herpes virus.

Smith T:

We use the monoclonals available in the shell vial assay for automatically identifying HSV type-1 and type-2. Outside of the epidemiologic interest at the present time, and perhaps from the standpoint of antiviral susceptibility, I don't think that type-1 versus type-2 is important in the routine medical practice. With the increasing use of acyclovir and the increasing reports of resistance of HSV to this and perhaps other agents, it may become more important. Most of us are familiar with the characteristic patterns of cytopathic effect, so that each and every strain doesn't necessarily have to be identified by particular antibody. Specific identification by reference antisera is not done for all organisms in any other area of microbiology. They use other criteria for identifying agents.

Amsterdam D:

I think there was an NCCLS preliminary bulletin that stated that it was mandated that one do that with antibody confirmation. That is the reason I asked that question.

Bukhari S (London School of Hygiene and Tropical Medicine, London, UK):

My question is about the application of DNA probes for detection of viruses with particular reference to CMV. The problem is that there are a lot of reports that there is cross-reaction of CMV with the host DNA. What is your impression about the specificity of the probe, because of the cross-reactivity of CMV with the human DNA, it could make it difficult to adopt it as a routine laboratory test.

Smith T:

Which probes are you referring to with regards to the cross-reaction?

Bukhari S:

Any DNA.

Smith T:

It is mandatory to utilize a number of positive and negative controls in the test. We have used the Enzo product for our *in situ* work, and the reagent does not react with uninfected cells.

Warford A (Kaiser Permanent, North Hollywood, CA):

With your large number of rotaviruses and your large bone marrow transplant patient population, are you seeing any rotaviruses in the adult bone marrow transplant patients?

Smith T:

No, most of ours occur in children. We did find RSV in the BAL of at least one patient. As patients become very immunosuppressed, you're going to see rotaviruses and others like RSV that commonly occur in infants and children, primarily, in other types of populations. Are you aware of any rotaviruses in adults.

Warford A:

Actually, almost 20% of our rotaviruses are in adults.

Peterson, E (University of California Irvine Medical Center, Orange, CA):

How do you combine conventional tube cultures with your shell vials. In other words, do you back up all you shell vials or are you selective?

Smith T:

We are selective, Ellena. For example, we use just two shell vials for urine specimens as an indication of first incidence of CMV infection. For blood, BALs and tissue specimens, we also inoculate tube cell cultures in addition to shell vials, as a back-up. In those kinds of specimens, unless you use a back-up, you will not recover the maximum amount of viruses. If I had one system, however, to use for CMV, I would definitely use shell vials. By using just shell vial alone and not using tube cell culture, you would miss 15% of the total CMV's detected. On the other hand, if you use tube cell cultures without shell vials, then you would miss 30%.

Peterson E:

How about HSV? Do you back those up?

Smith T:

We don't use a back-up system. I think that may become important. We are carrying out some studies now to really look at that situation much closer.

Kilbourne E (Mount Sinai School of Medicine, New York, NY):

I would like to make a general comment that is not directly related to viral diagnostics, because I'm not a viral diagnostician, per se. I happened to have been interviewed the other day by a man who is writing a book on the future of medicine in the year 2010. I, along with a number of so called experts, were asked to prognosticate what was going to be the turn and the pattern of diseases in the future. I scratched my head, mentally at least, and decided that more and more we're going to see endogenous latent reactable agents in the pattern. Your talk was of particular interest to me because it seems that 2010 is already here. Recognizing that you are a tertiary care center and the sampling is therefore askewed, I think it is interesting as a commentary of medical practice that this is already happening.

Needham C (Lahey Clinic Medical Center, Burlington, MA):

Tom, I have a couple of questions for you. We have actually been somewhat deterred by your data using shell vial isolation techniques from respiratory specimens for respiratory viruses. I think that the difference between your rates and those of others are striking. Can you give us any indication of what that difference might relate to?

Smith T:

The only respiratory virus that we looked at in detail was influenza virus. We really were not able to come up with the kind of results that we could use as a primary test, although others have found shell vials to be a sensitive assay system.

Needham C:

What accounts for that difference? I think you suggested 50-60% sensitivity versus others you have reported, maybe 90%. Do you think that's just the virus?

Smith T:

A number of technical variables are involved in shell vial techniques. A number of variables really have to be very carefully standardized. Use of fresh cells is one of the most important aspects. Monoclonal antibodies to early antigen products of replication are also important. There is always a learning curve involved in any new assay system, so that if you compare tubes to shell vials or if you compare A to B, you should be equally familiar with both techniques before you proceed with the comparison. Often, for example, in a very busy laboratory, three fourths of the time is spent with the conventional system and one fourth with the other assay. I think that those are the practical considerations that may reflect differences among laboratories.

Needham C:

Do you have any data regarding the isolation of CMV from just buffy coat preparation versus the concentration technique?

Smith T:

Several years ago, Bill Martin and others at UCLA found that the Ficoll Hypaque concentration technique was significantly more sensitive than just the buffy coat for the detection of CMV. They pointed out that, especially in the immunocompromised patient, leukopenia is a problem. Therefore, the Ficoll-Hypaque system will be much more efficient in recovering the limiting white cell fraction compared to buffy coat.

Lennette E (California Public Health Foundation, Berkeley, CA):

We're going to hear a great deal about sensitivity and specificity, and I think that there is a general misconception as to what sensitivity actually involves. You read paper after paper where test A is better than test B, and because test A gives you a titer of 16,000 and B gives you a titer of 32. If you go into the Manual of Clinical Microbiology, there is a chapter on this aspect, very simply written and right to the point. There is so much misperception that I have asked Wendy Strong to prepare a chapter on this for the next edition of Laboratory Diagnosis. The severity of the patient's illness and the intensity of antibody response has nothing to do with sensitivity. Sensitivity is, the test is positive in cases of the disease and negative in cases of no disease. It is a matter not of titer, but whether the test is plus or minus. It has to discriminate.

Smith T:

I totally agree.

Diaz F (Ottawa, Canada):

Some people suggested that DMSO, calcium or dexamethasone increases the sensitivity for CMV detection. Would you comment on that?

Smith T:

Yes. Pat West who is in the audience, did the original work on that. She finds a large difference in the reactivity of those biochemicals with cells from different sources. If you use MRC-5 cells, the effect of these chemicals may not be the same on cells from all sources. We have shown that the older the cells are when you receive them in the laboratory, the greater effect that these particular chemicals have in terms of the sensitivity of detecting virus. In other words, the more adverse the conditions, the better those biochemicals work. Age of the cell monolayers is a very important difference. Again, there are a lot of variables, but in our laboratory, we have not been able to demonstrate the efficacy of those particular reagents and biochemicals for detecting CMV and HSV.

Mendelson E (The Chaim Sheba Medical Center, Tel-Hashomer, Israel):

How important is the kind of antibody used in the shell vial system for detecting CMV? Do you use monoclonal antibody for early VCA or a mixture of antibodies.

Smith T:

It's very important. The original antibody that was developed by Dr. Gary Pearson that we use in our laboratory and that now is sold by du Pont has served us very well. We compared this antibody with the product from Syva. Both reagents reacted similarly for detection of CMV. It is important to work with an antibody directed to an immediate early antigen. The longer after inoculation that you wait for detection, the less intense is the fluorescence in the positive cells. Mixtures of monoclonals to both immediate early and late viral antigens is a wise approach.

Amsterdam D:

In reference to Dr. Lennette's comment, of course I agree with him, but there is another component to sensitivity that sometimes we use unappropriately and that is threshold sensitivity. The ability to detect the limiting number of replicating viral particles. Sometimes that effects the outcome of a positive or negative result. The question I was going to ask you Tom, was about rates of detection. In that comparative study that you mentioned in which you used transport medium as one on the indicators, what about the methods of handling within the laboratory? I don't remember the publication, but I know some laboratories use centrifugation, and that's been shown to actually diminish recovery. Do you know of any studies that have looked at both transport medium method of handling and maybe even aged cells?

Smith T:

Steve Lipson has shown that centrifugation of urine prior to inoculation, increased the detection of CMV in shell vials. Max Arens has found that centrifugation of HSV-containing specimens may reduce the infectivity of HSV. It really depends upon the virus and whether it is cell-associated. I really don't think centrifugation prior to inoculation is necessary, with the exception perhaps of stool or rectal specimens. We do not use centrifugation prior to inoculation in our laboratory and have not experienced undue contamination or toxicity.

Needham C:

I noticed that you did not include one technology in your presentation, for future directions, and that is flow cytometry. Do you think that there is a role for flow cytometry in viral detection?

Smith T:

I really can't comment on that, regarding viruses. We have used the flow cytometry for the detection of the trophozoites and cysts of *P. carinii*. I am appreciative of that type of technology, but I really don't feel qualified to comment on the use of it as a routine in virology.

EVALUATION OF ELECTRON MICROSCOPIC INFORMATION AVAILABLE FROM CLINICAL SAMPLES

Sara E. Miller

Duke University Medical Center
Durham, North Carolina, USA

INTRODUCTION

Much has been written on the procedures for using electron microscopy (EM) in diagnostic virology (Almeida, 1983; Kapikian et al. 1976; Miller, 1986, 1988, 1989; Oshiro and Miller, in press) and the viruses likely to be found in clinical samples (Hsiung, 1982; Lennette, 1985). Excellent atlases are also available for identifying virions (Doane and Anderson, 1987; Palmer and Martin, 1988). This information will not be repeated here. Rather, we will elaborate on characteristics of viruses that allow them to be recognized by EM, interpretation of EM observations, i.e., what do the results tell the physician, what tests are most appropriate for which specimens, and what action can be taken based on that information.

In the past, viral diagnosis was attempted primarily to assess the necessity of quarantine to prevent spread of the disease and to prevent the administration of unnecessary antibacterial therapy. Today with the availability of several antiviral drugs and many more in clinical trials, detecting viral diseases is becoming increasingly important. Indeed, because of the specificity of many of these drugs, it is important not only to determine that an illness is of a viral origin, but also to identify which virus is present. We are now in an era with respect to viral diseases comparable to that in the 1950's when antibiotics were beginning to be used for bacterial infections. Along with drug therapy, epidemiology is another major reason today to document viral infections.

EM is a valuable tool and in some cases may be the only one for discerning the presence of viruses; however, it can and should be used in conjunction with other virological tests. EM has several advantages. The main one is that it is rapid: (a) A direct negative stain can take as little as 20 min; even with the use of concentrating methods, it can be completed in 1-2 hr. (b) Thin sectioning usually takes about 24 hr but can be speeded up to 2-4 hr, which is certainly faster than any culturing procedures, and as fast as fluorescence staining. Another advantage is that for direct examination by negative staining, specific reagents such as antibodies, nucleic acid probes, or protein standards are not required. Use of these reagents necessitates an *a priori* notion of what virus may be present to narrow the reagents employed to a manageable number. A fur-

Medical Virology 10, Edited by L.M. de la Maza and
E.M. Peterson, Plenum Press, New York, 1991

ther advantage of EM is that infectious particles are not required, which is important if the sample has not been properly transported or in the case of viruses, such as the gastroenteritis viruses, that do not grow readily or at all in tissue culture. EM can also differentiate different viruses that produce similar cytopathology in tissue culture and can detect viral contaminants in cultures used for virus isolation. Because the test is a direct visualization of the agent, rather than a biochemical detection of viral components, e.g., enzyme immunoassays (EIAs) there is less chance of a false positive test.

One disadvantage of EM in diagnostic virology is that viruses within the same family most often have similar morphology and thus cannot be differentiated visually, e.g., herpes simplex virus (HSV) vs. varicella zoster virus (VZV). However, once the virus family is discerned, serology or immunoelectron microscopy (IEM) can further identify the virus. Another problem is that a fairly high number of virions (10^5-10^6/ml) must be present for any to be seen by negative staining, though several concentration methods are possible (Hayat and Miller, 1990). Other methods include clarification of debris by a low speed spin followed by ultracentrifugation, ultrafiltration, or precipitation. These drawbacks not withstanding, EM can be an important adjunct in viral diagnosis when used in conjunction with other methods.

SAMPLE COLLECTION

The collection and preparation of specimens for examination by techniques other than EM have been described in detail by Lennette (1985). To aid the physician in ordering EM virology tests, the following information will describe the kinds of samples that can be examined by EM, how much specimen should be collected, and how specimens should be sent or stored. The most frequent samples examined by EM are stools and liquids because they are more easily collected from the patient. Liquid specimens are listed in Table 1 along with an appropriate amount to send. It takes about 5 µl of liquid for a direct examination on one EM grid; however, the more sample available, the more likely it is that a viral agent could be seen after concentration procedures if it is present in low numbers.

Fecal specimens from cases of gastroenteritis are the most likely samples to be sent to the EM laboratory, not only because they are easily obtainable, but also because most of the viruses that cause gastroenteritis cannot be grown easily or at all in cell culture. Also, if one of the slide diagnostic kits for rotavirus or adenovirus is used, other viral pathogens will be missed. Stools should be sent in a cup or in the original diaper if the patient is an infant. Although a few hours at room temperature will not degrade these viruses, if there is a delay in transport, fecal samples should be refrigerated but not frozen. However, stools to be maintained for several months for epidemiological studies can either be kept at 4°C sealed tightly or under paraffin oil, or can be frozen at -70°C or in liquid nitrogen (but not in a regular refrigerator freezer).

Along with obvious liquids such as urine, cerebrospinal fluid (CSF), lavages, or synovial fluid, other specimens in this category include blister roofs that can be ground and extracted with water or volatile buffer and cytosmears on slides that can be rehydrated. If blisters do not contain enough fluid to be drawn into a syringe, they can be injected with 20-30 µl of sterile saline, and the saline can be drawn back into the syringe. Liquid specimens should be sent undiluted and unfixed in any kind of container including a syringe. Under no

TABLE 1. Amount of Various Specimens that Should be Sent to the EM Laboratory

Specimen	Amount
Blister fluid or blister roof extract	5-10 µl
Blood (not routinely examined by EM)	5-10 ml
Cerebrospinal fluid	1-5 ml
Pericardial fluid	1-5 ml
Pleural fluid, nasopharyngeal fluid, sputum (aspirates, lavages)	1-10 ml
Stool	0.5-10 ml
Tears	5-10 µl
Urine	5-10 µl
Tissue	1-3 mm^3

circumstances should they be placed into transport media as is common for delivery to the culture laboratory; this would not only dilute the viruses but also add proteins and salts that would obscure the field on the grid. As with stool, if liquid samples are to be delayed for several hours before they reach the EM lab, they should be refrigerated or shipped on wet ice but not frozen. Samples to be stored for long periods of time should be frozen at -70°C or in liquid nitrogen.

Sputum and other thick mucoid samples can be digested in the EM lab with a solution of 0.0065 M dithiothreitol in 0.1 M phosphate buffer, pH 7.0 (Stat-Pack Sputolysin Test; Behring Diagnostics, La Jolla, CA) or other reducing agents to make it manageable on the grid. A 1:1 ratio of reagent to sputum is incubated for 30-45 min at 25°C with frequent agitation until the viscosity of the sample is acceptable. After a low speed (1,000 x g) clarification, the supernatant is placed on a grid or concentrated and placed on a grid.

Blood can be examined after removing the cells in a low speed spin, and hepatitis B and parvovirus have been demonstrated by EM. However, blood is not routinely sent for EM. The number of different viral agents diagnosed from blood is small enough so that selection of specific reagents for serotyping is easy. Furthermore, serum contains proteins and lipids that are confusing by EM. Although human immunodeficiency virus type 1 (HIV-1) has been shown in thin sections of lymphocytes in early cases of acquired immunodeficiency syndrome (AIDS), it could never be identified by negative stain of serum. Thus, blood is best sent to the serology lab unless serological tests have been negative or a systemic infection by a particular agent in an immune compromised patient is suspected, and the physician wishes to test blood as well.

Tissue destined for embedment should be placed immediately upon removal into glutaraldehyde (2-5%, buffered) and transported to the EM laboratory or refrigerated until transport can be facilitated. This means that it is best to have vials of fixative available in the operating or autopsy room. If tissue must wait for a short time before being fixed, it should be kept moist with saline or buffer and cold, but never frozen. Tissue for homogenization and negative staining (see Sample Processing) should not be fixed, but maintained moist with a small amount of buffer or saline and transported on ice to the EM lab.

SAMPLE PROCESSING

A basic knowledge of specimen preparation will help the physician understand the time frame in which an answer can be expected. Liquid samples can be placed directly onto an EM grid with a support film, drained, negatively stained (Figure 1), and examined, all within 15-20 min. This direct procedure would be used for small amounts of specimens (e.g., tears and blister fluids) and as a preliminary examination for larger volumes. While rapid, this technique may not detect virions that are present in low numbers. For this reason the microscopist may use one of the several concentrating procedures that have been described in detail (Hayat and Miller, 1990). Briefly, large volumes can, after a low speed clarification (1,000 x g, 3-4 min), be ultracentrifuged (36,000 x g, 90 min). The pellet is resuspended in a small drop of water and then treated as in the direct method. Smaller volumes (100-200 µl) can be pelleted in a tube or directly onto a grid (100,000 x g, 30 min) in an Airfuge (Beckman, Fullerton, CA); this procedure requires less time. If performed in a 200 µl tube, it does not concentrate as highly as routine ultracentrifugation because less sample is pelleted. However, if 100 µl is spun directly onto a grid in the Beckman EM-90 rotor, the concentration may be as much as 10^4 X (Hammond et al. 1981). Other techniques such as agar diffusion, pseudoreplica technique, serum-in-agar, aggregation by antibody, and solid phase IEM, that require 20-30 min are described briefly in Figures 2-6. Polyacrylamide hydrogel can be used to concentrate liquid suspensions, and polyethylene glycol or ammonium sulfate

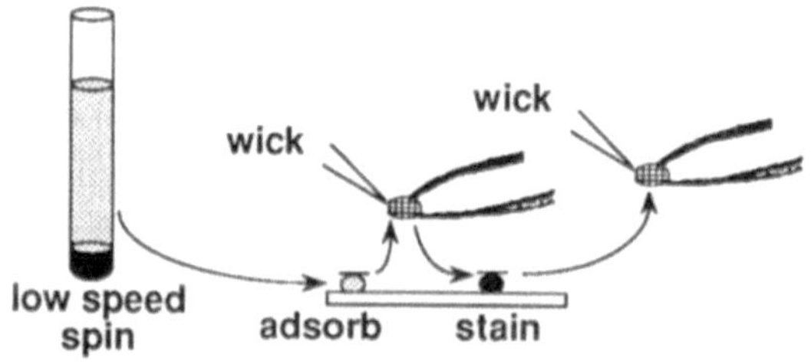

Figure 1. Direct negative stain. A water or volatile buffer extract of the sample is made, and large debris is removed by centrifugation at 1,000 x g for 3-4 min. A grid is incubated on a drop of the supernatant for 5-10 min. The grid is then drained on filter paper, placed on negative stain for 1-2 min, and drained (Miller, 1990a).

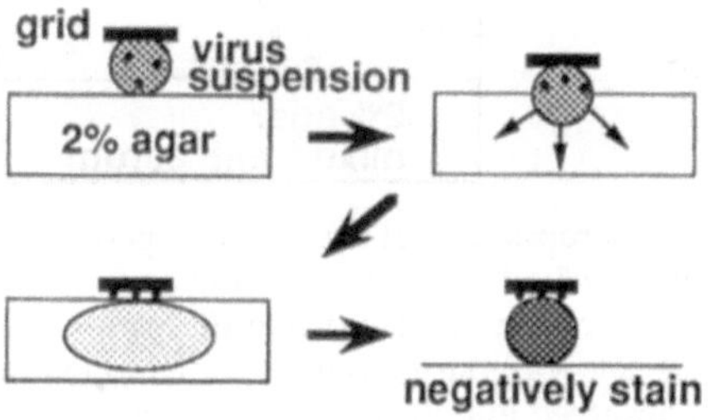

Figure 2. Agar diffusion. A drop of virus suspension is placed onto a block of 2% agar, and the liquid is allowed to diffuse into the agar (15-20 min), leaving the virus particles concentrated on the grid membrane. The grid is then negatively stained (Miller, 1990a).

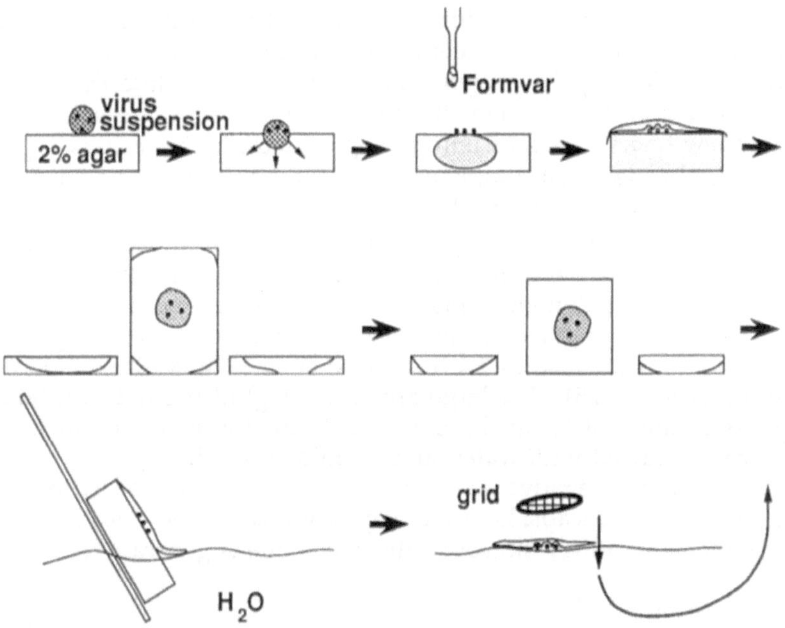

Figure 3. Pseudoreplica technique. After clarification at 1,000 x g, a virus suspension is placed onto 2% agar, and the liquid is allowed to be absorbed. A drop of Formvar (0.5%) is dropped onto the agar block. To remove excess Formvar that has run over the edge, the block is turned onto its side, and the two vertical edges are trimmed with a razor blade. The block is then turned 90°, and the other two edges are trimmed. This method of trimming prevents the Formvar from being pushed into the agar by the blade from the top and permits easier separation. The block is laid on a microscope slide, and the Formvar film is floated onto a water surface. It is then picked up on a grid in the same manner as coating grids with a plain Formvar film. (Miller 1990a; Hayat and Miller, 1990)

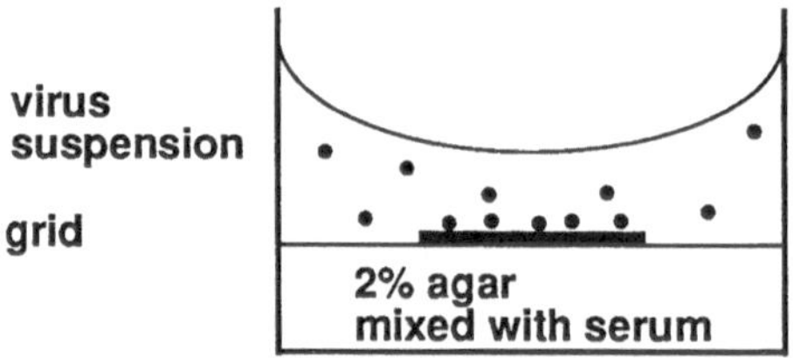

Figure 4. Serum-in-agar. Appropriate viral antiserum or pooled gamma globulin is mixed with cooled, but still molten, 2% agar and poured into a 96-welled plate. (Different concentrations or different antisera can be placed into the various wells.) A grid, film side up, is laid onto the surface, and virus suspension is added to the well. After the virus suspension has been almost absorbed into the agar (15-20 min), the grid is removed, drained, and negatively stained (Hayat and Miller, 1990; Miller, 1990a).

can be used to precipitate viruses from large volumes (e.g., several liters). These latter methods, also described in Hayat and Miller (1990), require several hours to overnight.

IEM is a term that is used variously to denote a concentrating step and/or a specific virus identification step. Immunological reactions have the advantage that they permit visualization of viruses at more dilute concentrations, and the disadvantage that they require an *a priori* notion of what virus might be present for selecting the proper antiserum to use. While these manipulations require a longer time than the direct method, they are still simple and take only a few hours. Various immunological techniques such as aggregation, coating, and gold labeling are explained in Figures 5, 7, and 8, and are described in detail in Hayat and Miller (1990).

Tissue samples usually require fixation, epoxy embedment, and ultrathin sectioning for virus examination. These procedures routinely take at least 24 hr for processing, although in the hands of an experienced microtomist who has the time and willingness to devote full attention to an emergency specimen, it can be accomplished in 2-4 hr (Doane et al. 1974; Miller and Lang, 1982; Miller and Nielsen, 1975). If a large amount (>1 g) of tissue is available, as in autopsy cases, some of it can be embedded and the rest ground in a tissue homogenizer, extracted with water, and examined rapidly by negative staining. Homogenization and negative staining works better for naked virions but also can demonstrate nucleocapsids of enveloped viruses if the nucleocapsids have a recognizable morphology; however, this technique is generally of low yield.

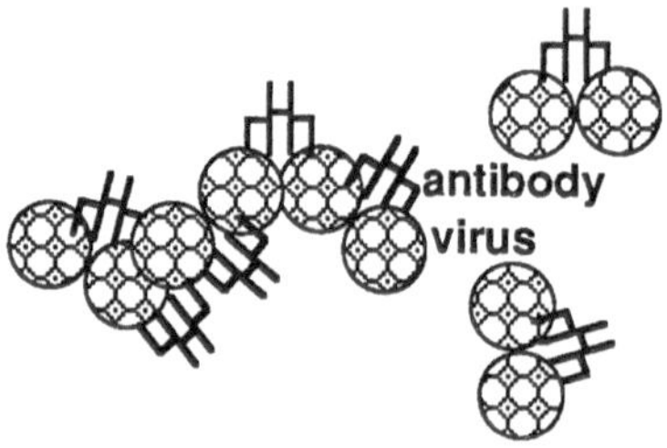

Figure 5. Virus aggregation by antibody. Clarified sample and antiserum are mixed (specific antibody, 1:100-1:1,000; pooled gamma globulin, 1:10-1:25) and incubated (overnight at 4°C or 2-3 hr at room temperature). The mixture is centrifuged at 17,000 x g for 1.5 h, and the pellet is resuspended in a small drop of water and negatively stained (Hayat and Miller, 1990; Miller, 1990a).

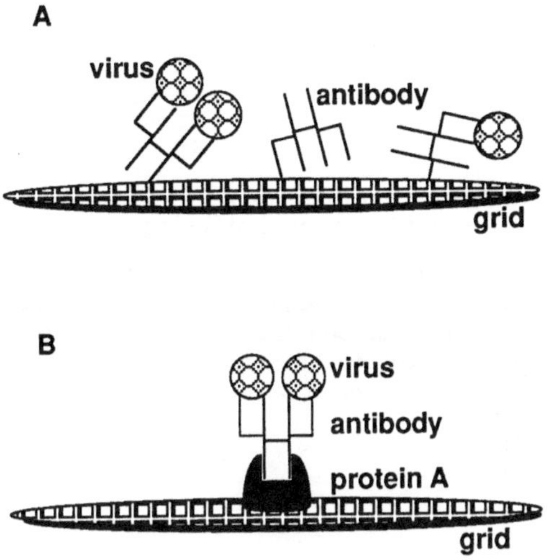

Figure 6. Solid phase immunoelectron microscopy (SPIEM). a. Antibody-coated grid. A grid is incubated for 30 min at room temperature on antibody (specific antibody, 1:1,000-1:2,000; pooled gamma globulin, 1:100-1:200). It is then washed on 20 drops of buffer and incubated for 30-60 min on a drop of virus suspension; finally, it is negatively stained. b. Protein A/antibody coated grid. First, the grid is incubated on a drop of protein A (10 μg/ml) for 10-15 min; it is then washed on drops of buffer and incubated on antibody and virus suspension as described in "a" above, except that the antibody concentration can be 10-100 times higher. This method attaches the antibody to the grid so that its virus-reactive sites are outward, thus preventing the antibody molecules from landing on the grid with their reactive sites down and blocking the binding of the virus. It also permits the use of more concentrated antiserum because the nonspecific inhibition of virus-binding by high protein concentration is overcome by the proper orientation of the virus-reactive sites (Hayat and Miller, 1990; Miller, 1990a).

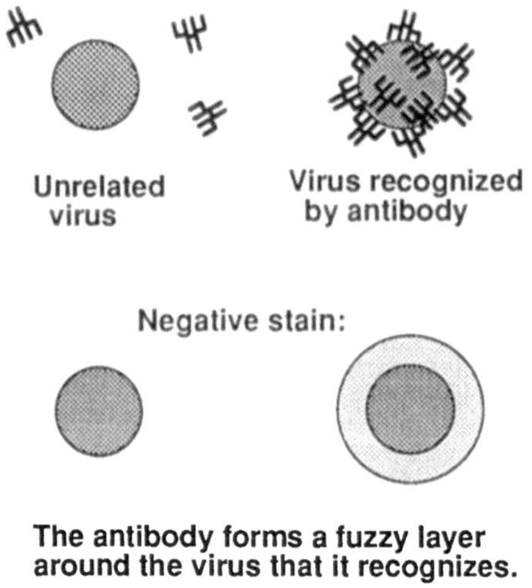

Figure 7. Virus serotyping by coating with antiserum. Viruses are attached to grids in any of the previously described methods. The grids are incubated on drops of specific antiserum (1:1,000-1:2,000) for 30 min at room temperature, washed in water, and negatively stained. If the antiserum has recognized the virus, a coat of fuzz will appear around the virus, permitting specific identification or serotyping. (Hayat and Miller, 1990).

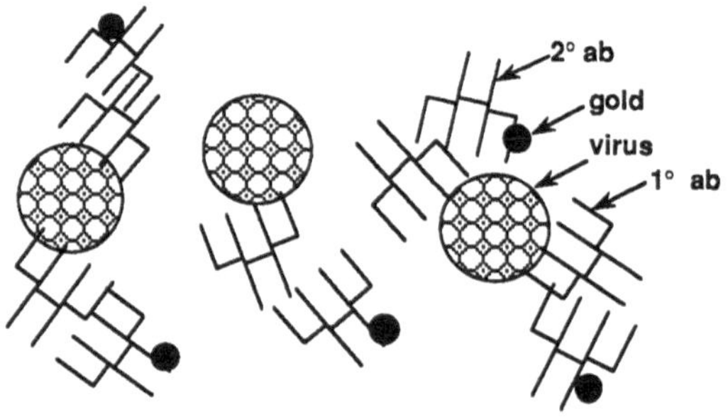

Figure 8. Immunogold labeling. Method 1: Virus is attached to a grid by one of the previously described methods. The grid is incubated with antiviral antibody (1:10-1:100) for 30 min at room temperature. After an extensive wash (at least 6 drops for 5 min each drop), the grid is incubated for 20 min at room temperature on a drop of secondary reagent (antibody against the primary antibody or protein A) (1:10-1:20) conjugated to 10 nm colloidal gold. It is then washed as before on buffer, rinsed in 3 drops of water, and negatively stained. Method 2: The virus suspension and primary antibody are incubated together 30 min at room temperature. The mixture is ultracentrifuged, and the pellet is resuspended in a small amount of buffer. Protein A- or antibody-gold is added and incubated for 20 min at room temperature. The suspension is again ultracentrifuged, and the pellet is resuspended in water and negatively stained (Hayat and Miller, 1990; Miller, 1990a).

A procedure for amplifying virus numbers in tissue is first to mince it and inoculate it into cell culture to permit a round or two of replication. Virions or forming virions can sometimes be seen in thin sections of the tissue culture before cytopathological effect is evident by light microscopy or fluorescence microscopy (Miller and Lang, 1982).

VIRUS IDENTIFICATION

To recognize the presence of a virus in a clinical sample, it is important for the electron microscopist to know the morphological characteristics by which viruses can be identified. A schematic of virus morphology as seen by negative staining is shown in Figure 9 and by thin sectioning in Figure 10. In negative stains, naked viruses are isometric and have a rigid outer coat that is not easily deformable; their size (20-80 nm) and the organization of their capsomers may render them identifiable (Figures 11, 12). Many of the small icosahedral viruses simply look like round fuzzy balls in patient samples and hence are called small round viruses (SRV) (Figure 13). Some of the small viruses have a rough surface; and a few can be identified in clinical samples (Figure 14). If not, they are referred to as small round structured viruses (SRSV). Enveloped viruses have a pliable membrane surrounding the core that is usually derived from cell membranes containing viral proteins; these virions are often pleomorphic due to drying artifacts. Occasionally enveloped virions have visible spikes on their outsides (Figure 15), but sometimes the spikes are so short that they are not easily discernable, and the virus cannot be distinguished from cellular debris. However, if the stain penetrates the membrane, and if the nucleocapsid has a characteristic shape, identification can be made even if the spikes are so short as to be indistinguishable (Figure 16). The nucleocapsid inside the membrane may be isometric like the naked viruses (Figure 16), helical (Figure 15), complex (Figure 17), or morphologically nondescript (Figure 18).

In thin sections, the location of viruses within the cells is important in identification. As a general rule, DNA viruses are usually seen in the nucleus (Figure 19), while RNA viruses are found in the cytoplasm (Figure 20). However, there are some exceptions. For example, poxviruses (DNA viruses) are constructed in the cytoplasm (Figure 21); herpesvirus nucleocapsids (DNA viruses) originate in the nucleus, but can make their way to the cytoplasm, enveloped or unenveloped (Figure 22); and nucleocapsids (Figure 15) of the measles virus-like agent sometimes found in subacute sclerosing panencephalitis (an RNA virus) have been seen in nuclei. Naked particles lyse their host cells to get out, and in late infection, unenveloped DNA virions can be seen in both the nucleus and cytoplasm. Naked viruses can sometimes be seen in matrices or paracrystalline arrays (Figure 19b). The nucleocapsids of enveloped viruses may bud into or out of cell membranes to obtain their outer covering (Figures 20, 22, 23, 24). They can be spherical (Figure 22), helical (Figure 20), or nondescript (Figure 23).

A few unusual viruses do not conform to the above generalities and deserve separate mention. Poxviruses are complex structures, and although enveloped, they have definite recognizable shape, either oval or brick-shaped (Figures 17 and 21). EM has been particularly useful in diagnosing viral skin lesions because poxviruses and herpesviruses can be rapidly and easily differen-

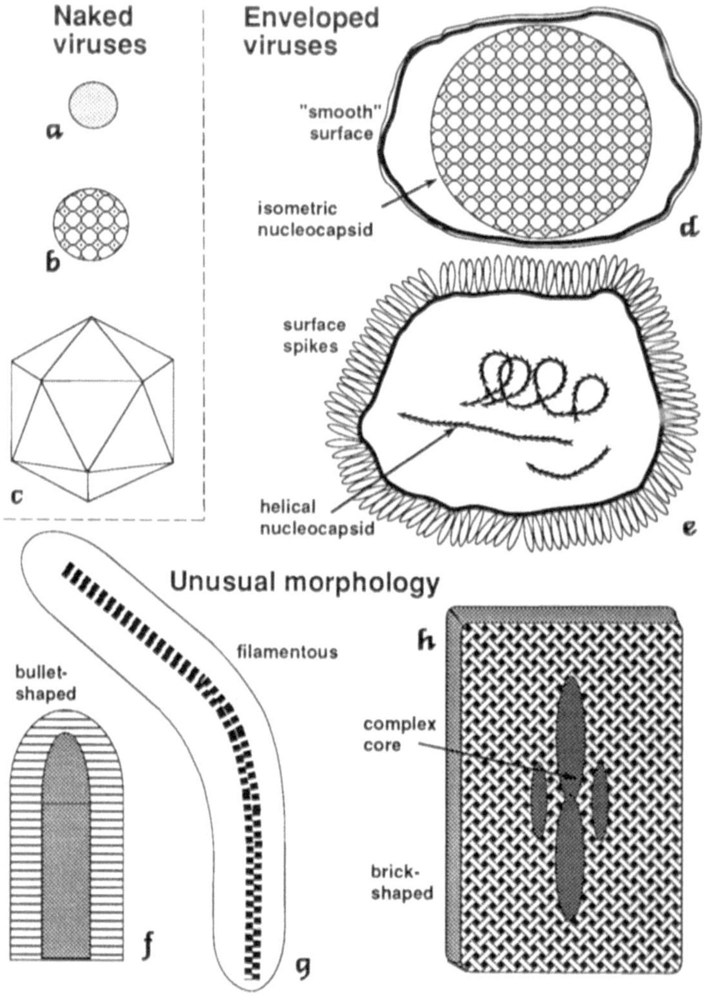

Figure 9. Schematic of the major morphological categories of viruses as seen by negative staining. a. small round featureless virus (e.g., parvovirus); b. small round structured virus (e.g., calicivirus); c. large virus with identifiable capsomeric structure (e.g., adenovirus); d. enveloped virus with short indistinguishable spikes that make the membrane appear smooth; this one is shown with a nucleocapsid of icosahedral symmetry (e.g., herpesvirus); some do not have a morphologically recognizable nucleocapsid; e. enveloped virus with spikes forming a fringe, and a helical nucleocapsid (e.g., measles virus); f. bullet-shaped virus (e.g., rabies virus); g. long filamentous virus (e.g., Marburg virus); h. brick-shaped virus (poxviruses) with a complex core (Miller, 1990a). The sizes are roughly to scale with the diameter of the nucleocapsid in *d* equal to 100 nm.

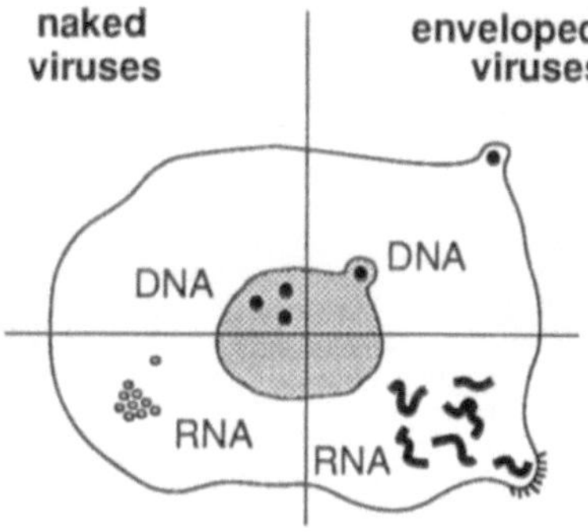

Figure 10. Schematic of viruses as seen by thin sectioning. Cellular location, shape of the nucleocapsid, and whether or not the virus is enveloped are important in identification (Miller, 1990a).

tiated by negatively staining blister fluid. Rhabdoviruses (Figure 24), e.g., rabies and LeDantec viruses, are also membraned, but routinely have a bullet shape where one end is rounded, and the other end is blunt. Though rhabdovirus infection is rare and thus the diagnostic opportunity is infrequent, the shape of the virion makes it easily recognizable in negative stains or thin sections. Finally, the filoviruses, though enveloped, are distinguishable by EM. Morphologically, they resemble the rhabdoviruses somewhat, except that they can be very long (sometimes up to 1,400 nm) filamentous virions with a diameter of 70-80 nm. In negative stains, they often are bent or curved and may appear in the shape of a shepherd's crook or a numeral 6. Fortunately, infection with filoviruses, e.g., Marburg and Ebola viruses, is rare.

Once the existence of a viral agent is determined, one can consult an atlas (Doane and Anderson, 1988; Palmer and Martin, 1989). Viruses can be identified morphologically with respect to families. Though specific characterization may rely on serological techniques, some important differentiations can be made. Finally, and very importantly, artifacts, cell components, and cell debris can resemble viral particles (Miller, 1986, 1989; Oshiro and Miller, in press). If an electron microscopist is unsure of a diagnosis, it is better to record the sample as negative or questionable. A false positive is potentially more detrimental than a false negative report because it might lead to the cessation of the search for an etiological agent.

DIAGNOSIS

Some viruses are usually restricted to certain organ systems, thus the origin of the specimen can yield clues to the identification of viral agents, and should be communicated to the microscopist. This information has been described in detail (Hayat and Miller, 1990; Miller, 1986; Oshiro and Miller, in press). Table 2 is a summary of likely viral pathogens in various systems that can be visualized by EM; diseases caused by viral agents have been described in detail elsewhere (White and Fenner, 1986). In the immunocompromised host,

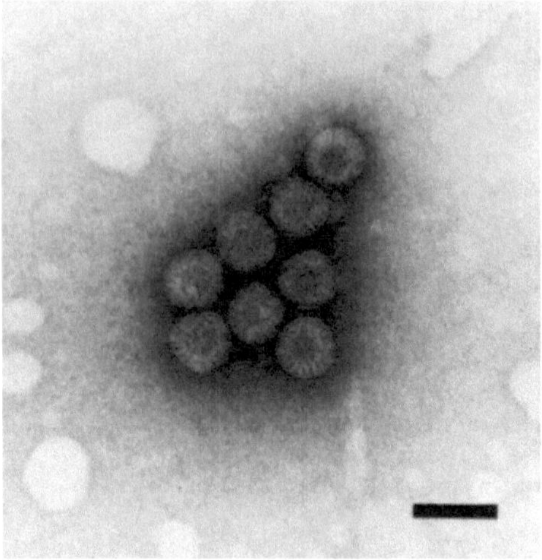

Figure 11. Rotavirus. This is one of the largest naked viruses infecting humans. Its size (70-75 nm) and the arrangement of its capsomers make it clearly identifiable in stool samples. Rotavirus is named from the fact that its capsomers appear like the spokes of a wheel. Group A rotavirus is responsible for 30-50% of the cases of infantile gastroenteritis, and infections with it are more prevalent in the winter months. Adult diarrhea rotavirus (ADRV) (Group B) can be seen, but is not more prevalent in children. Bar represents 100 nm.

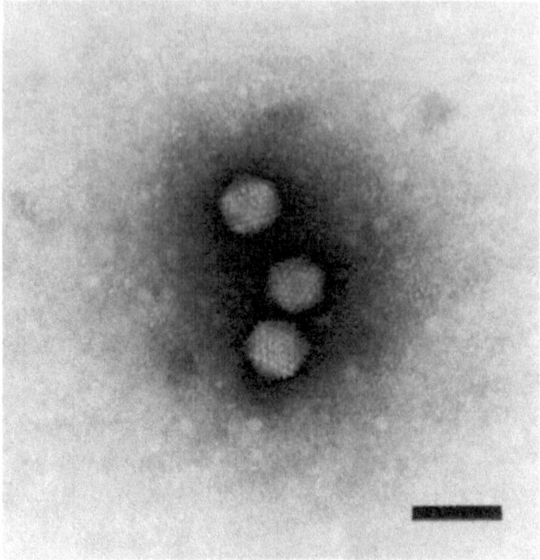

Figure 12. Adenovirus seen in stool but also in other parts of the body. Enteric adenoviruses cannot be distinguished morphologically from other strains. The 75-80 nm icosahedral virion has flat triangular facets made up of marble-shaped capsomers. Depending on its orientation on the grid film, its circumference may appear hexagonal unlike the spherical rotavirus. Bar represents 100 nm.

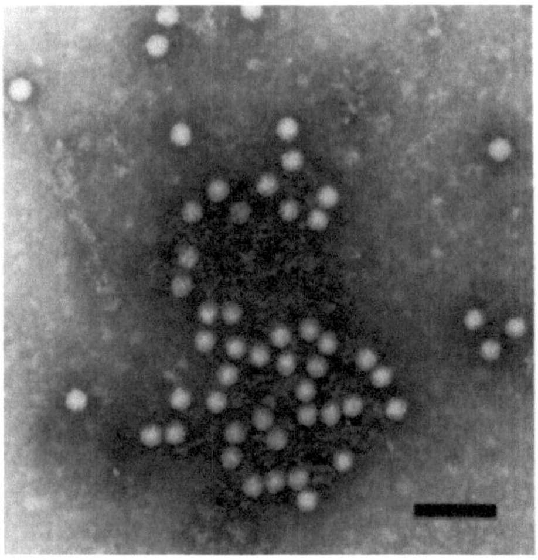

Figure 13. A small round virus without distinguishing morphological characteristics seen in stool from a patient with diarrhea. Bar represents 100 nm.

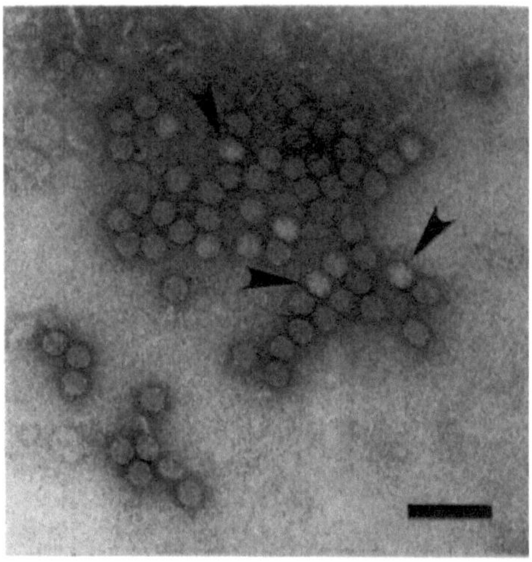

Figure 14. A small round structured virus seen in stool from a patient with diarrhea; it was later identified after printing the micrograph as astrovirus. Arrows denote three particles that show the star-shaped pattern. Bar represents 100 nm.

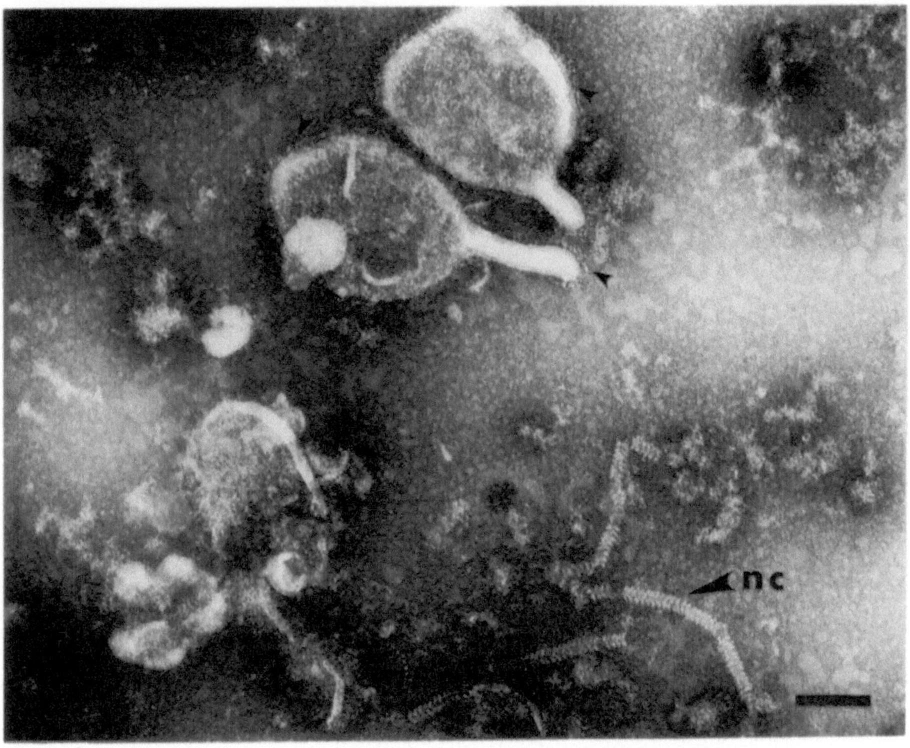

Figure 15. Measles virus. The complete particles are enveloped with spikes (small arrows) on the outside. The 18-nm helical nucleocapsids (nc) appear in a herringbone pattern, and may show circles (bottom center) if they become broken. Bar represents 100 nm.

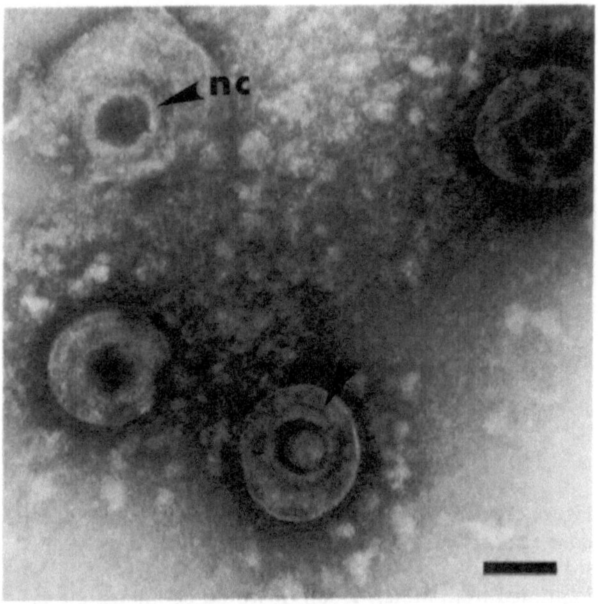

Figure 16. Herpes simplex virus. The enveloped particle has projections that are so short that they are not recognizable. Within the virion, 100-nm nucleocapsids (nc) are visible because the negative stain has penetrated the broken membrane. Some nucleocapsids here are damaged, and appear empty, but the one at the bottom center shows the nucleic acid core inside the nucleocapsid. Bar represents 100 nm.

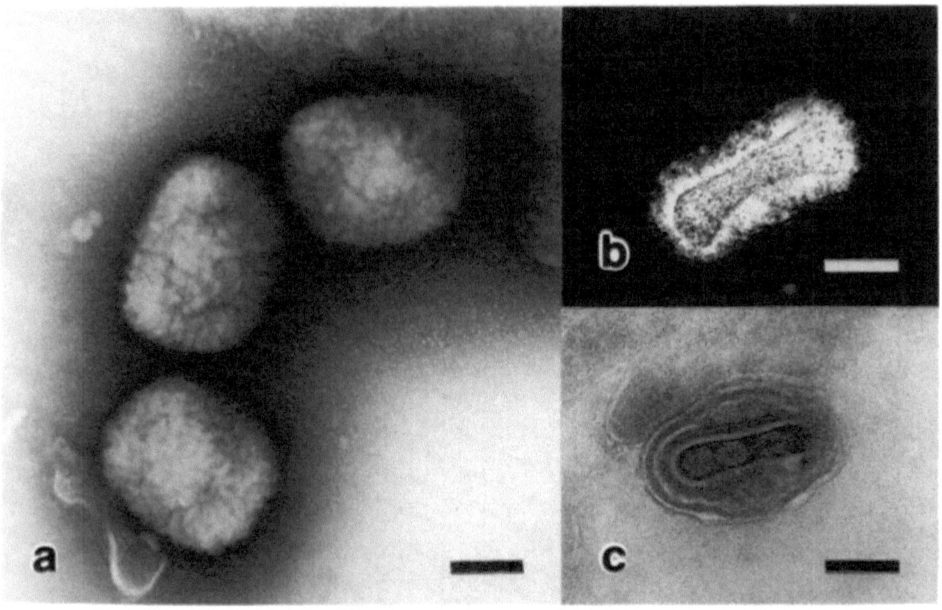

Figure 17. Vaccinia virus. a. Virions are brick-shaped or oval; the outer membrane surface is rough. b. The print is considerably overexposed to show the dump bell-shaped core that has been penetrated slightly by the negative stain (reprinted from Miller, 1986). c. An ultrathin frozen section of a vaccinia virus showing the complex core; contrasted with uranyl acetate. Bars represent 100 nm.

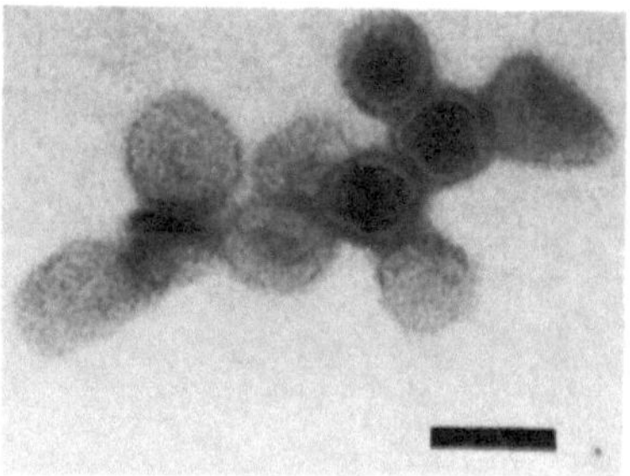

Figure 18. Rubella virus. An enveloped virus with projections so short as to be indistinguishable and a nucleocapsid that does not have identifying characteristics (reprinted from Miller, 1986). It would be impossible to identify this type of virus in an unknown sample because of its similarity to membranous cell debris. Bar represents 100 nm.

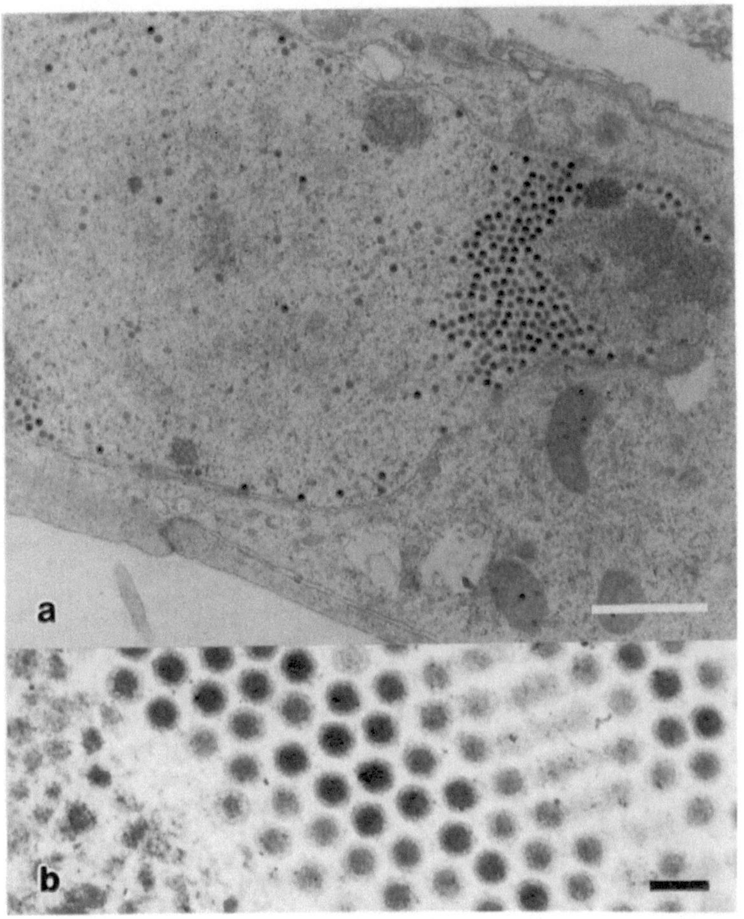

Figure 19. Adenovirus-infected cell. a. The DNA virions are formed in the nucleus. Late in infection, virions may also be seen in the cytoplasm as they lyse the host cell to get out. b. The icosahedral viruses can form paracrystalline arrays as seen here at high magnification. Bar in *a* represents 1 μm; bar in *b* represents 100 nm.

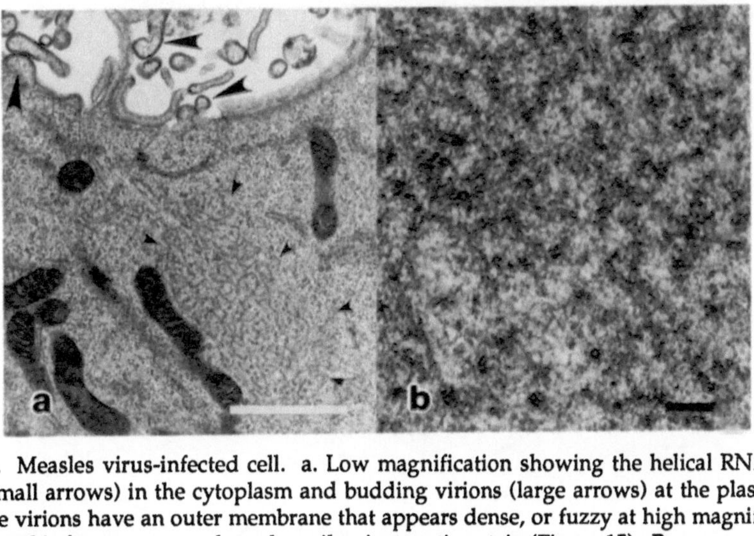

Figure 20. Measles virus-infected cell. a. Low magnification showing the helical RNA nucleocapsids (small arrows) in the cytoplasm and budding virions (large arrows) at the plasma membrane. The virions have an outer membrane that appears dense, or fuzzy at high magnification in thin section; this fuzz corresponds to the spikes in negative stain (Figure 15). Bar represents 1 μm. b. High magnification of the 18-nm worm-like nucleocapsids. Bar represents 100 nm.

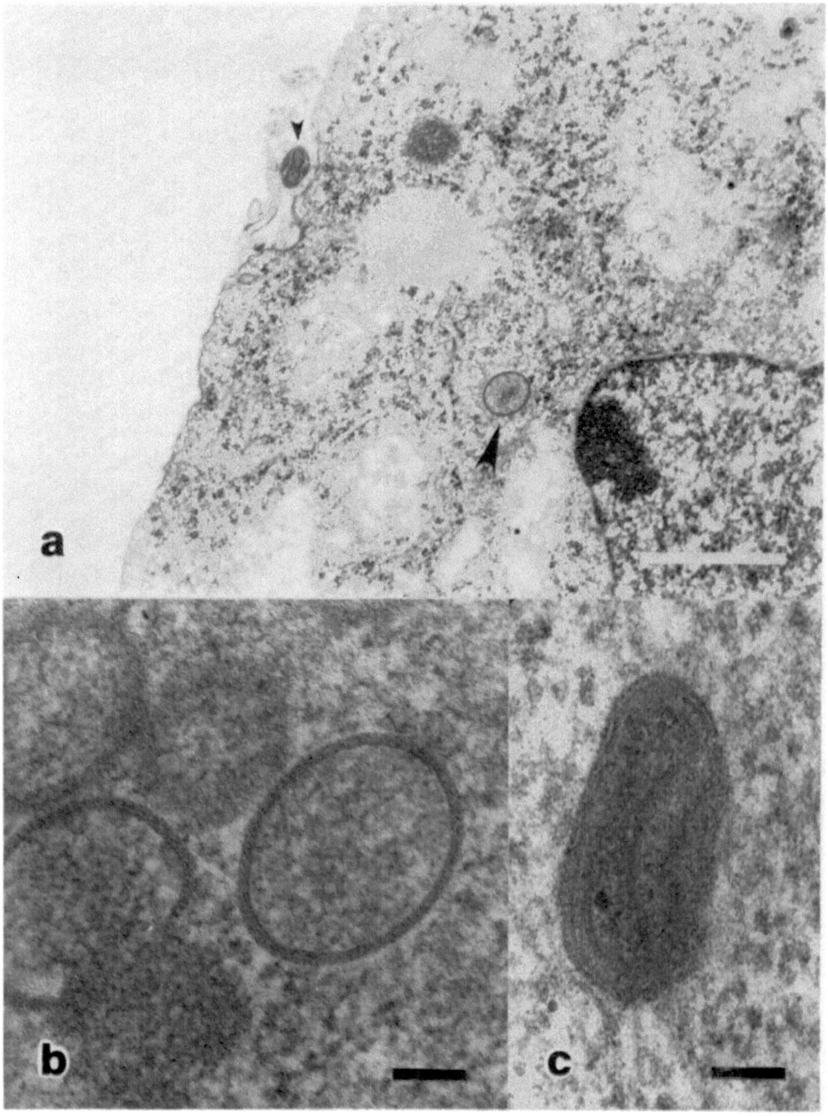

Figure 21. Poxvirus (vaccinia)-infected cell. a. Though the nucleic acid is DNA, this virus is constructed in the cytoplasm, and the envelope is created *de novo*, rather than by budding from cell membranes; immature particle (large arrow); mature particle (small arrow). Bar represents 1 μm. b. High magnification of immature particles. Bar represents 100 nm. c. High magnification of mature particle. Bar represents 100 nm.

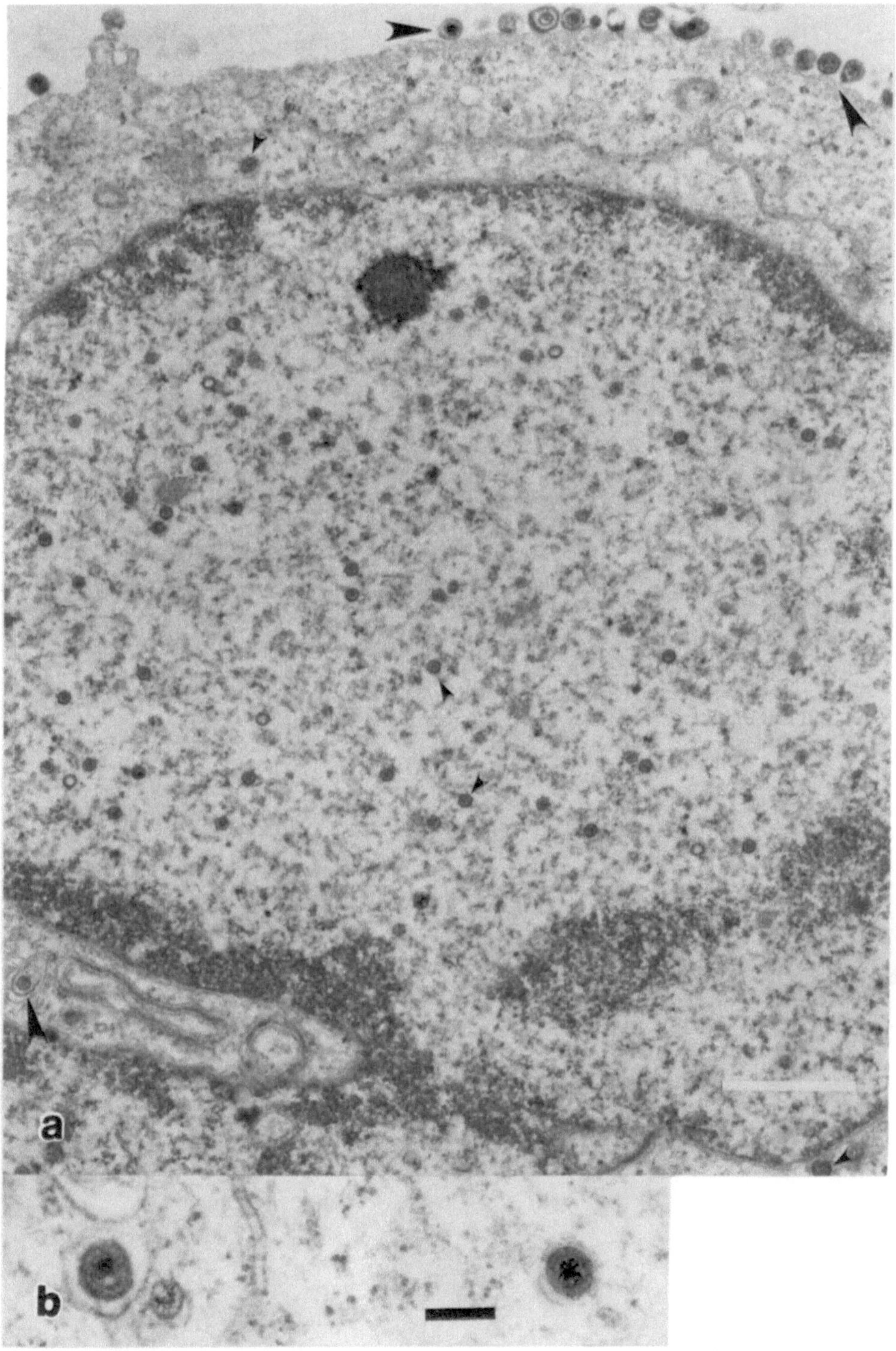

Figure 22. Herpes simplex virus-infected cell. a. Nucleocapsids (small arrows) are seen in the nucleus and cytoplasm; complete virions (large arrows) are seen in the cytoplasm and budding from the plasma membrane. The envelope can be obtained from the nuclear membrane, plasma membrane, or internal cell membranes. HSV often causes cells to over-produce membranes as seen in the lower left. Bar represents 1 μm. b. High magnification of cytoplasmic particles. Bar represents 100 nm.

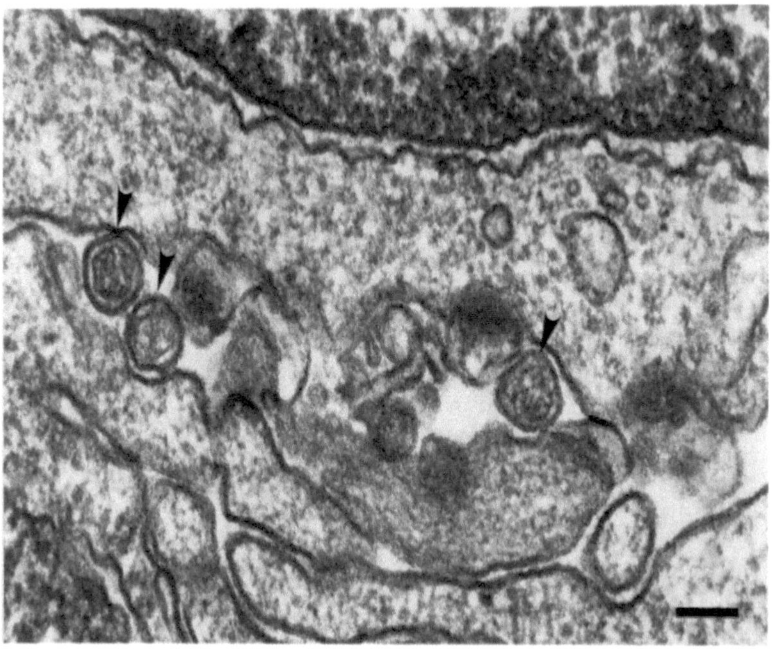

Figure 23. HTLV-1 (arrows), an RNA virus that buds from the cytoplasmic membrane to obtain its outer covering. Its nucleocapsid, although dense, and its short surface projections are not readily distinguishable amid cell debris in negative stains (reprinted from Miller, 1986). Bar represents 100 nm.

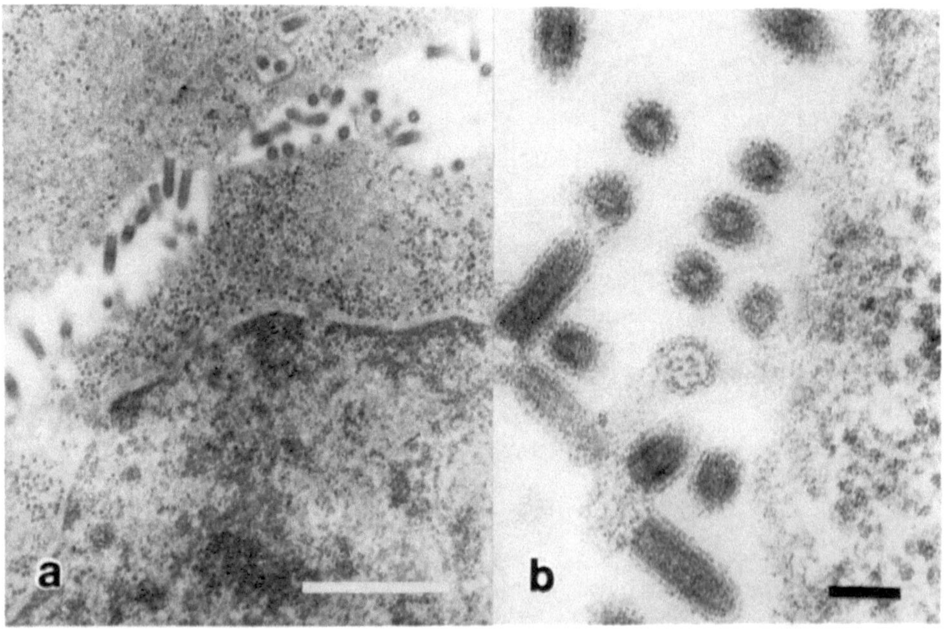

Figure 24. Cell infected with vesicular stomatitis virus, a rhabdovirus. In rabies-virus-infected cells, matrices of ribonucleoprotein form in the cytoplasm and can be identified by fluorescence microscopy as Negri bodies. Rhabdoviruses bud from the cytoplasmic membrane to form bullet-shaped virions shown at higher magnification in *b*. Bar in *a* represents 1 μm; bar in *b* represents 100 nm.

TABLE 2. Likely Viral Pathogens in Specimens Visualized by EM

BLISTERS	
	Herpes simplex virus/Varicella zoster virus
	Poxviruses (e.g., Orf/Vaccinia viruses)
CEREBROSPINAL FLUID	
	SRV (e.g., Coxsackie virus)
	Herpes simplex virus
	Measles virus
	Mumps virus
	Rubella virus
PLEURAL/PERICARDIAL FLUIDS	
	SRV (eg., Coxsackie virus)
SERUM (not usually processed by EM)	
	Hepatitis A virus
	Hepatitis B virus
	Non-A, non-B hepatitis viruses
	Parvovirus
	Filovirus (eg., Ebola virus)
SPUTUM/NASOPHARYNGEAL SECRETIONS	
	Respiratory syncytial virus
	Influenza/parainfluenza viruses
	SRV (e.g., rhinovirus, enterovirus)
	Adenovirus
	Mumps virus
	Cytomegalovirus
	Measles virus
	Reovirus
	Coronavirus
STOOL	
	Rotavirus
	Adenovirus
	Calicivirus
	Astrovirus
	SRV (small round viruses); featureless
	SRSV (small round structured viruses); rough surface (Norwalk virus-like)
	Coronavirus
SYNOVIAL FLUID	
	SRV (e.g., parvovirus, Coxsackie virus)
TEARS	
	Adenovirus
	SRV
	Herpes simplex virus
THROAT WASHES	
	Influenza/parainfluenza
	SRV (e.g., enterovirus)
	Adenovirus
	Mumps virus
	Measles virus
	Rabies virus
URINE	
	Cytomegalovirus
	Papovavirus

the systemic demarcation is not as distinct; however, certain types of viral infections are more prominent in these patients (Miller and Howell, 1988). In AIDS patients, the most commonly seen viral infection, and second in frequency only to *Pneumocystis carinii*, is cytomegalovirus (CMV). Other viruses in the herpes family are also frequent; these include HSV-1 and HSV-2, VZV and Epstein Barr virus (EBV), although the latter is not usually visualized by EM. Hepatitis B virus, polyomaviruses, and papillomaviruses are also common pathogens in these patients; adenoviruses and poxviruses are not uncommon. Coronaviruses, parvoviruses and human T-lymphotropic virus type 1 (HTLV-1) have also been reported.

EM RESULTS

Stool

For several reasons previously noted, EM is the method of choice for viral diagnosis from fecal specimens (Table 3). The recommendation from a recent workshop on gastroenteritis viruses at the Centers for Disease Control (Lew et al. submitted), for reporting results is that the agents listed in Table 4 be named when present. Their prevalence and seasonality as seen at 10 different EM viral laboratories in the US and Canada are also shown. Of 52,691 specimens examined, 16% were positive for a viral agent. Rotaviruses are the most prevalent, particularly in infants and especially in the winter months. Adenovirus and SRV tie for second in prevalence.

Quite often when a stool sample is received in the EM laboratory, the instructions on the requisition are for a "Rotazyme test" (Abbott Laboratories, Abbott Park, IL), as many physicians are aware of the commercially available slide kits for rotavirus detection (not all of which are named "Rotazyme", e.g., "Rotaclone", Cambridge Bioscience, Worcester, MA). There are also kits for adenoviruses. However, use of one of these kits would not detect any of the other potential agents. One laboratory in Canada with a heavy load screens samples with a rotavirus slide kit and performs EM on those that are negative (Lew et al. submitted). This procedure is sufficient for most specimens, but in the case of immunodeficient patients where a second virus may be present, it would miss anything but the one against which the antiserum was directed. We routinely perform direct EM on all stool specimens, and for those that are negative, we perform one of the concentrating procedures previously mentioned. Cell culture is inappropriate for fecal specimens because the viral agents found there either have not be propagated in tissue culture or grow slowly and with difficulty in the research but not the diagnostic laboratory.

The implications of a viral diagnosis depend on the virus. The presence of even a few rotaviruses in the stool of a patient with gastroenteritis is significant in the cause of disease. Rotavirus infection should be taken seriously, particularly in the very young; indeed in third world countries, it is responsible for many deaths. Fluid replacement with restoration of electrolyte balance should be initiated as soon as possible as dehydration can occur very rapidly. Further, since rotavirus is very contagious, stringent quarantine measures should be adopted to prevent spread throughout the nursery. In immunocompetent hosts, the infection is self limiting with supportive therapy, but in immunodeficient babies, therapy with immune globulin and ribavirin has been attempted.

TABLE 3. Procedures for Most Likely Virus Detection

Sample	Desired Diagnosis (presence or absence)	Technique
Stool	virus[a]	EM best/only
Urine	virus	EM, culture
Blisters	herpes vs. pox	EM, culture
Sputum	RSV	immunofluorescence
	virus	culture, EM
Nasopharyngeal secretions, lavages	virus	culture, EM
Pleural/ pericardial fluids	virus	culture, EM
CSF	virus	culture, EM less efficient
Tears	virus	culture, EM
Blood	hepatitis A, B	serology
	hepatitis, non-A/B	serology, EM, culture
	parvovirus	EM
Synovial fluid	virus	culture, EM
Tissue	papilloma	*in situ* hybridization
	EBV	*in situ* hybridization
	retroviruses	EIA, Western blot, *in situ* hybridization
	virus	culture, EM-thin sections

[a] Any virus

TABLE 4. Viruses Detected from Stool at Ten EM Viral Laboratories

Virus	Prevalence (%) Ave/Range[a]	Seasonality	Patient's age (yrs)
Rotavirus	48/26-83	Winter	< 1 (54 %) 1 - 4 (46 %)
Adenovirus	17/8-27	-	all
SRV/SRSV	10/0-40	-	infants (48%) older children & adults
Astrovirus	2/0-10	Winter	< 1 (64 %) 1 - 4 (36 %)
Calicivirus	1/0-4	-	≤ 1
Coronavirus	7/0-13[b]	-	?

[a] Range among the different centers
[b] One center reported 67%

From Lew et al. (submitted)

The detection of only a few adenoviruses in stool may not indicate intestinal infection. Adenoviruses can infect the respiratory tract, and enteric adenoviruses cannot be distinguished morphologically from respiratory adenoviruses that may have been swallowed. However, if there are numerous virions in a clinical setting of diarrhea, they probably are significant. In the case of cold or flu-like symptoms, the significance of intestinal adenoviruses is equivocal, but once such viruses are detected by EM, an immunological test (e.g., Adenoclone, Cambridge Bioscience, Worcester, MA) can be run to determine the strain if necessary. On the other hand, respiratory viruses can replicate in the intestinal mucosa and elicit pathologic effects; thus, differentiation between an enteric and other adenoviruses may be moot in the case of obvious diarrhea.

There is a problem with the interpretation of the presence of SRV in fecal samples. In North America, no one system of classification has been accepted for small isometric viruses. In Britain, Caul and Appleton (1982) define SRV as small (22-26 nm) round "featureless" viruses that do not induce an immune response and are not considered to be pathogenic. They seem to appear in stool after a diarrheal illness caused by one of the SRSV (Caul, personal communication). However, parvoviruses are 18-26 nm in diameter, and parvovirus-like agents have been seen by EM in feces of enteritis patients; these particles could be clumped with homologous serum (Paver et al. 1973). Furthermore, correct EM calibration is essential to classify one of these SRV as less than or more than 26 nm if the size is close to that. Thus, in this author's opinion, the issue of pathogenicity of SRV in stool has not yet been resolved. The SRSV include

Norwalk agent, mini-reovirus and other viruses whose surface is rough; most of these have been shown to be pathogenic. Astrovirus, calicivirus, and some small isometric viruses have been shown to cause important nosocomial infections (Blacklow, 1990; Lew et al., submitted; Riepenhoff-Talty, 1982). Since infection with these agents is often mild, and since precautions against the spread of diarrhea are most likely when the disease is severe, appropriate precautions may not be taken. This can result in higher treatment costs and possible severe ramifications for immunocompromised patients.

Coronaviruses and coronavirus-like particles (CVLP) have been reported in stool from both gastroenteritis patients and normal, asymptomatic individuals (MacNaughton and Davies, 1981). Furthermore, identification of these viruses is difficult, since they can be confused with membranous debris with spikes (e.g., mitochondrial membrane). A single or few isolated particles, unless they are diagnosed by an experienced microscopist, should not be considered conclusive. If there are numerous, similarly-sized particles present, their significance increases.

As with rotavirus, infection of the immunocompetent patient with any of the enteric viruses is self-limiting as long as homeostasis is maintained with supportive treatment. However, in the immunodeficient patient, diagnosis is a must for the administration of specific immune globulin and perhaps trial therapy with antiviral drugs. Little is known about the efficacy of drugs in this case, but there may be no other hope for some of these patients. The prospects for antiviral chemotherapy of gastrointestinal illnesses have been discussed by Babiuk et al. (1985).

Urine

Urine is another specimen for which EM can be very useful. Although viruses are not usually seen in urine during infections of other systems, they can be found there in certain cases. In congenital CMV infections, virion numbers are quite high in urine. Papovaviruses have been demonstrated in urine from pregnant women in the absence of disease and from kidney transplant patients. Parvoviruses have been seen in high numbers in urine from patients with aplastic anemia. Several different viruses have been isolated from urine in disseminated viral infections of immunocompromised hosts. In these instances, EM may be the method of choice because it is rapid and virus concentration in the sample is high. Papovaviruses and parvoviruses are not grown in the diagnostic culture laboratory, and growth of CMV requires 2-3 weeks. A fluorescence test is commercially available for CMV, but not for papova- or parvoviruses.

Respiratory Secretions/Pleural and Pericardial Fluids

In respiratory infections, if the physician suspects respiratory syncytial virus (RSV) (e.g., during a winter outbreak of lower respiratory tract illness), (s)he can request a fluorescent microscopy examination for that particular virus. However, if the fluorescence exam is negative, there would be no further information on the etiology, whereas EM may be able to demonstrate the presence of adenovirus or a rhinovirus. In the case of RSV, EM might be able to show the helical nucleocapsids, but would not be able to distinguish between RSV and the other paramyxoviruses. If EM had been performed first and showed helical nucleocapsids, the patient's symptoms and the epidemiology might suggest a differential diagnosis; then fluorescence microscopy or one of the enzyme immunoassays could confirm it. Fluorescence assays are also available for parainfluenza virus 1, 2, and 3; adenovirus; influenza virus A and B; HSV; VZV; and CMV. Rapid enzyme immunoassays are available for RSV, adenovirus, HSV, and VZV. Rhinoviruses, enteroviruses and parvoviruses have also been seen in throat washings and respiratory secretions. These agents appear by EM simply as SRV. Some do not grow in culture or require extended culture time (days to weeks) with multiple passages. EM is in some cases the only option for identifying such noncultivable agents. Since the ability of an agent to grow in tissue culture is generally not known *a priori*, inoculation of cultures and EM examination should both be undertaken. Drug therapy is available for some of these agents, a factor which increases the importance of viral identification.

The viruses most likely to be seen in pleural and pericardial fluids are those of the SRV variety (entero- and rhinoviruses). Other than demonstrating their presence and size, EM could not further differentiate them, unless a particular agent were suspected and antiserum against that agent were available.

Blister Fluid/Skin

For blisters, if the question is herpesvirus vs. poxvirus, the method of choice is EM; then fluorescence microscopy or IEM could further delineate the agent. If the symptoms (e.g., genital lesions vs. thoracic lesions) suggest a particular disease, EM could substantiate it. For example, in the first case, an EM finding of a herpesvirus would suggest HSV-2, (possibly HSV-1), while in the second case it would suggest VZV. Various poxvirus infections are seen in animal handlers, laboratory workers, and immunocompromised individuals. Differentiating these agents is important for therapy. Viruses have been demonstrated by EM, both with negative staining of scrapings and with thin sections (e.g., measlesvirus, papillomavirus). *In situ* hybridization (Chesselet, 1990; Metcalf et al. 1988; Norval and Bingham, 1987) is becoming more widely used for papilloma virus, particularly since the genome can be present in normal-appearing tissue at the lesion margin in the absence of complete virions. However, this procedure is not routinely available in all diagnostic virology laboratories.

Cerebrospinal Fluid

Viruses have been seen in CSF, but when present, they usually are not numerous; thus, culturing is the diagnostic method of choice. However, if several milliliters (2-5) are available, enough for culture and EM, it may be possible to demonstrate virus by EM after ultracentrifugation. Many enteroviuses cannot be readily grown, or require a long time and several blind passages. The EM diagnosis, if positive, would be of a small round virus of a given size; the size could narrow the possibilities, but precise identification could not be made directly. If a particular virus were suspected (e.g., poliovirus vaccine strain in an immunodeficient infant), and antiserum were available, identification could be made by IEM or by antibody neutralization in tissue culture.

Synovial Fluid

Parvovirus B19, rubellavirus, Coxsackie B virus, EBV, and arboviruses have been associated with arthritis. Parvovirus (which does not grow in culture) and Coxsackie viruses can be demonstrated by EM. Other viruses that are enveloped would be difficult to identify by EM because of their similarity to membranous debris, unless specific antiserum were available for IEM.

Blood

As previously mentioned, EM is generally not the method of choice for blood samples, but if serological tests are negative, EM as well as culture may be useful. Many different non-A, non-B hepatitis viruses have been described (referenced in Hayat and Miller, 1990), some of which may be demonstrated by EM. Serum samples including convalescent serum should be collected and preserved for IEM. Some of these agents have not been grown in tissue culture, and thus, may not be demonstrable by culturing. Also, parvovirus B19 (an SRV) has been seen by EM in blood from children with erythema infectiosum (fifth disease), in cases of aplastic crisis in sickle cell anemia, and in asymptomatic viremic volunteers (Cossart et al. 1975; Versteeg and Salimans, 1988). It does not grow in culture; thus, with exception of research procedures with molecular probes, EM and IEM are the procedures of choice for diagnosis.

Care must be taken when examining serum by negative staining because droplets of lipids and proteins present can appear as round spots and may resemble SRV. Clues to the viral nature of these structures are uniformity of size and shape and possibly a 3-dimensional appearance. Sometimes, but not always, stain collects around the underside of viruses producing a darker background immediately around the virion, while droplets often flatten out like a fried egg and appear simply as clear "holes" in the stain. Additionally, enveloped viruses such as *Retroviridae*, *Arenaviridae*, *Togaviridae*, and *Bunyaviridae*, that do not have long identifiable spikes or recognizable nucleocapsids cannot be distinguished from cell membrane debris by negative staining.

ANTIVIRAL DRUG THERAPY

One of the main reasons for pursuing a viral diagnosis today is the increasing availability of antiviral agents. Except for ribavirin, most currently available antiviral drugs do not have wide spectrum activity; accurate identification of viral infectious agents is therefore crucial for proper drug selection. The viruses for which some antiviral agent has been described are listed in Table 5. Detailed information on mechanism of action, licensed and investigational drugs, dosages, and research studies has been published (American Society of Hospital Pharmacists, 1990; De Clerq and Walker, 1987; Mills and Corey 1985, 1989; Olin, 1990). A brief summary of the more common ones follows.

Acyclovir

Acyclovir (ACV, Acyclovir sodium, acycloguanosine, Zovirax) is a well known anti-*Herpesviridae* drug. Suppressive therapy has been given orally in some patients with severe HSV infections to reduce the severity or frequency of symptomatic recurrences. The use of topical ACV in first episodes reduces the time for virus shedding, duration of pain and itching, and time for crusting and healing, but it is not as effective as oral and intravenous ACV. In the case of recurrent genital herpes, topical treatment is not effective, while oral ACV is, especially when started at the first sign of the prodrome; however, the benefit

TABLE 5. Human Viruses Against Which Drugs Are Available

CMV	Acyclovir	Ganciclovir[a]	Vidarabine[b]		Ribavirin[c]
EBV	Acyclovir				
HSV-1	Acyclovir	Ganciclovir[a]	Vidarabine		Ribavirin
HSV-2	Acyclovir	Ganciclovir[a]			Ribavirin
Herpes simae	Acyclovir	Ganciclovir[a]			
VZV	Acyclovir	Ganciclovir[a]	Vidarabine	Zidovudine[c]	
HIV-1				Zidovudine	Ribavirin
Arboviruses					Ribavirin[c]
Colorado tick fever virus					Ribavirin[c]
Crimean-Congo hemorrhagic fever virus					Ribavirin
Dengue fever virus				Amantadine[d]	
Encephalomyocarditis virus (EMC)					Ribavirin[c]
Hantaan virus (Korean hemorrhagic fever)					Ribavirin
Japanese encephalitis virus					Ribavirin[c]
Junin virus (Argentine hemorrhagic fever)					Ribavirin
Lassa fever virus					Ribavirin
Lymphocytic choriomeningitis virus				Amantadine[d]	
Machupo virus (Bolivian hemorrhagic fever)					Ribavirin
Pichinde virus					Ribavirin[c]
Rift Valley fever virus					Ribavirin
Rubella virus				Amantadine[d]	
Semliki Forest virus				Amantadine[d]	
Venzuelan equine encephalitis virus (VEE)					Ribavirin[c]
Yellow fever virus					Ribavirin[c]
Influenza A virus				Amantadine	Ribavirin
Influenza B virus				Amantadine[d]	Ribavirin
Influenza C virus				Amantadine[d]	
RSV				Amantadine[d]	Ribavirin
Parainfluenza virus				Amantadine[d]	Ribavirin
Measles virus					Ribavirin
SSPE virus					Ribavirin
Mumps virus					Ribavirin
Rhinoviruses virus					Ribavirin[c]
Coxsackie virus B1					Ribavirin
Enterovirus 72 (hepatitis A virus)					Ribavirin
Reovirus 1, 2, 3					Ribavirin
Rotavirus					Ribavirin[c]
Adenovirus					Ribavirin[c]
Hepatitis B virus					Ribavirin[c]
Poxvirus					Ribavirin[c]

[a] Principle use against CMV, but active against other *Herpesviridae*
[b] Less effective against HSV encephalitis than acyclovir
[c] *In vitro* activity. *In vivo* activity may not exist or may not have been shown.
d *In vitro* activity at high concentrations (> 10 μg/ml)

From American Society of Hospital Pharmacists, 1990a; Mills and Corey, 1985; Olin, 1990 is not as great as in first episodes. Because oral ACV is less expensive and easier to give than intravenous ACV, it is the therapy of choice for first episode genital herpes. ACV appears to be more effective than vidarabine for treating HSV encephalitis and also for treating chickenpox in the immunocompromised host.

Both oral and intravenous ACV are effective in speeding healing and shortening the period of acute VZV infections when started within 48-72 hr after onset of symptoms. It may also reduce the occurrence of post-herpetic neuralgia. The efficacy of parenteral ACV in CMV treatment is unclear, although it has produced improvement in some immunocompromised patients with CMV pneumonia. ACV inhibits EBV replication in active production *in vitro* but has no effect on the episomal EBV DNA. It has been shown to reduce virus shedding in acute infectious mononucleosis but does not substantially affect the course of the disease. It has been used against EBV infection in AIDS patients to reduce the clinical progression of hairy leukoplakia, and to treat several patients with fever, interstitial pneumonitis, panleukopenia, and high levels of anti-EBV antibodies. However, it has had little effect in nonimmunocompromised hosts, including attempts to treat chronic fatigue syndrome. Variations in therapeutic efficacy with respect to virus type, lesion location, and mode of drug delivery are discussed in detail by Mertz (1989). ACV has been used concomitantly with AZT without increased toxicity, and preliminary *in vitro* data suggest that it may potentiate the antiretroviral activity of AZT.

ACV is a purine nucleoside analog that is preferentially absorbed by herpesvirus-infected cells and converted to the triphosphate form that is a toxic analog of deoxyguanosine triphosphate. It also interferes with HSV DNA polymerase by forming a complex with the enzyme and the DNA template. Without the 3' OH group for the 5'- to 3'-phosphodiester linkage, it acts as a chain terminator. It is inactive against the other DNA viruses such as vaccinia virus and adenovirus 5 as well as several RNA viruses against which it has been tested.

Several other antiherpesviral drugs are being examined. FIAC (2'-fluoro-5-iodoarabinosyl-cytosine), a pyrimidine analog, is very active *in vitro* against herpesviruses. Clinical trials in the immunocompromised host show dramatic decreases in time to last lesions, crusting, pain, and dissemination; however, there is mild liver toxicity. BVDU (bromovinyldeoxyuridine) has a much increased activity over ACV in VZV infections. BW-A515U is a prodrug of ACV that produces high levels of plasma ACV after oral administration. BVaraU (1-b-D-arabinofuranosyl-E-5[2-bromovinyl] uracil), FMAU (fluoromethylarabinosyluridine), and FIAW (fluoroiodoarabinosyluridine) are active *in vitro* against herpesviruses and resemble ACV in their mechanism of action. All are nucleoside analogs.

Amantadine Hydrochloride

Amantadine hydrochloride (amantadine HCl) (Symadine, Symmetrel) is used primarily in the prophylaxis of influenza A after exposure of high risk persons; however, early immunization is still the method of choice for prevention. If administered 24-48 hr after onset of influenza A, amantadine reduces the duration of the disease and provides a more rapid return to activities and improvement in lung function. It is not effective *in vivo* against influenza strains other than type A nor parainfluenza viruses, rhinoviruses, adenoviruses, RSV, and others.

Amantadine HCl is a synthetic cyclic primary amine that is administered orally. Its mechanism of action is not completely understood, but it appears to inhibit membrane-associated events such as penetration of the virus into the host cell and release of infectious viral nucleic acid.

Rimantidine is a derivative of amantadine that is used in the United Soviet Socialist Republic against influenza A and may become available in the United States soon. Animal studies show it to be somewhat more effective than amantadine.

Ansamycin

Ansamycin (rifabutine), a derivative of the antimycobacterial rifamysin binds retroviral reverse transcriptase *in vitro*. Clinical trials in AIDS patients have not been rewarding.

Antimoniotungstate

Antimoniotungstate (HPA-23) has anti-reverse transcriptase activity in a number of animal retroviruses both *in vitro* and *in vivo* and has been tested in AIDS patients. Although a transient reduction in peripheral blood virus was seen in some patients, others had no improvement. The drug causes thrombocytopenia, liver dysfunction and fever.

Foscarnet

Foscarnet (phosphonoformate trisodium, PFA) is active against all herpesviruses; it may be considered for ACV-resistant herpesvirus isolates and for CMV infection in AIDS patients on AZT. It is being tested as a topical anti-HSV drug and as a systemic anti-CMV drug. Clinical trials of intravenous foscarnet have shown some nephrotoxicity and abnormalities of serum calcium and phosphorus levels.

Foscarnet and the related phosphonoacetic acid (PAA) are pyrophosphate analogs. They interfere with the viral polymerase of some DNA and RNA viruses. The reverse transcriptase of retroviruses is inhibited by PFA but not PPA.

Ganciclovir

Ganciclovir sodium (BW B759U, BIOLF-62, DHPG sodium, GCV sodium, Cytovene, nordeoxyguanosine) is active *in vitro* against all herpesviruses, but its principle use is against CMV retinitis in immunocompromised patients; it is being investigated in gastrointestinal CMV infections in AIDS patients and bone marrow transplant recipients. Because of high toxicity, the use of ganciclovir should be weighed against its side effects including mutagenicity, carcinogenicity, adverse reproductive potential, renal and hematological toxicity, psychosis, and phlebitis. Ganciclovir should not be used in immunocompetent hosts, and combined use with AZT results in profound, prolonged neutropenia. A synergistic effect on CMV is seen with ganciclovir and interferon α or β.

Ganciclovir is a purine nucleoside analog of guanine that interferes with DNA synthesis by competing with deoxyguanosine. Its increased activity over

ACV in CMV inhibition is thought to result from slower catabolism of ganciclovir triphosphate by intracellular phosphatases.

Ribavirin

Ribavirin (ICN-1229, Virazole, RTCA, Tribavirin) has the widest spectrum of all the antiviral drugs. Its principle use is against RSV in severe lower respiratory tract infections, and it has been used in influenza A and B infections. Although not currently included in U.S. Food and Drug Administration approved labeling, it has been effective against Lassa and Crimean-Congo fevers as well as measles virus, HSV-1 and 2, enterovirus 72 (hepatitis A), and adenovirus, and is under investigation for management of HIV-1 infection. However, it is antagonistic to AZT and should not be used concomitantly. It is inhibitory *in vitro* to a number of different RNA and DNA viruses including various *Arenaviridae, Bunyaviridae, Orthomyxoviridae, Paramyxoviridae, Reoviridae, Retroviridae, Togaviridae, Adenoviridae, Herpesviridae, and Poxviridae*. In some cases, though, the virus inhibition *in vivo* is less than that *in vitro*.

Ribavirin interferes with viral RNA and DNA synthesis and subsequently, protein synthesis by a mechanism that, like those of many other antiviral drugs, is not completely known yet. It exhibits a greater effect on viral DNA and RNA synthesis than cellular nucleic acid synthesis except in a few cases such as HSV-1 and enterovirus 72. Its activity depends on its intracellular conversion to ribavirin-5'-triphosphate and -monophosphate; the triphosphate competes with ATP and GTP for viral RNA polymerase and inhibits the enzymes responsible for capping the 5' viral mRNA with guanosine. The monophosphate inhibits IMP dehydrogenase, the enzyme that synthesizes GTP. Ribavirin inhibits phosphorylation of thymidine but unlike ACV is not readily incorporated into DNA or RNA.

Selenazole and tiazofurin are selenium- and sulfur-containing nucleoside analogs related to ribavirin. *In vitro*, they are active against most of the viruses that are sensitive to ribavirin, and in some cases, their interaction with ribavirin is additive or synergistic. Their efficacies and toxicities *in vivo*, particularly that of the selenium compound, have not been established.

Suramin

Suramin (Germanin), is one of the longest known drugs that has been considered as an antiviral agent. It has been used for many years against African trypanosomiasis and onchocerciasis. Suramin, a sodium salt derivative of napthalenetrisulfonic acid, is active *in vitro* against retroviruses and was used in clinical trials against HIV-1. However, it causes considerable side effects including adrenal, urinary, and hepatic abnormalities; fevers; rashes; as well as rare idiosyncratic reactions. It is no longer seriously considered as a therapy in AIDS.

Vidarabine

Vidarabine (adenine arabinoside, Ara-A) has been used to treat HSV encephalitis, resulting in reduced mortality, but appears to be less effective than ACV; comparative studies are underway. It has been used to treat shingles and chickenpox in immunocompromised patients but is ineffective in treatment of

VZV encephalitis. In both immunocompromised and immunocompetent hosts, vidarabine has been used to treat CMV infections.

Vidarabine is a purine nucleoside produced from fermentation by a streptomycete. Its mode of action is unclear, but it may block viral DNA polymerase; it is not significantly incorporated into the viral DNA.

Adenine arabinoside monophosphate (ara AMP) inhibits HBV replication and can be given by intramuscular injection. Its efficacy in clearing one of the internal antigens of hepatitis B (HBeAg) from chronically infected individuals has been variable.

Zidovudine

Zidovudine (azidothymidine, AZT, Compound S) is probably the most widely known antiviral agent due to its popularity in the AIDS press. Studies have shown that when initiated early, AZT reduces the progression of the disease; further study is necessary to determine whether length of survival is increased if the drug is administered to symptomless HIV-positive individuals. In addition to HIV-1, it is active against some other mammalian retroviruses and has *in vitro* activity against EBV.

AZT is a synthetic thymidine analog that, when incorporated into DNA, results in the inability to form phosphodiester linkages. *In vitro* it interferes with viral reverse transcriptase. The U.S. Food and Drug Administration has designated it as an orphan drug for management of HIV-1. An orphan drug is one whose cost of research and production surpasses the revenue expected from sales, but for humanitarian reasons, the producer makes it available in return for certain monetary advantages (such as tax breaks).

Other Considerations for Antiviral Therapy

In addition to antiviral agents, drugs that afford various forms of symptomatic relief are occasionally employed in the treatment of viral illness. Awareness of the manifestations of different viral infections can influence symptomatic therapy. For example RSV and parainfluenza virus precipitate the release of histamines into nasopharyngeal secretions and thus contribute to wheezing. However, this is not true in rhinovirus or coronavirus infections. Therefore, H1 antihistamines would not be effective therapy for common colds. On the other hand, rhinoviruses cause the production of high kinin levels in secretions, which suggests that antiinflammatory drugs might be a useful therapy.

Obviously, the best antiviral therapy is prevention with vaccines (American Society of Hospital Pharmacists, 1990c; Jordan, 1988; Quinnan, 1990), but unfortunately they are not available for all viruses. A few antisera are available for passive immunization when exposure has not permitted vaccination (American Society of Hospital Pharmacists, 1990b); but again, not all viral infections can be treated in this manner. Considerable research is in progress on vaccine production, and this subject is too vast to be covered here. We have focused on a synopsis of drugs currently available or presently in clinical trials. Other possibilities are stimulation of the host response to viral infections with immunomodulators such as interferons, interleukins, thymic humoral factors, cyclosporin A, ampligen, or isoprinosine (Mills and Corey, 1989), alone or in combination with drug therapy. Some antiretroviral drugs have *in vitro* synergistic interaction with others or with interferon or soluble CD4; preliminary studies suggest that alternating therapy or combination therapy may decrease toxicity or increase efficacy. However, immunotherapy could be counterproductive, for example, if it produces activated cells that viruses use for replication or activates latent virus. Investigational drugs active *in vitro* against viruses must be thoroughly studied to determine such things as their toxicity *in vivo*, their ability to produce adequate serum levels, and their interaction with other therapeutic agents.

SUMMARY

EM is a very advantageous procedure for diagnostic virology and should be used in conjunction with culture and other specialized techniques that require specific probes. It is a very rapid procedure and is particularly good for viral agents that cannot be cultured and for specimens that are liquid. If specific antiserum is available, viruses can be serotyped by EM.

The morphological characteristics that should be noted for virus identification by negative staining are whether the viruses are naked or enveloped, and if enveloped whether they have recognizable spikes. Also, the shape of the nucleocapsid in enveloped virions is important. In thin sections the virus location in infected cells is a clue to the nucleic acid type. Other characteristics just mentioned for negative stains are important in thin sections as well.

Specimens to be examined by negative staining (liquids and stools) should be sent to the EM laboratory unfixed and undiluted; tissue samples should be fixed immediately upon removal in buffered glutaraldehyde. Specimens for EM should never be frozen in a regular refrigerator freezer. Long term storage of stool specimens can be accomplished in liquid nitrogen, at -70°C, or in sealed containers at 4°C. Tissue specimens, if they are not processed through epoxy resin, should be stored at 4°C in glutaraldehyde.

Other procedures for the laboratory diagnosis of viral infections are reviewed by Spector and Lancz (1986), Yolken (1990), and White and Fenner (1986), and rapid laboratory techniques are discussed by Henshaw (1988; submitted).

The current armamentarium of antiviral drug therapy includes ACV, amantadine, ganciclovir, ribavirin, vidarabine, and AZT; many other agents are being developed. The availability of such therapeutic options makes the diagnosis of viral infections increasingly important.

REFERENCES

Almeida JD (1983) Uses and abuses of diagnostic electron microscopy. Curr Top Microbiol 104:147-158.

American Society of Hospital Pharmacists (1990a) Antivirals. In AHFS Drug Information. Bethesda: American Society of Hospital Pharmacists, Inc., pp. 353-388.

American Society of Hospital Pharmacists (1990b) Serums. In AHFS Drug Information. Bethesda: American Society of Hospital Pharmacists, Inc., pp. 1880-1920.

American Society of Hospital Pharmacists (1990c) Vaccines. In AHFS Drug Information. Bethesda: American Society of Hospital Pharmacists, Inc., pp. 1921-1073.

Babiuk LA, Sabara MI and Frenchick P (1988) Prospectives for the treatment of gastrointestinal tract virus. In De Clercq E (ed), Clinical Use of Antiviral Drugs. Boston: Martinus Nijhoff Publishing, pp. 341-359.

Blacklow NR (1990) Medical virology of small round gastroenteritis viruses. In de la Maza LM and Peterson EM (eds), Medical Virology, Vol. 9, New York: Plenum Press, pp. 111-128.

Caul EO and Appleton H (1982) The electron microscopical and physical characteristics of small round human fecal viruses: an interim scheme for classification. J Med Virol 9:257-265.

Chesselet M-F (1990) *In Situ* Hybridization Histochemistry. Boca Raton: CRC Press, Inc..

Cossart YE, Field AM, Cant B and Widdows D (1975) Parvovirus-like particles in human sera. Lancet 1:72.

De Clerq E and Walker RT (eds) (1987) Antiviral Drug Development. A Multidisciplinary Approach. New York: Plenum Press.

Doane FW and Anderson N (1987) Electron Microscopy in Diagnostic Virology. New York: Cambridge University Press.

Hammond GW, Hazelton PR, Chuang I and Klisko B (1981) Improved detection of viruses by electron microscopy after direct ultracentrifuge preparation of specimens. J Clin Microbiol 14:210-221.

Hayat MA and Miller SE (1990) Negative Staining. New York: McGraw-Hill Publishing Co.

Henshaw NG (1988) Identification of viruses by methods other than electron microscopy. ASM News 54:482-485.

Henshaw NG (submitted) Diagnostic virology. In Joklik WK, Willet HP, Wilfert CM and Amos DB (eds), Zinsser Microbiology. New York: Appleton-Century-Crofts.

Hsiung GD (1982) Diagnostic Virology Illustrated by Light and Electron Microscopy. New Haven: Yale University Press.

Jordan WS Jr. (1988) Program for accelerated development of new viral vaccines. Prog Med Virol 35:1-20.

Kapikian AZ, Dienstag JL and Purcell RH (1976) Immune electron microscopy as a method for the detection, identification, and characterization of agents not cultivable in an in vitro system. In Rose NR and Friedman H (eds), Manual of Clinical Immunology. Washington: American Society for Microbiology, pp 467-480.

Lennette DA (1985) Collection and preparation of specimens for virological examination. In Lennette EH, Balows A, Hausler WJ, Jr, Shadomy HJ (eds) Manual of Clinical Microbiology, 4th Ed., Washington, D.C., American Society for Microbiology, pp. 687-693.

Lew J, Glass R, Petric M, LeBaron C, Hammond G, Miller SE, Robinson C, Boutillier J, Talty M, Payne C, Franklin R, Oshiro L and Jaqua M-J (submitted) Six year retrospective surveillance of gastroenteritis viruses identified at ten electron microscopy centers in the United States and Canada.

MacNaughton MR and Davies HA (1981) Human enteric coronaviruses. Arch Virol 70:301-313.

Mertz GJ (1989) Mucocutaneous herpes simplex virus infections: prophylaxis and treatment. In Mills J and Corey L (eds), Antiviral Chemotherapy. New Directions for Clinical Application and Research. New York: Elsevier Science Publishing Co., Inc., pp. 1-22.

Metcalf TG, Jiang X, Estes MK and Melnick JL (1988) Nucleic acid probes and molecular hybridization for detection of viruses in environmental samples. Prog Med Virol 35:186-214.

Miller SE (1986) Detection and identification of viruses by electron microscopy. J Electron Microsc Tech 4:265-301.

Miller SE (1988) Diagnostic virology by electron microscopy. ASM News 54:475-481.

Miller SE (1989) Electron microscopy in rapid viral diagnosis. EMSA Bull 19:53-59.

Miller SE (1990a) Electron microscopy in diagnostic virology. I. Specimen handling and procedures for identifying viruses in samples. In Mukherjee TM (ed), National Training Course on the Application of Electron Microscopy in Biomedical Research and Diagnosis of Human Diseases. Adelaide, So. Australia: World Health Organization, pp. 100-105.

Miller SE (1990b) Electron microscopy in diagnostic virology. II. Virus morphology and confusing structures. In: Mukherjee TM (ed), National Training Course on the Application of Electron Microscopy in Biomedical Research and Diagnosis of Human Diseases. Adelaide, So. Australia: World Health Organization, pp. 106-114.

Miller SE and Howell DN (1988) Viral infections in the acquired immunodeficiency syndrome. J Electron Microsc Tech 8: 41-78.

Miller SE and Lang DJ (1982) Rapid diagnosis of herpes simplex infection: amplification for electron microscopy by short term in vitro replication. J Infect 4:37-41.

Miller SE and Nielsen RA (1979) Rapid embedding: tissue to TEM in less than 2 hours. Proc. Southeastern Electron Microsc Soc 2:27.

Mills J and Corey L (1985) Antiviral Chemotherapy. New Directions for Clinical Application and Research, Vol. 1. New York: Elsevier Science Publishing Co., Inc.

Mills J and Corey L (1989) Antiviral Chemotherapy. New Directions for Clinical Application and Research, Vol. 2. New York: Elsevier Science Publishing Co., Inc.

Norval M and Bingham RW (1987) Advances in the use of nucleic acid probes in diagnosis of viral diseases of man. Arch Virol 97:151-165.

Olin BR (ed) (1990) Antiviral agents. Drug Facts and Comparisons. St. Louis: JB Lippincott Co.

Oshiro LS and Miller SE (in press). Electron microscopy. In Lennette EH (ed), Laboratory Diagnosis of Viral Infections, 2nd Ed. New York: Marcel Dekker.

Palmer EL and Martin ML (1988) Electron Microscopy in Viral Diagnosis. Boca Raton: CRC Press, Inc.

Paver WK, Caul EO, Ashley CR and Clarke SKR (1973) A small virus in human faeces. Lancet 1: 237-240.

Quinnan GV Jr (1990) Immunization against viral diseases. In Galasso GJ, Whitley RJ and Merigan TC (eds), Antiviral Agents and Viral Diseases of Man. New York: Raven Press, pp. 727-969.

Riepenhoff-Talty M, Saif LJ, Barret HJ, Suzuki H and Orga PL (1982) Potential spectrum of etiological agents of viral enteritis in hospitalized infants. J Clin Microbiol 17:352-356.

Specter S and Lancz GJ (1986) Clinical Virology Manual. New York: Elsevier Science Publishing Co., Inc.

Versteeg J and Salimans MM (1988) The human parvoviruses. In de la Maza LM and Peterson EM (eds), Medical Virology, Vol VII. New York: Elsevier Science Publishers, pp. 33-56.
Yolken RH (1990) Laboratory diagnosis of viral infections. In Galasso GJ, Whitley RJ and Merigan TC (eds), Antiviral Agents and Viral Diseases of Man. New York: Raven Press, pp. 141-181.
White DO and Fenner FJ (1986) Medical Virology. Orlando: Academic Press, Inc.

DISCUSSION

de la Maza L (University of California Irvine Medical Center, Orange, CA):

Are you offering these types of diagnostic tests on a stat basis? If somebody comes with a specimen, are you willing to perform the test in the middle of the night, with a turn-around time of 20 minutes?

Miller S:

Absolutely, and I've been called at home. We have gotten to know the pediatricians, so that when they order a test and want it stat, they call me up and say, "how do I get there." The problem with some of the larger hospitals (we are a tertiary hospital) is that there is a single collecting laboratory. The nurse collects the sample, and it gets shuffled off to Central Collection along with all the other things that get sent there: the bacteria, the fungi, the parasites, and everything else. Then they sort everything, and the courier picks it up. If you have just missed the last courier for the day, it can sit around somewhere and get lost for a while. If they physician wants something stat, (s)he calls up, and I say, "I'm in room 339 Jones; collect the sample and send it over by medical student." We pretty much can stop what we're doing and perform the test. Again, the "quick and dirty" direct method is the one that is 20-30 minutes. If we do the direct procedure and it's negative, then we start playing with the sample and will run one of the concentration procedures that takes more time. Almost always, we use the Airfuge, and then sometimes we use some of these other techniques as well.

DETECTION OF VIRAL SEQUENCES IN FORMALIN FIXED, PARAFFIN EMBEDDED TISSUES FROM HIV-1 INFECTED PATIENTS USING THE PCR

Darryl Shibata

University of Southern California School of Medicine
Los Angeles, California, USA

INTRODUCTION

The development of new techniques can further the elucidation of the biology of many diseases. The invention of the polymerase chain reaction (PCR) in the 1980's by Mullis and co-workers at the Cetus Corporation (Saiki et al. 1985) has revolutionized the field of molecular virology. The ability of the PCR to detect as few as one target present among 100,000 to 1,000,000 human cells has permitted the direct examination of *in vivo* viral infections. Prior to the PCR, extremely low numbers of virus were often "invisible" by many techniques. Culture techniques were generally the most sensitive method of direct viral detection. However, it was theoretically possible that some latent viruses would not grow in culture. Immunohistochemistry and *in situ* hybridization techniques can detect the expression of viral antigens or RNA, or the presence of multiple viral genomes. Unfortunately, these techniques suffer from various drawbacks. In particular, latent viral infection with small numbers of viral genomes (generally less than 50 copies per cell) would escape detection. The Southern blot technique can detect viral DNA regardless of its state of activation, but generally needs at least one viral genome present per 100 cells for detection.

Therefore, there existed a vast "sensitivity" gap in the techniques for viral detection. For many viruses, especially the herpes family, lifelong persistent infection was known to exist but direct detection of the latent *in vivo* infection was extremely difficult. Because of this "sensitivity" gap, many different theories on viral persistence and burden could be postulated but not proven.

The PCR, with its markedly increased sensitivity, can directly measure the low viral levels present in many human infections. Human immunodeficiency virus type 1 (HIV-1), Epstein-Barr virus (EBV), and human cytomegalovirus (CMV) are three persistent infections which are commonly simultaneously present and have been difficult to detect *in vivo*. The PCR has facilitated study of these important human pathogens.

Medical Virology 10, Edited by L.M. de la Maza and
E.M. Peterson, Plenum Press, New York, 1991

PCR ON PARAFFIN EMBEDDED TISSUES

The PCR, since it replicates only a small segment of DNA (typically 200-1,000 base pairs or less), does not require high molecular weight DNA as a substrate. Very degraded DNA templates can be utilized. Indeed, the PCR has been used to sequence very ancient specimens (Paabo, 1989). Because of this ability, many unconventional sources of DNA may be analyzed.

Most medical tissue specimens are fixed in buffered formalin, embedded in paraffin, and then sectioned for microscopic examination. This technique has been performed for many decades and will continue as the most common method of tissue preservation and analysis for the near future. Therefore, techniques which can utilize this common source of human tissues would be most valuable for both retrospective and prospective studies.

The DNA present in formalin fixed, paraffin embedded tissues can be extracted. In some cases (Dubeau et al. 1986; Goelz et al. 1985), the DNA is intact enough for restriction enzyme analysis, although in many cases the DNA is too size degraded for the Southern blot. The PCR, however, can utilize the size degraded DNA as an amplification substrate (Imparin et al. 1987). Because only very small amounts of DNA are needed, a single thin (5-10 μm) section similar in thickness to those cut for microscopic examination can be used (Shibata et al. 1988a) The analysis of a single section has the advantage that the DNA in the lesion examined can be defined by a "mirror" adjacent section stained for histological examination. In addition, the bulk of the fixed tissue can be saved for future studies or for confirmation of the original PCR results.

The extraction of the DNA from the paraffin blocks is extremely quick (Shibata et al. 1989a; Wright and Manos, 1990). A single section is placed in a microfuge tube, and then deparaffinized with an organic solvent (xylene or octane). The tissue is then incubated with a proteinase K solution overnight at 37°C or several hours at 50°C. The aqueous DNA containing phase can be directly added to the PCR tube after boiling to inactivate the proteinase K. Although further purification of the DNA is possible, this procedure usually yields an excellent PCR substrate and its simplicity helps avoid possible problems with specimen contamination.

The best amplification is obtained from paraffin tissues fixed in 10% buffered formalin or ethanol. Other fixatives are generally not as suitable for analysis (Greer et al. in press; Rogers et al. 1990). Prolonged fixation (greater than 2-3 days) should be avoided. The PCR on the DNA extracted from the fixed sections is generally not as efficient compared to purified DNA preparations. Therefore, additional PCR cycles (typically 30-50 cycles) are generally required to obtain comparable sensitivity. Studies on mixtures of fixed and paraffin cell lines have demonstrated that a sensitivity of at least one target in 1,000 cells (Shibata et al. 1988b). The PCR products are generally not as "clean" as those produced from high molecular weight DNA. Electrophoresis of the PCR products will often reveal multiple non-specific bands or a background of non-specific amplification.

Despite these problems, PCR is generally more sensitive than other current methods of nucleic acid analysis on fixed tissues. The analysis of archival fixed tissues allows retrospective epidemiologic studies. For example, human papillomavirus types 16 and 18 were amplified by PCR from fixed cervical cancer specimens from the 1940's (Shibata et al. 1988d), and HIV-1 has been detected in the fixed autopsy tissues of an English seaman who died in 1959 (Corbitt et al. 1990). Also, since tissues are routinely fixed in formalin and paraffin embedded, a large variety of specimens, including very rare or unusual lesions can be easily collected. As a minor but very important experimental point, the fixed tissues are biologically safe.

HIV-1 INFECTIONS

The investigation of latent human viral infections is hampered by the low numbers of virus present and the inability to readily obtain many human tissues. The detection of these latent viruses is most easily accomplished by measuring the host immune response to the infection. Antibodies produced against viral antigens can be detected in the vast majority of latent infections. The direct detection of infected cells would be invaluable for the further study of latency.

HIV-1 is a retrovirus which causes immunodeficiency, resulting in acquired immunodeficiency syndrome (AIDS) and death (Fauci, 1988). HIV-1 infects CD 4 positive cells which are commonly T-helper lymphocytes or monocytes. Its RNA genome is replicated by the viral reverse transcriptase into a DNA proviral copy. Thereafter, the virus many be latent with little or no expression or active with viral replication and cell death.

Direct detection of HIV-1 was initially difficult leading to the impression that very low numbers of HIV-1 were present. Indeed, this difficulty was taken as evidence that HIV-1 may not be the cause of AIDS (Duesberg, 1988). With improved cell culture techniques such as the recognition of the inhibition caused by CD 8 positive cells (Walker et al. 1986), HIV-1 could be cultured from the majority of antibody positive individuals (Ho et al. 1989). Southern blot analysis of tissues from HIV-1 infected individuals could only detect HIV-1 from a minority of specimens (Shaw et al. 1984). *In situ* nucleic acid hybridization techniques on peripheral blood and lymph nodes could demonstrate HIV-1 in only a small minority of lymphocytes, approximately one in 10,000 to 100,000 cells (Biberfeld et al. 1986; Harper et al. 1986). These data provided a paradox. How could HIV-1 cause such a profound immunodeficiency if only a minority of T-helper cells were infected?

This impression of an extremely low HIV-1 burden appears to be incorrect and most likely reflects the limitations of the techniques employed (Ho et al. 1989; Schnittman et al. 1989; Shibata et al. 1989). The Southern blot has a sensitivity of approximately one HIV-1 provirus in 100 cells, and therefore, specimens with virus present in lesser amount will be falsely negative. The PCR, with its higher sensitivity, is a better technique for the detection of HIV-1 provirus.

PCR detection of HIV-1 was first reported by Kwok et al. (1987). PCR could detect HIV-1 from the majority of peripheral blood specimens from HIV-1 infected individuals (Ou et al. 1988). The number of infected peripheral blood mononuclear cells could be estimated by PCR after serial dilutions of purified population of intact CD 4 positive cells (Psallidopoulos et al. 1989; Schnittman et al. 1989). It was found that HIV-1 could be amplified from as few as 100 CD 4 cells, implying that at least one in one hundred of these cells were HIV-1 infected. The numbers of cells expressing HIV-1 was approximately ten-fold less. The proportion of HIV-1 infected cells was dependent on the stage of the HIV-1 infection. Individuals with AIDS had the highest levels of infection (at least 1%) while asymptomatic HIV-1 infected individuals had lower levels of infection (0.003 to 1%, mode 0.2%) of their CD 4 cells. Other investigators (Simmonds et al. 1990) also found similar levels of HIV-1 infected peripheral blood mononuclear cells and by comparing the PCR on serial dilutions of extracted DNA versus intact cells, estimated that most of the infected cells contained a single copy of provirus. Similar findings of relatively high levels of HIV-1 in the plasma and peripheral blood mononuclear cells of AIDS patients with lower levels in asymptomatic patients were documented by culture (Ho et al. 1989).

The CD 4 positive or T-helper cells present in the peripheral blood represent a mobile element of the HIV-1 infection and are a small minority (1%) of the total body lymphoid population. The majority of lymphoid cells do not circulate but remain for variable amounts of time in lymph nodes. Part of the spectrum of HIV-1 infection includes enlarged lymph nodes, commonly designated as persistent generalized lymphadenopathy (PGL). The histology of these lymph nodes consists of florid follicular and interfollicular B-cell hyperplasia (Ewing et al. 1985). The etiology of PGL is unknown. With progression of the disease, PGL tends to regress with follicular atrophy (Turner et al. 1987). HIV-1 provirus could detected from only 21% of fresh PGL biopsies by the Southern blot (Shaw et al. 1984) and *in situ* hybridization could identify HIV-1 in all PGL biopsies but only one in 10,000 to 100,000 cells appeared infected. The numbers of infected cells appeared to decrease with progressive disease (Biberfeld et al. 1986). In contrast, using, using PCR, HIV-1 could be detected from all fixed PGL biopsies (Shibata et al. 1989). PCR after serial ten-fold limiting dilutions of the DNA extracted from these PGL biopsies demonstrated that the number of proviral copies was approximately one per 100 to 10,000 cells. The PGL biopsies from patients with or without AIDS had similar numbers of HIV-1 provirus. Biopsies with follicular involution also had similar numbers of HIV-1 provirus as compared to biopsies with florid follicular hyperplasia. Because multiple proviral copies may be present in one cell (Shaw et al. 1984), the proportion of infected lymph node cells cannot be estimated by this analysis. However, serial dilutions of intact lymph node cells also gave similar results (Shibata, unpublished observations), suggesting that the proportion of infected lymph node cells is approximately 0.01 to 1%.

These PCR studies demonstrate that the extent of HIV-1 latent infection is greater than previously estimated. Although the majority of lymphoid cells are uninfected, it is difficult to place these findings in their proper perspective since the extent of other latent viral infections have not been previously measured. The study of other latent viral infections by similar methods would help improve out understanding of the significance of the HIV-1 data.

CMV INFECTIONS

CMV is a member of the herpes virus family. CMV infections are common with over half of the adults infected in most populations. The acute infection may be relatively asymptomatic or present as acute heterophil negative infectious mononucleosis. After the initial infection, there is a lifelong persistence of CMV reflecting a balance between viral replication and host immunity. In most infected individuals, no further CMV disease occurs, although CMV can be cultured from the urine and cervix of asymptomatic individuals. Latent infection is usually detected by the presence of serum antibodies against CMV antigens. Latent CMV infections can reactivate in immunodeficient hosts and produce significant morbidity and mortality.

Detection of CMV infected cells during the latent asymptomatic infection has been difficult. Indeed, the cell type which is the reservoir for the latent CMV infection has not been identified with any certainty as CMV can infect virtually any cell type including granulocytes, monocytes, and lymphocytes (Braun and Reiser, 1986; Einhorn and Ost, 1984; Rice et al. 1984; Schrier et al. 1985). Cultures of peripheral blood for CMV are negative in asymptomatic seropositive individuals. Similarly, studies of normal kidneys have not identified CMV infected cells. Despite these negative studies, CMV must be present in peripheral blood since transfusion from a CMV seropositive donor is associated with a risk of seroconversion in the seronegative recipient (Tegtmeier, 1985). CMV must also be present in the kidney since transplantation of a kidney from a seropositive individual to a seronegative recipient is associated with CMV infection (Gann et al. 1988).

It was a theoretical possibility since CMV is a latent virus which is present in most individuals, that the PCR would have little utility since it would provide information similar to serologic assays - CMV would be detected in all individuals with antibodies against CMV. This has not been the case. For example, CMV was not detected by PCR in the peripheral blood of normal CMV seropositive individuals, but could be easily detected in CMV culture positive blood specimens from immunocompromised individuals (Shibata, 1988c) or the urine from neonates (Demmler et al. 1988).

The lack of detectable CMV in the blood of "normal" CMV seronegative or seropositive individuals by the PCR prompted a more intense search for the site of CMV latency. Fixed autopsy tissues provide a readily available source of virtually any tissue for analysis. However, CMV could not be readily detected by the PCR from any autopsy tissue, including the brain, or lymph nodes. Some very faint CMV PCR amplification signals could be detected in the lung or adrenals of some autopsies (Shibata, unpublished data), but these patients had been ill for long periods before death and may have been immunocompromised. The difficulty in demonstrating convincing CMV PCR amplification in the "normal" autopsies then lead to the analysis of autopsies and lymph node biopsies from HIV-1 infected individuals.

CMV could not be detected in the fixed PGL biopsies, even in patients with AIDS. The tissues from nine AIDS autopsies were amplified for both HIV-1 and CMV sequences. Both CMV and HIV-1 could be detected in selected tissues. For CMV, two general patterns emerged which could be correlated with histologic findings (Shibata and Klatt, 1989). If characteristic Cowdry type A inclusions typical for CMV infected cells were absent in the entire autopsy, most of the tissues were CMV PCR negative with the exception of the lung and adrenals. It appears that detectable CMV reactivation in the immunocompromised host first occurs in the adrenal glands or lung. Interestingly, histologic CMV inclusions are also most often present in these two organs in AIDS autopsies (Klatt and Shibata, 1989).

In contrast, if Cowdry type A inclusions were present in any tissue, then all tissues (skin, colon, heart, lymph node, liver, spleen, ovary, testes, prostate, cervix) were positive for CMV, probably reflecting CMV viremia at the time of death since the CMV PCR would detect both infected cells and cell free virions. This study demonstrates the value of autopsy studies in which the entire body can be studies with molecular techniques ("molecular autopsy"). In the clinical setting, CMV PCR of isolated tissue biopsies may not be meaningful since in the presence of CMV viremia, all tissues may be positive and not specifically infected. Further investigation will be necessary to extend these findings to the tissue biopsies from living patients.

In these same autopsy specimens, HIV-1 could be detected in all lymph nodes, half of the spleens, and in the single thymus examined. All the lymph nodes were depleted of follicles, reflecting the histology (Ewing et al. 1985) and profound immunodeficiency usually present at autopsy of AIDS patients. HIV-1 was also detected in some brain specimens. HIV-1 provirus was detected in a minority of other organs and usually in tissues with chronic inflammation. Although HIV-1 was more often detected in tissues with chronic inflammation and likely present in mononuclear cells, the presence of HIV-1 could not be reliably predicted by morphology. The HIV-1 PCR positive tissues had variable numbers of mononuclear cells and less than half of the tissues with chronic inflammation had detectable HIV-1 provirus. These findings suggest that HIV-1 infected cells are commonly present in lymphoid tissue and variably present at sites of inflammation. Unlike CMV, massive dissemination of HIV-1 was not detected although the PCR would not detect circulating HIV-1 virions since they contain RNA.

EBV INFECTIONS

EBV is another member of the herpes family (Thorley-Lawson, 1988). EBV infections are widespread in most populations and most HIV-1 infected individuals are also EBV infected. EBV usually infects B-lymphoid or nasopharyngeal cells, although T-cells can also be infected. The common clinical manifestation of EBV related disease is the acute infection which can be asymptomatic or result in infectious mononucleosis. Except in immunodeficient individuals, acute EBV infection seldom produces significant morbidity. Similar to CMV infections, after the acute infection, a persistent lifelong infection reflecting a balance between viral replication and host immunity is established. In the majority of individuals, the latent infection remains asymptomatic, although in some cases, EBV containing cells can become neoplastic. Familiar examples are African Burkitt's lymphoma and nasopharyngeal carcinoma. In addition, EBV related B-cell lymphoproliferations can occur in immunocompromised individuals (transplant and HIV-1 infected patients). EBV is thought to play a critical role in the etiology of these neoplasms, but the exact mechanism is unknown.

The numbers of EBV infected cells appears to be quite small during a normal latent infection. Using culture techniques, relatively large numbers of EBV infected cells can be found in the peripheral blood during an acute infection. Afterwards, with the onset of host immunity, the number of EBV infected cells declines to approximately one EBV infected cell per one million mononuclear cells (Rocci et al. 1977). Using nucleic acid hybridization techniques (Southern blot and *in situ*), EBV infected cells can be easily detected during acute infections and in EBV related proliferations (Niedobitek et al. 1989; Weiss and Movahed, 1989), but the detection of EBV infected cells during a normal latent infection has been generally unsuccessful (Staal et al. 1989; Weiss and Movahed, 1989).

Similarly, the detection of EBV by PCR in the blood and fixed lymph node biopsies of latently infected normal individuals has also been generally unsuccessful. Using PCR on purified DNA, EBV was detected from a minority of blood and lymph node specimens from normal individuals, but could be detected in a larger proportion of specimens from patients with Sjorgren's syndrome (Saito et al. 1989). EBV could also not be detected from ten formalin fixed, paraffin embedded lymph node biopsies or 20 spleens from normal individuals (Peiper et al. 1990). In our laboratory, it has been difficult to detect EBV in the peripheral blood or fixed lymph node biopsies from normal individuals (Shibata, unpublished data). In contrast, one study could detect EBV in the peripheral blood from 54% of healthy normal individuals and from 86% of HIV-1 infected individuals (Gopal, 1990). These differences in the ability to detect EBV in normal individuals may reflect differences in techniques and sensitivity since culture studies have demonstrated that very low levels of EBV infected cells do circulate. Apparently, for most investigators, the numbers of EBV infected cells present during normal latent infections are usually too small for PCR detection.

EBV can be amplified from throat washings obtained from about 25% of normal individuals (Gopal et al. 1990; Sixbey et al. 1990) and from some biopsies of tonsils or the nasopharynx (Shibata, unpublished data). In addition, EBV has been detected from EBV related neoplasms such as nasopharyngeal carcinoma and high grade non-Hodgkin's lymphomas, and infectious mononucleosis using PCR (Peiper et al. 1990).

HIV-1 infected individuals may have varying degrees of immunodeficiency which impair the suppression of latent EBV infections. The levels of EBV infected peripheral blood mononuclear cells increases ten-fold in HIV-1 infected individuals to approximately one in 100,000 cells as detected by culture techniques (Brix et al. 1986). HIV-1 infected individuals are also at increased risk for the development of high grade B-cell non-Hodgkin's lymphoma (Levine et al. 1984). Approximately one half of these lymphoma contain EBV sequences. Monoclonal or oligoclonal immunoglobulin gene rearrangements can be detected in approximately 20% of PGL biopsies (Pelicci et al. 1986), suggesting that the PGL lymphoid hyperplasia may be a prodrome for lymphoma. In addition, reactivation of EBV may also occur in the PGL biopsies, accounting for the B-cell hyperplasia and an increased risk of EBV containing lymphoma.

EBV can be detected by the PCR from approximately one third of the fixed PGL biopsies from HIV-1 infected individuals without AIDS (Shibata et al. in press). EBV was also detected by *in situ* hybridization in the same biopsies, supporting the PCR data. The numbers of EBV genomes appeared small by PCR dilution analysis and no more than 1% of the cells were positive by *in situ* hybridization. The presence of EBV in the PGL biopsies was associated with an increased risk for the development or the concurrent presence of EBV-related lymphoma. The EBV infected cells detected in these PGL biopsies associated with lymphoma may represent the background of EBV-related lymphoproliferation which can lead to the development of lymphoma or may represent histologically occult involvement with EBV containing lymphoma cells.

LATENT VIRAL INFECTIONS - HIV-1, EBV, AND CMV

Latent viral infections may be best understood in the context of comparisons between these common human infections. The HIV-1 epidemic has focused much attention and research on latent viral infections. The PCR has allowed the sensitive detection of very low numbers of actively or latently infected cells.

CMV and EBV represent two herpes viruses whose association with humanity is worldwide and probably ancient. With a few exceptions the majority of individuals can be infected and survive the acute infection with minimal morbidity or mortality. Despite the host immune response, the viruses are able to persist throughout the lifetime of the individuals. Reactivation of the infections can occur in the setting of immunodeficiency and produce considerable morbidity and mortality. In addition, EBV infection may directly contribute to the development of certain neoplasms. In the majority of individuals, these two viruses remain forever latent at levels which represent a balance between viral replication and host immunity. Studies using the PCR have demonstrated that this level is remarkably low and in most cases, these viruses have been extremely difficult to detect from tissues of normal individuals.

In contrast, HIV-1 infections have emerged acutely in the 1980's into the current epidemic. HIV-1 infections have not yet spread throughout the world and are largely confined to specific social groups. Similar to the herpes viruses, HIV-1 maintains a persistent and apparently lifelong infection. HIV-1, after five to ten years, inevitably kills its host through the destruction of its immune system. Unlike the herpes viruses, the host does not appear to respond adequately to the initial infection. HIV-1 provirus can be easily detected in the tissues and blood of asymptomatic and symptomatic individuals.

In particular, the enlarged PGL lymph nodes commonly present in HIV-1 infected individuals contain without exception relatively large amounts of provirus. In contrast, despite some degree of immunodeficiency, CMV and EBV are not present in similar numbers in the majority of such hyperplastic lymph nodes. The etiology of PGL is unknown, but apparently for the majority of individuals, it is not secondary to an increase of EBV or CMV infected cells. HIV-1 can stimulate B-cell proliferation (Schnittman et al. 1986) and may be the direct cause of the PGL hyperplasia. Certain growth factors produced by HIV-1 infected monocytes and T-lymphocytes, such as IL-6 (Groopman et al. 1989; Nakakima et al. 1989), may be operative in this regard.

In conclusion, by comparing EBV, CMV and HIV-1 infections, it can be seen that although the numbers of HIV-1 infected cells are small, other latent viral infections are characterized by even lower levels of infections. HIV-1 therefore never established a true latency analogous to EBV or CMV since significant numbers of HIV-1 infected cells are not eliminated by the host immune response. The differences in the biology of the viruses (retrovirus versus herpes virus) and their target cells, as well as the longer historical adaptation between humanity and the herpes viruses may account for these differences.

REFERENCES

Biberfeld P, Chayt KJ, Marselle LM,, Biberfeld G, Gallo RC, Harper ME (1986) HTLV-III expression in infected lymph nodes and relevance to pathogenesis of lymphadenopathy. Am J Pathol 125:436-432.

Birx DL, Redfield RR, Tosato G (1986) Defective regulation of Epstein-Barr virus infection in patients with acquired immunodeficiency syndrome (AIDS) or AIDS-related disorders. N Engl J Med 314:874-879.

Braun RW, Reiser HC (1986) Replication of human cytomegalovirus in human peripheral blood T cells. J Virol 60:29-36.

Corbitt G, Bailey AS, Williams G (1990) HIV infection in Manchester, 1959. Lancet 336:51.

Demmler GJ, Buffone GJ, Schimbor CM, May RA (1988) Detection of cytomegalovirus in urine from newborns using polymerase chain reaction DNA amplification. J Infect Dis 158:1177-1184.

Dubeau L, Chandler LA, Gralow JR, Nichols PW, Jones PA (1986) Southern blot analysis of DNA extracted from formalin-fixed pathology specimens. Cancer Res 46:2964-2969.

Duesberg P (1988) HIV is not the cause of AIDS. Science 241:514.

Einhorn L, Ost A (1984) Cytomegalovirus infection of human blood cells. J Infect Dis 149:207-214.

Ewing EP, Chandler FW, Spira TJ, Byrnes RK, Chan WC (1985) Primary lymph node pathology in AIDS and AIDS-related lymphadenopathy. Arch Pathol Lab Med 109:977-981.

Fauci AS (1988) The human immunodeficiency virus: infectivity and mechanisms of pathogenesis. Science 239:617-622.

Gann JW, Ahlmen J, Svalander C, Olding L, Oldstone MBA, Nelson JA (1988) Inflammatory cells in transplanted kidneys are infected with human cytomegalovirus. Am J Pathol 132:239-248.

Goelz SE, Hamilton SR, Volgelstein B (1985) Purification of DNA from formaldehyde-fixed and paraffin-embedded human tissue. Biochem Biophys Res Commun 130:118-126.

Gopal MR, Thompson BJ, Fox J, Tedder RS, Honess (1990) Detection by PCR of HHV-6 and EBV DNA in blood and oropharynx of healthy adults and HIV-seropositives. Lancet 1:1598-1599.

Greer CE, Peterson SL, Kiviat NB, Manos MM (in press) PCR amplification from paraffin-embedded tissues: effects of fixative and fixation time. Amer J Clin Pathol.

Groopman JE, Molina JH, Scadden DT (1989) Hematopoietic growth factors. Biology and clinical applications. N Engl J Med 321:1449-1459.

Harper ME, Marselle LM, Gallo RC, Wong-Staal F (1986) Detection of lymphocytes expressing human T-lymphotropic virus type III in lymph nodes and peripheral blood from infected individuals by *in situ* hybridization. Proc Natl Acad Sci 83:772-776.

Ho DD, Moudgil T, Alam M (1989) Quantitation of human immunodeficiency virus type I in the blood of infected persons. N Engl J Med 321:1621-1625.

Imparin CC, Saiki RK, Erlich HA, Teplitz RL (1987) Analysis of DNA extracted from formalin-fixed, paraffin-embedded tissues by enzymatic amplification and hybridization with sequence-specific oligonucleotides. Biochem Biophys Res Commun 142:710-716.

Kwok S, Mack DH, Mullis KB, Poiesz B, Ehrlich G, Blair D, Friedman-Kein A, Sninsky JJ (1987) Identification of human immunodeficiency virus sequences by using *in vitro* enzymatic amplification and oligomer cleavage detection. J Virol 61: 1690-1694.

Levine AM, Meyer PR, Begandy MK, Parker JW, Taylor CR, Irwin L, Lukes RJ (1984) Development of B-cell lymphoma in homosexual men. Ann Intern Med 100:7-13.

Nakajimak, K, Martinez-Maza O, Hirano T, Breen EC, Nishanian PG, Salazar-Gonzales JF, Fahey JL, Kishimoto T (1989) Induction of IL-6 production by HIV. J Immunol 142:531-516.

Niedobitek G, Hamilton-Dutoit S, Herbst H, Finn T, Vetner M, Pallesen G, Stein H (1989) Identification of Epstein-Barr virus infected cells in tonsils of acute infectious mononucleosis by *in situ* hybridization. Hum Pathol 20:796-799.

Ou CY, Kwok S, Mitchell SW, Mock OH, Sninsky JJ, Krebs JW, Feorino P, Warfield O, Schochetman G (1988) DNA amplification for direct detection of HIV-1 in DNA of peripheral blood mononuclear cells. Science 239:295-297.

Paabo S (1989) Ancient DNA: extraction, characterization, molecular cloning, and enzymatic amplification. Proc Natl Acad Sci 86:1939-1943.

Pelicci P-G, Knowles DM, Arlin ZA, Wieczopek R, Lucin P, Dina D, Basilico C, Dalla-Favera R (1986) Multiple monoclonal B cell expansions and c-myc

oncogene rearrangements in acquired immune deficiency syndrome-related lymphoproliferative disorders. J Exp Med 164:2049-2060.
Psallidopoulos MC, Schnittman SM, Thompson LM, Baseler M, Fauci AS, Lane HC, Sulzman NP (1989) Integrated proviral HIV type 1 is present in CD4+ peripheral blood lymphocytes in healthy seropositive individuals. J Virol 63:4626-4631.
Rice GPA, Schrier RD, Oldstone, MBA (1984) Cytomegalovirus infects human lymphocytes and monocytes: virus expression is restricted to immediate-early gene products. Proc Natl Acad Sci USA 81:6134-6138.
Rochi G, de Felici A, Ragona G, Heinz A (1977) Quantitative evaluation of Epstein-Barr-virus-infected mononuclear peripheral blood leukocytes in infectious mononucleocsis. N Engl J Med 296:1332-1334.
Rogers BB, Alpert LC, Hine EAS, Buffone GJ (1990) Analysis of DNA in fresh and fixed tissue by the polymerase chain reaction. Am J Pathol 136:541-548.
Saiki RK, Scharf S, Faloona F, Mullis KB, Horn GT, Erlich HA, Arnheim N (1985) Enzymatic amplification of β-globin genomic sequences and restriction site analysis for diagnosis of sickle cell anemia. Science 230:1350-1354.
Saito I, Servenius B, Compton T, Fox RI (1989) Detection of Epstein-Barr virus DNA by polymerase chain reaction in blood and tissue biopsies from patients with Sjogren's syndrome. J Exp Med 169:2191-2198.
Schnittman SM, Lane HC, Higgins SE, Folks T, Fauci AS (1986) Direct polyclonal activation of human B lymphocytes by the acquired immune deficiency syndrome virus. Science 233:1084-1086.
Schnittman SM, Psallidopoulos MC, Lane HC, Thompson L, Baseler M, Massari F, Fox CH, Salzman NP, Fauci AS (1989) The reservoir for HIV-1 in human peripheral blood is a T cell that maintains expression of CD4. Science 245:305-308.
Schrier RD, Nelson JA, Oldstone MBA (1985) Detection of human cytomegalovirus in peripheral blood lymphocytes in a natural infection. Science 30:1058-1051.
Shaw GM, Hahn BH, Arya SK, Groopman JE, Gallo RC, Wong-Staal F (1984) Molecular characterization of human T-cell leukemia (lymphotrophic) virus type III in the acquired immune deficiency syndrome. Science 226:1165-1171.
Shibata D, Arnheim N, Martin WJ (1988a) Detection of human papilloma virus in paraffin-embedded tissues using the polymerase chain reaction. J Exp Med 167:225-230.
Shibata D, Fu YS, Gupta JW, Shah KV, Arnheim W, Martin WJ (1988b) The detection of human pappillomavirus in normal and dysplastic tissue by the polymerase chain reaction. Lab Invest 59:555-559.
Shibata D, Martin WJ, Appleman M, Cousey D, Leedom J, Arnheim N (1988c) Detection of cytomegalovirus DNA in peripheral blood of patients infected with human immunodeficiency virus. J Infect Dis 158:1185-1192.
Shibata D, Martin WJ, Arnheim N (1988d) Analysis of DNA sequences in forty-year-old paraffin-embedded thin-tissue sections: a bridge between molecular biology and classical histology. Cancer Res 48:4564-4566.
Shibata D, Brynes RK, Nathwani BN, Kwok S, Sninsky J, Arnheim N (1989) Human immunodeficiency viral DNA is readily found in lymph nodes biopsies from seropositive individuals: analysis of fixed tissue using the polymerase chain reaction. Am J Pathol 135:697-702.

Shibata D, Klatt EC (1989) Analysis of human immunodeficiency virus and cytomegalovirus infection by polymerase chain reaction in the acquired immunodeficiency syndrome. Arch Pathol Lab Med 113:1239-1244.

Shibata D, Weiss LM, Nathwani BN, Brynes RK, Levine AM (in press) Epstein-Barr virus in benign lymph node biopsies from individuals infected with the human immunodeficiency virus is associated with the concurrent or subsequent development of non-Hodgkin's lymphoma.

Simmonds P, Balfe P, Peutherer JF, Ludlam CA, Bishop JO, Brown AJ (1990) Human immunodeficiency virus-infected individuals contain provirus in small numbers of peripheral mononuclear cells and at low copy number. J Virol 64:864-872.

Sixby JW, Shirly P, Chesney PJ, Buntin DM, Resnick L (1989) Detection of a second widespread strain of Epstein-Barr virus. Lancet 2:761-765.

Staal SP, Ambinder R, Beschorner WE, Hayward GS, Mann R (1989) A survey of Epstein-Barr virus DNA in lymphoid tissue. Am J Clin Pathol 91:1-5.

Tegtmeier GE (1985) Cytomegalovirus and blood transfusion. In Dodd RY, Barker LF (eds) Infection, Immunity and Blood Transfusion. New York, Alan R. Liss, pp. 175-199.

Thorley-Lawson DA (1988) Basic virologic aspects of Epstein-Barr virus infection. Sem Hematol 25:247-260.

Turner RR, Levine AM, Gill PS, Parker JW, Meyer PR (1987) Progressive histopathologic abnormalities in the persistent generalized lymphadenopathy syndrome. Am J Surg Pathol 11:625-632.

Walker CM, Moody DJ, Stites DP, Levy JA (1986) CD8+ lymphocytes can control HIV infection *in vitro* by suppressing virus replication. Science 234:1563-1566.

Weiss LM, Movahed LA (1989) *In situ* demonstration of Epstein-Barr Viral genomes in viral-associated B cell lymphoproliferations. Am J Pathol 134:651-659.

Wright DK, Manos MM (1990) Sample preparation from paraffin-embedded tissues. In Innis MA, Gelfand DH, Sninsky JJ, White TJ (eds) PCR Protocols: A guide to Methods and Applications. San Diego, Academic Press, pp. 153-158.

DISCUSSION

Peterson E (University of California Irvine Medical Center, Orange, CA):

On the seronegative, PCR positive HIV patients, what do you do in your laboratory to assure that it is a true HIV signal?

Shibata D:

We have not encountered this situation except in one case. In this case we couldn't get the patient back for re-biopsy or follow-up, so what we did was analyze several paraffin blocks from his lymph node biopsy. All the blocks, repeated multiple times, were still HIV positive. The HIV present was a tremendous amount and was very easy to detect.

Peterson E:

Do you chart your results? What are you doing with your results as far as treatment and history of the patient, etc?

Shibata D:

There has been some problem in the sense that our Chairman talks about licensed tests versus non-licensed tests. What I do now is verbally communicate, because I think there can be a large misinterpretation of what these tests mean. I don't do any PCR testing unless I actually talk to a clinician before hand so that they understand what the results might mean.

Peterson E:

Do those go on the chart from the Pathology laboratory, or is it just a verbal communication that it's positive?

Shibata D:

There is another laboratory at our institution that does put results in the chart. I mainly do research, so I don't generally chart these results. I think I would, though, if they said I should.

de la Maza L (University of California Irvine Medical Center, Orange, CA):

One of the big problems with PCR is cross contamination. When you are cutting sections in a microtome, I assume that people don't take much care in cleaning up, and the technologists are processing thousands of specimens. How are you handling this problem?

Shibata D:

Contamination is a real problem in PCR and probably is preventing the wide-spread application in clinical medicine in the sense that you can go wrong very fast. Handling microtomes, it turns out, that the contamination doesn't seem to be much of a problem because when one takes these blocks, one has to produce a flat surface, so when one starts cutting the blocks, one cuts a lot of wax before you finally get to the tissue. That tends to self clean the blade. You just have to take certain precautions like washing your hands in between. You do try and keep the area clean. You instruct the technician in special techniques. I generally don't give it to any histotech and tell her to cut it. I have specially trained histotechs which usually cut multiple sections so we can confirm results from independent isolations.

DETECTION OF VIRAL NUCLEIC ACIDS BY Qβ REPLICASE AMPLIFICATION

Cynthia G. Pritchard and James E. Stefano

GENE-TRAK Systems
Framingham, Massachusetts, USA

INTRODUCTION

The need for clinical assays of exquisite sensitivity is well established. For example, the ability to detect reliably a single infected cell present in a sample is critical in screening blood supplies for human immunodeficiency virus type 1 (HIV-1). Direct detection of nucleic acid sequences specific for the pathogen circumvents several problems associated with other assay methods. First, isolation of the organism responsible for an infection is not always possible. Second, antigen levels may be extremely low in the initial stages of infection. Finally, indirect detection is complicated by antibody titers that may be low and that may fluctuate during the course of infection. A number of methods for direct detection of low levels of particular nucleic acids have been described. Prominent among these are: the polymerase chain reaction (PCR) (Saiki et al. 1988) the transcription-based amplification system (TAS) (Kwoh et al. 1989); and the ligase chain reaction (LCR) (Backman and Wang, 1989). We have explored an alternative method of signal amplification which employs Qβ replicase to exponentially amplify probe molecules that have been bound to target nucleic acids.

Qβ replicase is a four subunit RNA-dependent RNA polymerase that is assembled in *Escherichia coli* either during natural infection by bacteriophage Qβ (Haruna and Spiegelman, 1965a) or upon induction of recombinant *E. coli* bearing cDNA copies of the 65 kilodalton (kD) phage-encoded subunit on expression plasmids (Mills, 1988) (Figure 1). With an additional host-encoded protein (host factor), the enzyme is highly template selective, replicating only the 4.2 kilobase RNA genome of the phage (Haruna and Spiegelman, 1965b). The replicase plus host factor uses the viral (+) strand as a template to direct the synthesis of a complementary (-) strand. Since both the parent and progeny RNA molecules can serve as templates for the synthesis of additional (+) and (-) strands, an exponential increase is observed in the number of strands, until there are enough strands to saturate the available enzyme molecules (Spiegelman et al. 1968; Weissman et al. 1968). Thereafter, synthesis proceeds in a linear fashion.

Medical Virology 10, Edited by L.M. de la Maza and
E.M. Peterson, Plenum Press, New York, 1991

The enzyme also is capable of replicating a limited number of small RNA molecules between 80 and 300 nucleotides in length (Kacian et al. 1972; Mills et al. 1975; Schaffner et al. 1977), many of which were isolated originally from *in vitro* Qβ replicase reactions to which template had not been added. Replication of these "variant" RNAs does not require host factor, and many are excellent templates for Qβ replicase, which copies them in an exponential manner similar to Qβ phage RNA. The best studied of these RNA molecules is midivariant (MDV-1), a 221 nucleotide RNA (Figure 2). In our hands, each round of replication of MDV-1 takes about 20 sec. A single molecule of this RNA will yield 10^{12} progeny strands in approximately 13 min, an amount of RNA which is detectable by a variety of simple means such as fluorescence.

Lizardi and colleagues (1988) found that additional sequence could be inserted into MDV-1 RNA, producing molecules that replicate faithfully in an exponential manner, with rates similar to those of the parent molecule. These observations suggested that such Qβ templates could be useful as amplifiable reporter probes in hybridization assays. Before these assays could be proven feasible, however, a number of technical obstacles dealing with assay background had to be overcome. 1) Unlike target amplification techniques such as PCR, the sensitivity of assays in which a hybridized probe is amplified will be limited by non-specifically bound reporter probes which may be carried through the hybridization process. These non-hybridized probes will be replicated and will generate a signal in the absence of target nucleic acid (Lomeli et al. 1989). We have employed a powerful method of background reduction called reversible target capture (RTC) to decrease levels of non-specifically bound reporter probe (Hunsacker et al. 1989; Morrissey and Collins, 1989; Morrissey et al. 1989). 2) All exquisitely sensitive assay methods have the common problem of detecting low level contaminants from the environment, including false positives caused by cross-contamination of specimens. For assays with an amplification step, stringent controls for production of reagents,

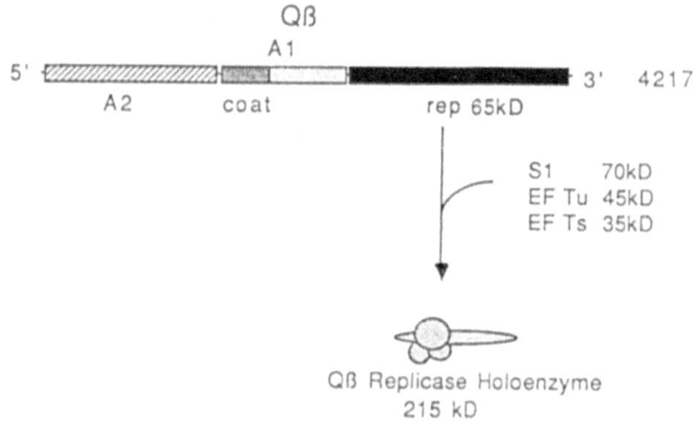

Figure 1. Schematic representation of the RNA genome of Qβ phage. The proteins for which each segment of the genome codes are indicated. The 215 kD Qβ replicase holoenzyme is composed of four proteins. Three are from the *E. coli* host: ribosomal protein subunit S1, and two protein synthesis elongation factors, EF-Tu and EF-Ts. Only one subunit is coded for by the phage RNA (65 kD).

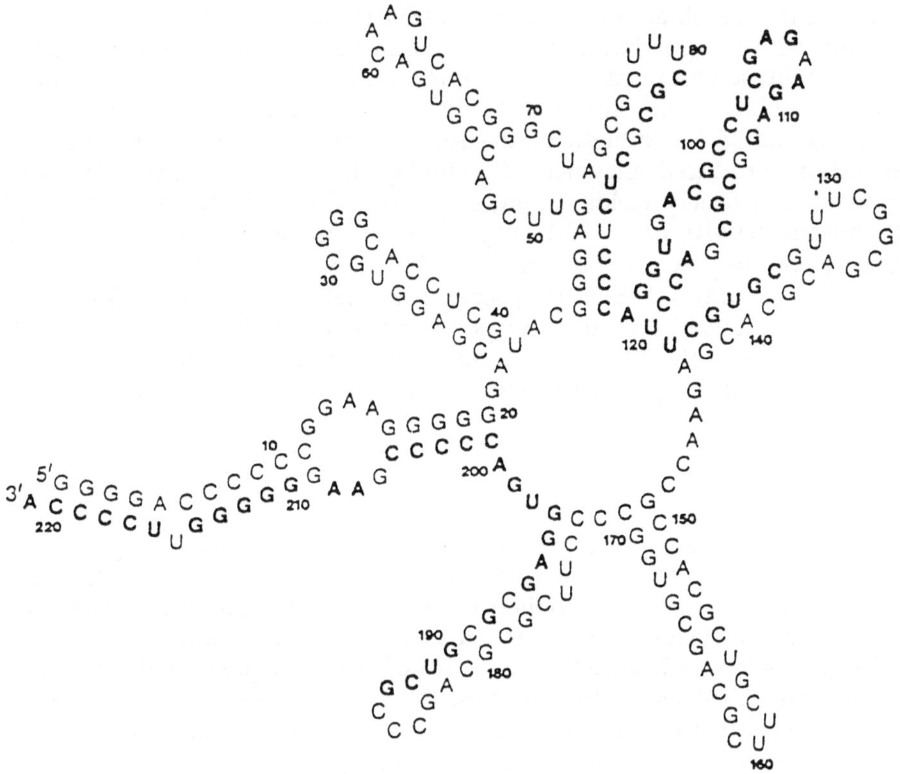

Figure 2. MDV-1 (+) RNA folded into the secondary structure predicted to be most stable by computer analysis (Zucker and Stiegler, 1981).

hardware, and disposables are required to keep all materials free from extraneous amplifiable entities, whether primer/target complexes, previously amplified products, or MDV probe molecules. Ideally, to eliminate completely the possibility of release of probes or amplified products into the environment, a closed assay system should be employed. 3) Assay sensitivity also would be reduced by replication of endogenous MDV-1 RNA which was known to contaminate all early preparations of Qβ replicase. However, recent advances in the purification of Qβ replicase have yielded enzyme preparations free of such contaminating RNA molecules (Robert DiFrancesco, in preparation). We have employed such preparations of Qβ replicase in conjunction with reversible target capture to develop assays for the direct detection of viral nucleic acids. The result is a sensitive, nonisotopic assay that can directly detect DNA or RNA targets and can be completed in less than three hours.

MATERIALS AND METHODS

Enzymes/Nucleotides

Bacteriophage T7 RNA polymerase (EC 2.7.7.6) was purchased from Promega (Madison, WI). Calf thymus terminal deoxynucleotidyl transferase

(EC 2.7.7.31) was obtained from Supertechs (Bethesda, MD). Ribonucleoside triphosphates were purchased from Pharmacia LKB Biotechnology (Piscataway, NJ). Qβ replicase was purified by the method of DiFrancesco (1989) from either of two recombinant *E. coli* clones (Mills, 1988; L. Burg, unpublished results) containing the 65 kD subunit of the replicase in expression plasmids. Both preparations produced enzymes with similar characteristics (data not shown). This method routinely yields 0.4 mg of purified Qβ replicase per gram of cells, with specific activity of >1500 U/mg. One unit is defined as 1 nmol of GMP incorporated into poly C for 10 min at 37°C. Analysis of the purified enzyme by electrophoresis through SDS polyacrylamide gels stained with Coommassie blue or silver showed only the four polypeptides of the holoenzyme and a low MW protein, likely the 12 kD host factor. Qβ replicase purified by this method does not synthesize a product in the absence of template under standard replication conditions.

Capture Probes

Single stranded capture probes complementary to selected regions of the target nucleic acid were prepared using β-cyanoethyl phosphoramidite chemistry on a 380A Synthesizer (Applied Biosystems, Foster City, CA). Four different deoxyoligonucleotides, complementary to nucleotide numbers 4,498 - 4,536, 4,675 - 4,698, 4,699 - 4,738, and 4,822 - 4,861 of HIV-1 (sequence data from Bionet System) were used as capture probes. A tail of about 150 deoxyadenylate residues was added to the 3' end of each of these by incubation with terminal deoxynucleotidyl transferase (Nelson and Brutlag, 1979).

MDV RNA Reporter Probe

Recombinant MDV-1 RNA containing a sequence complementary to nucleotides 4,739 - 4,778 of HIV-1 RNA (Bionet numbering) was synthesized *in vitro* by transcription from a recombinant plasmid. Complementary oligonucleotides were synthesized, annealed, and cloned between the SmaI and EcoRI sites of the fal-st MDV cDNA construct described by Lizardi et al. (1988), which is used to produce an MDV RNA with a structured internal probe specific for *Plasmodium falciparum*. We chose an existing MDV cDNA construct with an internal insert so that reaction products could be distinguished easily by size from wild-type MDV-1 when analyzed by gel electrophoresis. When the recom-

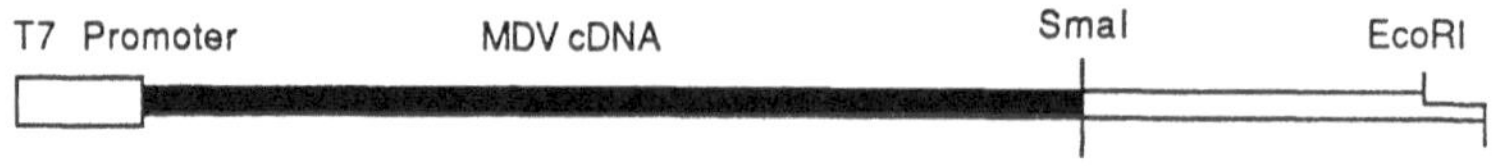

Figure 3. Structure of the plasmid from which MDV probes were transcribed. The heavy black line represents MDV-1 cDNA. When the plasmid is cleaved with SmaI and incubated *in vitro* with T7 RNA polymerase, the resulting transcripts consist of the sequence of MDV-1 (+) RNA. This plasmid serves as vector for the construction of recombinant RNA probes.

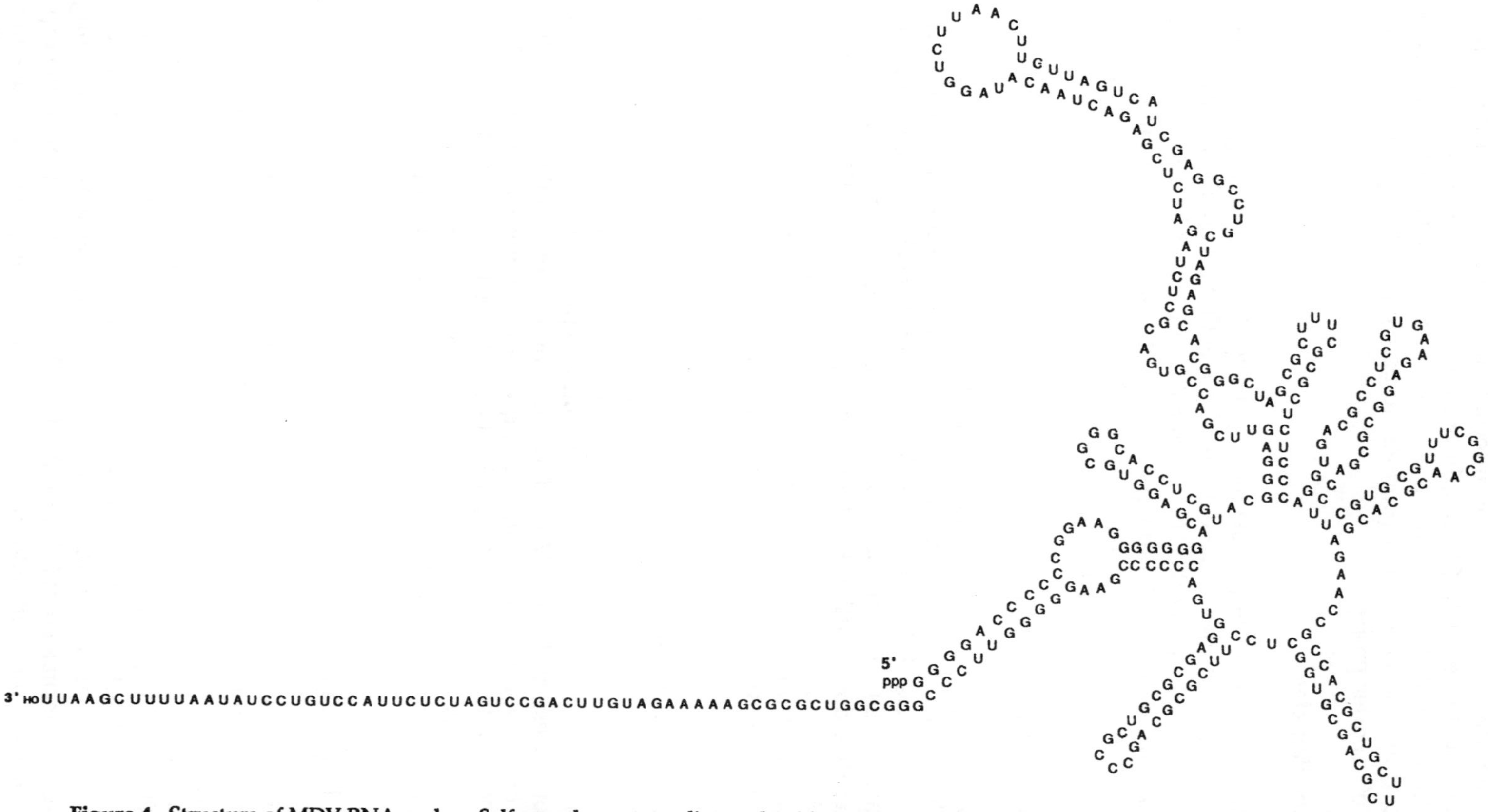

Figure 4. Structure of MDV RNA probe. Self complementary oligonucleotides were synthesized corresponding to the HIV-1 target region from nucleotides 4,739 - 4,778. These were annealed and cloned into pMDV fal-st (Lizardi et al. 1988) cut with SmaI and EcoRI. When cleaved with EcoRI and transcribed by T7 RNA polymerase, an RNA containing an internal (fal-st) insert and a 3' HIV-1 specific probe extension was generated.

binant plasmid was cut with EcoRI and transcribed by T7 RNA polymerase (Figure 3), an RNA containing an internal insert and a 3' probe extension was generated (Figure 4) (Stefano, in preparation). The 3' probe site yields more efficient hybridization than internal inserts (data not shown). The stable internal insert allows the reaction products to be identified as products of the replication of reporter probe.

Paramagnetic Particles

Sub-micron ferric oxide paramagnetic particles (Advanced Magnetics, Inc., Cambridge, MA) were derivatized with oligo d(T) as described previously (Morrissey et al. 1989) and blocked for 4 h at 68°C in 100 mM Tris, 4% BSA, 0.5% Sarkosyl, 10 mM EDTA, 0.05% Bronopol, 0.01% antifoam (Dow-Corning FG-10), pH 7.8. The final suspension of particles was 0.125% (w/v) solids with a binding capacity of 300 pmol of $d(A)_{50}$ per mg.

Sample Preparation/Hybridization

Two model systems were used to optimize these assays for HIV-1: 1) lysates of T-lymphoblastoid cells containing various amounts of added synthetic HIV-1 RNA (see below) and 2) HIV-1-infected H9 cells (Biotech Research Labs, Rockville, MD). T-lymphoblastoid cells were prepared from whole blood of HIV-1 negative donors using Ficoll gradients (Carter et al. 1987). A synthetic target RNA was generated by transcription of a cloned fragment of the *pol* region of an HIV-1 provirus with SP6 RNA polymerase as described previously (Pelligrino et al. 1987). This RNA corresponds to nucleotides 2,094 - 5,158 of HIV-1 RNA (Bionet numbering) (Figure 5).

To assay for HIV-1 RNA, T-lymphoblastoid cells or HIV-infected H9 cells were lysed in a buffer containing 5 M guanidine thiocyanate (GuSCN), 10% dextran sulfate, 100 mM EDTA, 200 mM Tris, pH 8.0, 1% sarkosyl. Twenty microliter aliquots of T-cell lysate containing various amounts of the synthetic HIV *pol* RNA or H9 cell lysate were added to 1.5 ml polypropylene tubes (Micronics, Flow Laboratories, McLean, VA). Fifty microliters of a mixture of 10^{11} molecules of MDV RNA probe and 2.5 ng of each capture probe (in 1.5 M GuSCN, 100 mM Tris pH 7.8, 10 mM EDTA, 0.5% sarkosyl, 0.5% BSA, 0.01% antifoam), was added to each tube. The mixtures were incubated at 37°C for 30 min to allow formation of ternary hybrids between target, capture and reporter probes.

Reversible Target Capture

Fifty microliters of a suspension of oligo d(T)-derivatized paramagnetic particles was added to each tube and incubated for 5 min to capture the ternary hybrids (Figure 6). The particles were washed to remove MDV RNA probe that was not bound to target. To each tube, 200 µl of wash buffer containing 1.5 M GuSCN, 100 mM Tris pH 7.8, 10 mM EDTA, 0.5% sarkosyl, 0.5% bovine serum albumin (Sigma, Fraction V), 0.01% antifoam was added. The tubes in the rack were mixed on a platform vortex (SMI, American Dade, Miami, FL) at setting six for 30 sec and incubated at 37°C for 2 min. The particles containing the captured hybrids were collected onto the sides of the tubes by placing the rack in a magnetic separator device (GENE-TRAK Systems). The used washed buffer was aspirated, and the particles washed two additional times in the same manner. To release the ternary complexes from the paramagnetic particles, without disruption of the hybrid complexes themselves, 50 µl of a buffer containing 3.25 M GuSCN, 100 mM Tris pH 7.8, 65 mM EDTA, 0.5% BSA, 0.5% Sarkosyl was added to each tube. The suspensions were mixed and incubated at 37°C for 5 min. This treatment selectively dissociates the dA:dT hybrids which bind the hybrid complexes to the magnetic particles. The spent particles were collected in the magnetic separator and the supernatants were transferred to tubes containing 50 µl of fresh paramagnetic particles suspended in buffer without guanidine. The mixtures were incubated for 5 min at 37°C to recapture the ternary hybrids onto the fresh particles. The particles were washed three times as above and the hybrids were eluted and recaptured onto a third aliquot of fresh particles. After washing the final set of particles three times with buffer containing GuSCN, they were washed three additional times with 200 µl of 50 mM Tris pH 8.0, 1 mM EDTA, 300 mM potassium chloride, 0.1% NP-40. The hybrid complexes were eluted in 50 µl of a buffer containing 50 mM Tris pH 8.0, 1 mM EDTA, 0.1% NP-40. Aliquots of the final eluates were used to prime Qβ amplification reactions.

Qβ Amplification

Qβ amplification reactions were monitored in one of two ways, termed end point assays and "real time" kinetic assays. For end point amplification assays, 20 µl of the final eluate was added to 30 µl of an ice-cold reaction mixture containing Qβ replicase. The final concentrations of the components of the re-

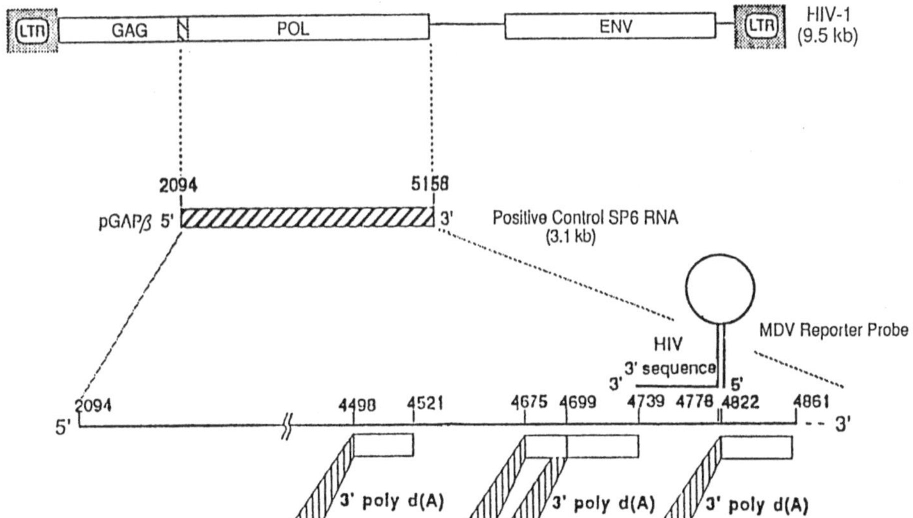

Figure 5. Schematic representation of the HIV-1 genome, positive control RNA, and position of the capture and reporter probes used in the reversible target capture assay. Nucleotide positions correspond to a consensus sequence for HIV-1 (Bionet). The synthetic target RNA was generated by transcription of a cloned restriction fragment of the *pol* region of an HIV-1 provirus with SP6 RNA polymerase (Pelligrino et al. 1987).

action were 100 mM Tris pH 7.5, 15 mM $MgCl_2$, 400 μM each ATP, CTP, GTP, UTP, and 1 μg of purified Qβ replicase (specific activity of 2,000 units/mg). The tubes were warmed to 37°C to initiate the reaction. After various periods of incubation, subsets of the replicate reactions were stopped by adding 25 μl of 80 mM EDTA containing 3.2 μg/ml propidium iodide. Fifty microliters of the completed reactions were transferred to a microtiter plate, which was photographed over a 365 nm ultraviolet transilluminator using a Wratten #27 filter and Polaroid type 667 film.

For real time kinetic assays, 50 μl aliquots of buffer containing various amounts of MDV RNA reporter probes were added on ice to 50 μl of reaction mixtures containing 2 μg/ml propidium iodide in addition to the standard components listed above. The mixtures were warmed to 37°C and the course of the reaction was followed in a Fluoroskan II microtiter plate reader (Dynatech, Burlington, MA) using a 480 nm excitation filter (14 nm band pass) and a 610 nm emission filter (6 nm band pass). We used the Titertek Automate software (Flow) for data collection.

RESULTS

We assayed for the presence of HIV-1 RNA in T-lymphoblast lysate using a replicatable RNA reporter probe complementary to a conserved region of the *pol* gene. Figure 5 shows the hybridization sites. Two of the four capture probes hybridize to the target adjacent to the reporter probe site. The third

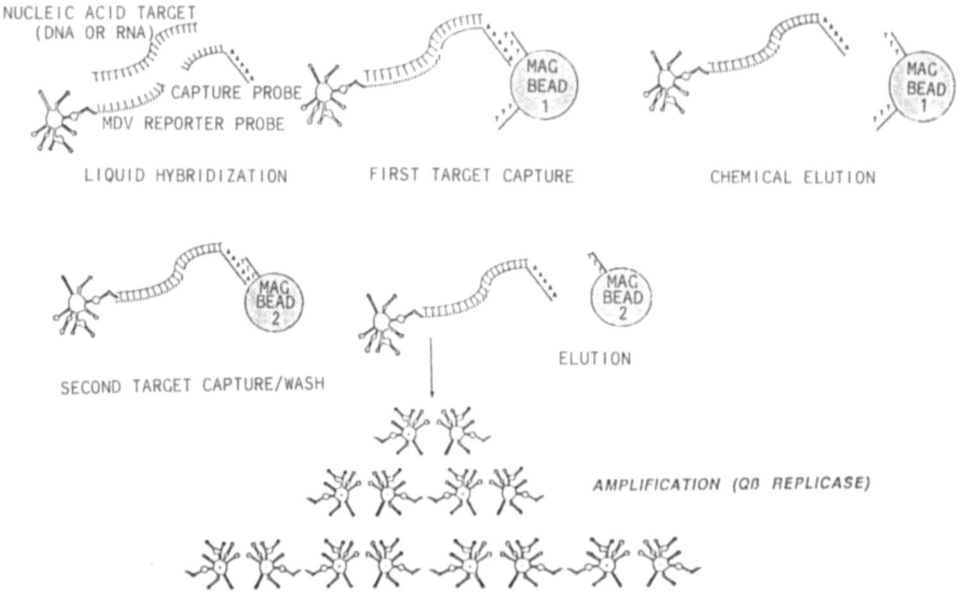

Figure 6. Schematic representation of reversible target capture with Qβ amplification. Mag bead = paramagnetic sub-micron particle. Only two cycles of RTC are represented. The capture probes have deoxyadenosine residues on their 3' ends, which mediate binding to the particles. Chemical elution indicates release of the hybrids from the particles with 3.25 M GuSCN.

anneals to a site about 40 nucleotides 3' to the reporter probe, and the last to a site 200 nucleotides 5' to the reporter probe. An SP6 transcript of a cloned portion of an HIV-1 provirus isolate designated pGAPβ containing all four of these sites was used as a positive control (Pelligrino et al. 1987).

Cells were lysed by vigorous vortex mixing in a buffer containing 5 M guanidine thiocyanate. This chaotrope also inhibits nucleases and promotes rapid solubilization of the cell matrix (Thompson and Gillespie, 1987). The hybridization and background reduction method that we employed is illustrated in Figure 6. After hybridization of the capture and reporter probes to complementary regions of the target, the hybrid complexes were captured onto oligo d(T)-derivatized particles via the poly d(A) tails on the capture probes. These particles are paramagnetic, so that they remain dispersed unless placed in a magnetic field. This allows efficient washing, release, and capture. The small size of the particles produces a large surface area, yielding binding kinetics similar to those for solution hybridization. The particles are suspended in buffer without GuSCN, so their addition lowers the GuSCN concentration, allowing hybridization of the poly d(A) extensions on the capture probes to the oligo d(T) on the particles. If target is present in the sample, it will be bound via the capture probe to the particles, and it will be carrying reporter probe which has hybridized to it. The particles were washed extensively to remove probe that was not specifically bound to target.

The ternary hybrids were released from the particles by adding a concentration of GuSCN sufficient to dissociate the poly d(A)-oligo d(T) hybrid but not the hybrids formed between the target HIV-1 RNA and the capture and reporter probes (Figure 6). The spent particles, which nonspecifically retain some reporter probes, were discarded.

The eluates containing the released hybrid complexes were added to an equal volume of fresh paramagnetic particles and incubated for 5 min, thus effecting recapture. Control experiments indicated that approximately 80% of an isotopically labeled target bound to the particles and could be recaptured following release (data not shown).

After three cycles of washing of the second set of particles, the released ternary hybrids were recaptured onto a final set of magnetic particles. The final set of particles was washed with another buffer to remove any GuSCN, which inhibits Qβ replicase. The hybrid complexes were eluted from the particles in a low ionic strength buffer compatible with the amplification reaction. An aliquot of each eluate was added to a well that contained complete reaction mixture on ice. The reactions were initiated by incubating the mixtures at 37°C. Each target nucleic acid molecule present in the final eluate should have an MDV probe hybridized to it. Qβ replicase will use neither the target nucleic acid nor the capture probes in the eluate as template, but will amplify any MDV probe molecules that are present. The Qβ replicase reactions were stopped by addition of EDTA and the product was visualized by addition of an intercalating dye whose fluorescence increases upon binding to the MDV RNA synthesized in the reaction.

An end point assay for HIV-1 RNA employing reversible target capture with a replicatable RNA reporter probe complementary to a conserved region of the pol gene could detect 1 fg (600 molecules) of HIV-1 RNA (Figure 7). After 8 min of amplification, product RNA molecules were detectable as fluorescence in the wells corresponding to the highest levels of target RNA. Increasing the time of replication yielded fluorescence in reactions corresponding to lower target levels. After 14 min, fluorescence was observed in the wells corresponding to 1 fg of input target RNA, whereas no signal was observed in the reactions initiated from samples that contained no target (Figure 7). After 16 min of replication, all the reactions began to show some fluorescence.

Similar responses were observed in assays of HIV 1-infected H9 cells, in which signals were easily distinguished above background in reactions corresponding to one infected cell, even after 18 min of amplification (Figure 8).

Electrophoresis of a portion of each of the reactions represented in Figure 7 demonstrated that the RNA products corresponded in size to that expected from the replication of the probe, i.e., they were distinguishable from MDV-1 RNA and other variant RNAs that might have been associated with the enzyme.

The sensitivity of these assays is limited by potential for replication of probe molecules which are non-specifically carried through the RTC process, probably by adsorption to the paramagnetic particles. By comparison with Qβ reactions initiated with known amounts of RNA probe, we estimate that this "background" corresponds approximately to 100 probe molecules remaining after three cycles of RTC. Thus each cycle of reversible target capture appears to reduce the background by approximately 1,000-fold.

To decrease the amount of sample required, simplify the manipulation of Qβ amplification reactions, and allow quantitation, a method for real time kinetic analysis was developed. This method is identical to that described for the end point assay, except that the reaction mixture contained an intercalating dye at a level which slightly slows, but does not otherwise change the reaction properties (J. Stefano, unpublished results). The sensitivity of a real time kinetic assay for HIV-1 RNA was identical to that of an end point assay, i.e., one fg or 600 target molecules (Figure 9). As in the end point assay, background signal limits the sensitivity of the real time assay, producing fluorescence after about 22 min of amplification.

One potential advantage of the real time kinetic assay is the ability to yield quantitative data regarding target levels based on the time course of the reactions. In these assays, the time at which a signal is detected at some predeter-

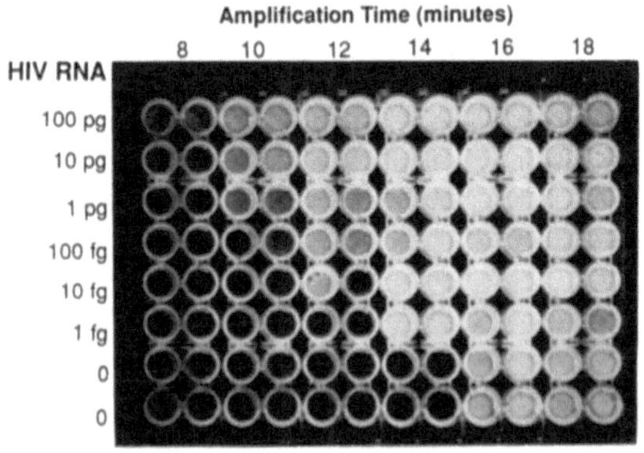

Figure 7. Results of an end point Qβ amplified assay for HIV-1 RNA. Duplicate sample of the final eluates were added to chilled reaction mixtures and warmed to 37°C to initiate the amplification reactions. At the indicated times, reactions were stopped by addition of EDTA and the product was visualized by addition of an intercalating dye.

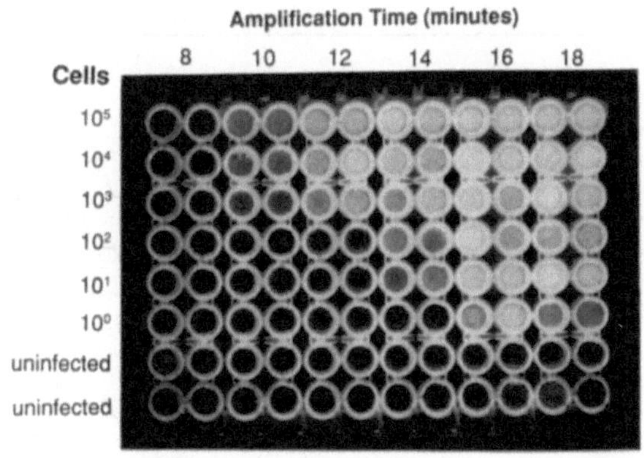

Figure 8. Results of an end point Qβ amplified assay for HIV-1-infected H9 cells. The assay was performed as described in the preceding figure, except that aliquots of infected H9 cell lysate served as the source of HIV-1 RNA. Dilutions of the lysate were made in uninfected T-lymphoblast cell lysate. The equivalent number of HIV-1-infected cells added to each hybridization mixture is shown.

mined point above the baseline is in inverse proportion to the logarithm of the number of variant molecules placed into the reaction. Above the background level, the number of MDV RNA probes is equal to the number of target nucleic acid molecules to which these are hybridized. Therefore, by comparison of the kinetics of parallel reactions initiated with known amounts of probe, a direct measure of target level can be obtained.

DISCUSSION

Assays for target nucleic acid which employ amplification of a hybridizable MDV RNA reporter probe rather than amplification of the target nucleic acid can take advantage of the fact that intervening steps such as thermal dena-denaturation and primer binding are not required. Assay conditions can be modified slightly from those described above to allow the direct detection of DNA targets with MDV RNA probes, so conversion of a DNA target into RNA by reverse transcription is not required. These characteristics shorten the time required for assay and amplification. Reversible target capture purifies target nucleic acids from cellular debris and other potential inhibitors of the amplification reaction. This also allows amplification to occur under ideal conditions rather than under compromise conditions dictated by annealing characteristics of primers as in target amplification systems. Moreover, since only MDV RNA and not the probe region is replicated, conditions for amplification in all assays are identical and do not have to be optimized separately.

However, signal amplification assays must include highly selective processes to eliminate non-hybridized probes which otherwise also would produce signal (background) even in the absence of target nucleic acids. We have used reversible target capture on paramagnetic particles to reduce this background to minimal levels before using Qβ replicase to amplify an HIV-1 specific MDV RNA probe. As reported here, 600 molecules of synthetic target RNA containing the *pol* region sequence were detectable after about 12 min of amplification.

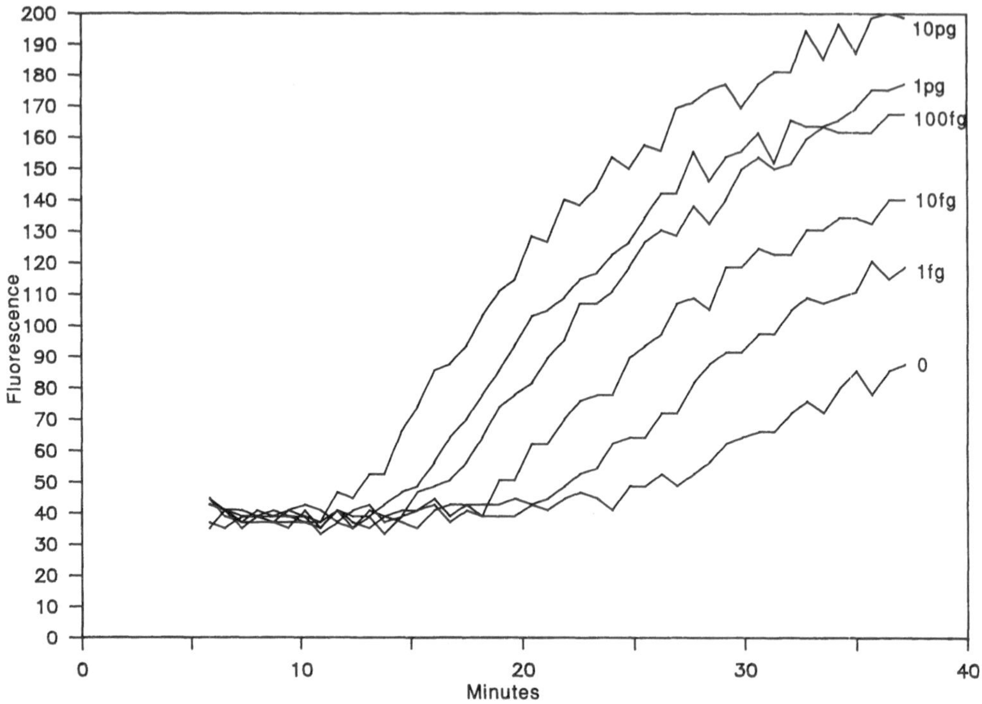

Figure 9. Kinetic Qβ replicase amplified assay for HIV-1 RNA. An aliquot of each eluate was added on ice to reaction mixtures containing intercalating dye in addition to the standard components. Mixtures were warmed to 37°C and the course of the reaction was followed in a Fluoroscan II microtiter plate reader. The amounts of synthetic HIV-1 *pol* RNA added to the initial hybridization mixture are indicated.

As determined from the analogous assay with an isotopically labeled probe, typically 5,000 molecules of HIV-1 RNA are associated with a single T lymphoblast infected with HIV-1 in culture (Ma et al. 1990). Provided that RNA levels in T-lymphoblastoid cells collected from blood are similar, this assay should prove useful for direct detection of HIV-1 in clinical samples. Recently, significant levels of HIV-1 RNA were detected in plasma unaccompanied by detectable levels of p24 antigen (Hewlett et al. 1988). Such RNAs could be detected directly by our approach (G. Radcliffe, unpublished data). Efforts to extend this assay to clinical samples are currently in progress.

This assay requires a number of repetitive manipulations to separate the target-bound probe from non-hybridized probe. In spite of this limitation, the total assay including final detection typically requires only 2-3 h. Furthermore, since the assay can be performed in a batch mode, multiple (up to 96) samples can be tested in the same time period. By comparison, the PCR usually requires a sample preparation step(s) before the thermal cycling process (Saiki et al. 1988). Additional post amplification steps such as hybridization usually also are necessary to verify the sequence of the amplified product DNA. Although the Qβ amplification method lacks a comparable ability to verify that the product RNA arose from replication of a target-bound probe molecule, it has a significant advantage in that the conditions under which hybridization and amplification occur may be selected to optimize independently the selectivity and efficiency of each. For example, the initial hybridization may be carried out under conditions which can discriminate single-base differences in targets. Moreover, since the sequence of capture and reporter probes may be independently varied and both must hybridize to generate a signal, target specificity may reside in either or both probes.

Lomeli et al. (1989) recently reported a similar assay in which amplified MDV product was detected instead by incorporation of ^{32}P-labeled nucleotides. They were able to distinguish only 10^5 target molecules above background. This is about a six-fold improvement in sensitivity over a non-amplified version of the assay employing an isotopically labeled detector probe (Gillespie et al. 1989). In contrast, we observed a 1,000-fold greater sensitivity than the isotopic method. This difference may be due to our use of an enzyme preparation which is free of endogenous RNA template. We also observed that the alternative insertion site of the probe sequence into the MDV-1 molecules that we employed yields more efficient hybridization (Stefano, in preparation). However, even with these improvements and the use of highly stringent RTC conditions, some non-hybridized probe remains which replicates and limits the ultimate sensitivity of our assay. Further improvements in RTC technology and probe design are anticipated to aid in the elimination of this background. For example, "smart" RNA reporter probes, which replicate only if target-associated (Lizardi et al. 1988), should improve the assay sensitivity.

REFERENCES

Backman K and Wang C (1989) Method for detecting a target nucleic acid sequence. European Patent Application #0/320308.

Carter W, Brodsky I, Pellegrino M, Henriques H, Parenti D, Schulof R, Robinson W, Volsky D, Paxton H, Kariko K, Suhadolnik R, Strayer D, Lewin M, Einck L, Simon G, Scheib R, Monefiori D, Mitchell W, Paul D, Meyer W, Reichenbach N and Gillespie D (1987) Clinical, immunological, and virological effects of ampligen, a mismatched double-stranded RNA, in patients with AIDS or AIDS-related complex. Lancet 6: 1286-1292.

DiFrancesco R (1989) Purification of Qβ replicase. US Patent Application Number 07/364,306.

Eoyang L and August J (1971) Qβ RNA polymerase from phage Qβ-infected *E. coli*. In: Cantoni G and Davis D (eds.), Procedures in Nucleic Acid Research New York: Harper and Row. Vol. 2: 829-842.

Gillespie D, Thompson J, and Solomon R (1989) Probes for quantitating subpicogram amounts of HIV-1 RNA by molecular hybridization. Mol Cell Probes 3: 73-86.

Haruna I and Spiegelman S (1965a) Specific template requirements of RNA replicases. Proc Natl Acad Sci USA 54: 579-587.

Haruna I, and Spiegelman S (1965b) Recognition of size and sequence by an RNA replicase. Proc Natl Acad Sci USA 54: 1189-1193.

Hewlett I, Gregg R, Ou C, Hawthorne C, Mayner R, Schumacher R, Schochetman G and Epstein J (1988) Detection in plasma of HIV-1 specific DNA and RNA by polymerase chain reaction before and after seroconversion. J Clin Immunoassay 11: 161-166.

Hunsaker W, Badry H, Lombardo M and Collins M (1989) Nucleic acid hybridization assays employing dA-tailed capture probes II. Advanced multiple capture methods. Anal Biochem 181: 360-370.

Kacian D, Mills D, Kramer F and Spiegelman S (1972) A replicating RNA molecule suitable for detailed analysis of extracellular evolution and recombination. Proc Natl Acad Sci USA 69: 3038-3042.

Kwoh D, Davis G, Whitfield K, Chappelle H, DiMichele L and Gingeras T (1989) Transcription-based amplification system and detection of amplified human immunodeficiency type 1 with a bead-based sandwich hybridization format. Proc Natl Acad Sci USA 86: 1173-1177.

Lizardi P, Guerra C, Lomeli H, Tussie-Luna I and Kramer F (1988) Exponential amplification of recombinant-RNA hybridization probes. Biotechnology 6: 1197-1202.

Lomeli H, Tyagi S, Pritchard C, Lizardi P and Kramer F (1989) Quantitative assays based on the use of replicatable hybridization probes. Clin Chem 35: 1826-1830.

Ma X, Sakai K, Sinangil F, Golub E, Volsky D (1990) Interaction of a non-cytopathic human immunodeficiency virus type 1 (HIV-1) with target cells: efficient virus entry followed by delayed expression of its RNA and protein. Virology 176:184-194..

Mills D (1988) Engineered recombinant messenger RNA can be replicated and expressed inside bacterial cells by an RNA bacteriophage replicase. J Mol Biol 200: 489-493.

Mills D, Kramer F, Dobkin C, Nishihara T and Spiegelman S (1975) Nucleotide sequence of microvariant RNA: another small replicating molecule. Proc Natl Acad Sci USA 72:4252-4256.

Morrissey D and Collins M (1989) Nucleic acid hybridization assays employing dA-tailed capture probes. Single capture methods. Mol Cell Probes 3: 189-207.

Morrissey D, Lombardo M, Eldredge J, Kearny K, Groody P and Collins M (1989) Nucleic acid hybridization assays employing dA-tailed capture probes. I. Multiple capture methods. Anal Biochem 181: 345-359.

Nelson T and Brutlag D (1979) Addition of homopolymers to the 3' ends of duplex DNA with terminal transferase. Methods in Enzymology 68: 41-47.

Pelligrino M, Lewin M, Meyer W, Lanciotti R, Bhaduri-Hauck L, Volsky D, Sakai K, Folks T and Gillespie D (1989) A sensitive solution hybridization technique for detecting RNA in cells. Biotechniques 5: 452-459.

Saiki R, Gelfand D, Stoffel S, Scharf S, Higuchi R, Horn G, Mullis K and Erlich H (1988) Primer-directed enzymatic amplification of DNA with a thermostable DNA polymerase. Science 239: 487-491.

Schaffner W, Ruegg, K and Weissman C (1977) Nanovariant RNAs: nucleotide sequence and interaction with bacteriophage Qβ replicase. J Mol Biol 117: 877-907.

Speigelman S, Pace N, Mills D, Levisohn R, Eikhom T, Taylor M, Peterson R and Bishop D (1968) The mechanism of RNA replication. Cold Spring Harbor Symp Quant Biol 33: 101-124.

Thompson J and Gillespie D (1987) Molecular hybridization with RNA probes in concentrated solutions of guandine thiocyanate. Anal Biochem 163: 281-291.

Weissman C, Feix G and Slor H (1968) *In vitro* synthesis of phage RNA: the nature of the intermediates. Cold Spring Harbor Symp Quant Biol 33: 83-100.

Zucker M and Stiegler P (1981) Optimal computer folding of large RNA sequences using thermodynamics and auxiliary information. Nuc Acids Res 9: 133-148.

DISCUSSION

de la Maza L (University of California Irvine Medical Center, Orange, CA):

You are telling us that you are going to come up with this fantastic piece of equipment where we just need a drop of the patient's urine, and come back two hours later and we have the answer in our pocket. How far are we from that point?

Pritchard C:

I would say that it's at least two years, because we realized fairly early in the development of these assays that while we could do them, we could not do them reproducibly because of the environmental contamination rate. The only way practically to do these assays in the clinical lab, or for that matter, anywhere, is to do them in a completely closed assay format. We've been working with an engineering company that had developed a system very similar to the du Pont ACA packs, and it's based on Surlyn chemistry. We had a prototype instrument on which we were doing the assays in an open format, but the launch was delayed because we had to develop the Surlyn packs in which to perform the chemistry and an instrument to perform the assays in the closed format. The target date now is late 1992, unfortunately, not tomorrow.

Peterson E (University of California Irvine Medical Center, Orange, CA):

You had mentioned one fungal probe, and we don't have any, that I know of readily available in the labs. What types of things are you aiming at?

Pritchard C:

We have a long list of organisms. We have a large in-house probe group that develops specific probes for various organisms. We are targeting a variety of organisms, from all different sorts of bacteria to viruses and Chlamydia. We are also targeting things like generic bacterial and fungal probes for determination of fungemias and bacteremias in blood.

Needham C (Lahey Clinic Medical Center, Burlington, MS):

First, I would like to say congratulations. That was a very elegant presentation with very elegant genetic manipulations. I would like to tell you that I was recently trying to explain this technology to my colleagues in my department, and I was not anywhere nearly as eloquent as you were. I have a couple of questions. One, the last maneuver that you developed, the ribozyme release strategy, does that work efficiently in a background matrix that is more or less independent of the specimen type?

Pritchard C:

That is a very interesting point and I'm glad you brought it up. Ultimately, if the ribozyme system works as efficiently as we would like, you could envision doing only one round of reversible target capture and release, and have a wide enough "window" to obtain the ultimate sensitivity that you needed. Right now, we're still doing three rounds of reversible target capture to reduce the background as much as possible. I can't envision ever being able to do ribozyme release in a dirty sample; in other words, add the probes, hybridize, ribozyme release the target bound probe and amplify. For one thing, you would still have all the components from the patient sample matrix that could interfere with the subsequent Qβ amplification reaction, and also with the ribozyme reaction. For example, magnesium on the sample could allow ribozyme cleavage. However, you can envision doing only one round of RTC to get rid of all the cell matrix, and other interfering components. It would be a very fast assay in that case.

Needham C:

My last is really more a comment than a question. I think that one of the more interesting facets of the amplification technology is what does it mean to find one gene of something, and clearly if it's HIV, at least presently our understanding is that it means disease. Obviously, in a lot of other clinical microbiology, we based a lot of our decision making on infection versus disease in terms of the number of organisms present. So, I think that the quantitation issue is going to remain an important one, even on the clinical side.

de la Maza L:

How critical is the specimen that you put in the reaction? It's nice to work with purified RNA and stuff like that, but when you put a drop of urine or a drop of sputum, then things get a little haywire. Do you have to process the specimen?

Pritchard C:

Actually, that's the nice thing about it. Right now assays for CMV just require mixing an equal volume of sample urine with the 5 M guanidine buffer. The same is true of plasma samples for HIV, just a 50/50 mixture of sample with buffer works quite well. With these types of samples, no long, involved sample processing is required. We do envision some sample processing for stool, although we've had some success with assays for Campylobacter in which the stool was solubilized directly in 5 M guanidine buffer. We may have to concentrate the samples. If you're looking for one infected cell in 10 ml of blood, there will be a concentration step required.

DETERMINANTS OF MEASLES MORTALITY: HOST OR TRANSMISSION FACTORS?

Peter Aaby

Institute of Anthropology
University of Copenhagen
Copenhagen, Denmark

INTRODUCTION

The WHO estimates that measles kills 1.5-2 million annually in the developing countries, making it the major killer among vaccine preventable diseases (Aaby et al. 1987a). Though the accuracy of this estimate can be discussed, there is no doubt that measles poses a major public health hazard in developing countries. Furthermore, there is increasing evidence that measles infection is connected with long-term effects on morbidity and mortality which have not been considered when evaluating its importance (Aaby and Clemens, 1989). It is therefore likely that measles presents a larger health problem than usually understood. Though measles has often been considered "the simplest of all infectious diseases" (Maxcy, 1948), it may well be the single pathogen connected with the largest number of childhood deaths.

Since our understanding of the causes of severe infection has many implications for preventive health policies, there are good reasons to understand why some children become more ill than others. Variations in severity of infection have usually been explained in terms of the care provided or of host factors influencing the individual's immunological capacity; these latter factors would include malnutrition (Morley, 1973; Walsh, 1983), age at infection (Reves, 1985; Walsh, 1983), genetic susceptibility (Black et al. 1977), sex-linked immune differences and chronic conditions (Hertz et al. 1976).

Malnutrition is no doubt the most widely accepted explanation of why infections like measles, which today are harmless in the industrialized world, still kill many children in some parts of Africa, Asia and Latin America (Morley, 1973; Walsh, 1983). Low age at infection has also been considered a major determinant of the level of mortality. Thus, the decrease in case fatality rate (CFR) in the industrialized countries has been explained as a result of a presumed rise in the age at infection (Reves, 1985). The host factor approach is based on the premise that there is something distinctive about the individuals who contract severe or fatal infection. Specific disease interventions may therefore have a limited impact on survival. Hence, some have claimed that children prevented from dying of measles are the "weak" ones who presum-

Medical Virology 10, Edited by L.M. de la Maza and
E.M. Peterson, Plenum Press, New York, 1991

ably are more likely to die of other infections (Mosley, 1985; The Kasongo Project Team, 1981).

It is my contention that the host factor approach is inadequate for explaining the known variation in acute measles mortality (Aaby, 1989). Since intensive exposure increases severity of measles (Aaby et al. 1984a), it is suggested that the causes of severe disease may lie in the process of transmission of infection rather than in the host itself. From this perspective, children prevented from dying of measles due to intensive exposure should not be more likely to die of other infections. Specific disease intervention may therefore have a strong impact on general survival (Aaby et al. 1981a, 1987a).

The present communication discusses the explanatory potential and implications of these contrasting approaches to the understanding of severe measles infection. Most observations are based on longitudinal community studies of measles infection in Guinea-Bissau and several other African communities. Furthermore, historical studies from European communities (Aaby et al. 1986a; Pfeilsticker, 1863) and records from the infectious disease hospital in Copenhagen (Blegdamshospitalet) in the period 1915 to 1925 (Aaby, 1988a, 1990) have been used to examine whether the experience from Africa also applies to the period in the industrialized countries when measles mortality was very high. The case fatality ratio for all ages in the hospital in Copenhagen in 1915 to 1925 was 13% (284/2,208) (Aaby, 1988a).

MEASLES MORTALITY

Acute Mortality - the Size of the Problem

The EPI estimate (1987) of 2 million measles deaths annually is based on the assumption that of the 83.3 million children in developing countries surviving annually to the age of one year only 24% were protected by vaccination (Aaby et al. 1987a). The remaining children all get measles and the CFR of these children is 3.2% on the average. The CFR has been assessed to be 3-4% throughout Africa and Asia, and 2% in the major countries in South America. It is very difficult to evaluate the CFR in many of these areas due to lack of reliable registration. The most reliable estimates of CFRs from measles should come from areas with prospective studies of child mortality. The available data concerning CFRs in such community studies are summarized in Table 1. With few exceptions, the CFRs in these studies exceed the EPI estimate, at least in the beginning of a project period. For Africa, the difference between official estimates and observed CFRs is glaring. For West and Central Africa, mortality in the community may well be 2 to 3 times higher than the estimated 3%. Improved estimates of acute measles mortality would not necessarily raise the global estimate, since the CFRs in Asia and Latin America could be lower. However, the precise size of the problem of acute deaths from measles is uncertain.

Delayed Impact of Measles on Mortality

Several studies from West Africa and India-Bangladesh have indicated that children previously infected with measles have significant excess morbidity (Bhaskaram et al. 1984; Shahid et al. 1983) and mortality compared with community controls. Hull et al. (1983) reported an outbreak of measles in a vil-

TABLE 1. Measles CFR in Prospective Community Studies

Country	Immunization	CFR (no. of measles cases) 0-4 years	All ages
Africa Rural			
Guinea-Bissau	NI [a]	34%(101)	24%(162)
Gambia	NI	22%(259)	
Senegal	NI	20%(44)	13%(68)
Senegal	NI	18%(537)	
Senegal	NI	10%(966)	7%(1,500)
Kenya	I [b] (20%)	8%(331)	6%(424)
Nigeria	NI	7%(222)	
Africa Urban			
Guinea-Bissau	NI	21%(356)	17%(459)
Guinea-Bissau	I	15%(124)	14%(161)
Zaire	I	6%(1,069)	
Other areas - Rural			
Guatemala	NI	5%(292)	4%(449)
Guatemala	NI	4%(231)	3%(276)
India	NI	3%(72)	1%(181)
Bangladesh	NI	4%(510)	4%(896)
Bangladesh	NI	2%(3,458)	
Other areas - Urban			
India	NI	0%(318)	

[a] NI=no immunization
[b] I=immunization program (% coverage)
From Aaby et al. 1987

lage in the Gambia which they re-visited 3 and 9 months later to assess the impact of infection. As indicated in Table 2, after acute infection former measles patients had a significantly higher risk of dying compared to community controls (odds ratio (OR)=10.0; 95% confidence interval (CI): 4.8-20.9). The excess mortality seemed particularly high for children below one year of age. The results from Gambia could be confounded by background factors distinguishing cases from controls. For example, it seems clear that deaths occurred mainly in domestic compounds where many children lived close together (Hull, 1988). The risk of measles infection as well as mortality in general may have been greater in larger compounds. Parental attitudes could play a role, since many of the controls had been immunized (Hull et al. 1983). However, the relative risk was particularly high for children under one year of age where very few of the controls had been immunized prior to the outbreak and the difference in mortality in general was so large that it is unlikely to be due solely to confounding between the risk of measles infection and the determinants of child mortality in general.

TABLE 2. Mortality During 9 Months of Follow-up for Measles Patients and Community Controls - The Gambia

Age at infection	Mortality of measles cases Acute	1-9 mos later	Mortality of controls 0-9 mos later
3-11 mos	18%(2/11)	56%(5/9)	3%(3/94)
1-2 yrs	9%(3/35)	13%(4/32)	2%(3/190)
3-4 yrs	6%(2/31)	7%(2/29)	1%(1/182)
5-6 yrs	0%(0/36)	6%(2/36)	1%(2/188)

Based on Hull et al. 1983

Studies from Nigeria and Burkina Faso have likewise found significantly higher risk of dying for cases in the months following measles compared with controls (Osagie, 1986; van de Walle, 1986). In Bissau, we found excess mortality as well in the second year after measles infection. The small children under two years of age who had measles during an epidemic of measles in the beginning of 1979 had a mortality of 5.9% (7/118) during 1980 compared to only 1.3% (3/237) among the children from the same community who had not had measles (relative risk (RR)=4.7; 95% CI: 1.4-15.7) (Aaby et al. 1984c).

Since few studies have assessed the delayed impact of measles infection, it is difficult to determine its risk factors, its impact relative to the acute mortality and its contribution to overall mortality. However, the experience from West Africa suggests that delayed mortality may account for as much as acute measles mortality. More studies in this area are clearly needed.

VARIATION IN MEASLES MORTALITY

Though the total impact of measles infection on mortality may be uncertain, some major contrasts in severity of infection have been established by epidemiological research. An adequate explanation of severe measles should be able to account for the major contrasts in mortality. The most important variations in measles mortality may be described as follows:

Geographical Variation

Measles mortality is higher in Africa than in Asia or Latin America (Table 1) (Aaby, 1988b). Within Africa, measles mortality is higher in West Africa than in East Africa, as is child mortality generally.

Rural-Urban Differences

Within developing countries, age specific CFRs are higher in rural areas than in cities (Aaby, 1988b).

Institutions and Overcrowding

When measles was severe in the industrialized world, several studies documented that measles was particularly severe in institutions grouping many susceptible individuals; e.g., orphanages, child institutions, emigrant passenger ships, military and refugee camps (Aaby, 1988b; Aaby et al. 1984a). For example, during the First World War, an epidemic broke out in a German refugee camp where people were living in barracks with 200-300 individuals. In the first two months, the CFR was 10% (5/50) but this increased during the following months to 45% (286/628) (Reder, 1918). Apparently, measles becomes more severe when several individuals become ill simultaneously. The same experience has been made more recently in developing countries where several community studies have shown mortality to be higher in houses with multiple cases than in families with only a single case of measles (Aaby, 1988b).

Virgin-soil Epidemics

Very high CFRs have usually been encountered in virgin-soil epidemics, i.e. outbreaks in communities where no one has previously had measles and all are susceptibles. For example, when measles first hit Fiji in 1875, it killed 25% of the population of 80,000 (Morley, 1973). Similarly high CFRs have occurred in many other virgin-soil epidemics (Aaby, 1985).

Decline in Mortality

In the industrialized world, a major decline in measles mortality occurred in the beginning of this century (Reves, 1985). In England, where developments are best described, the decline took place between 1890 and 1940, just before the advent of antibiotics. At the turn of the century, several studies based on outbreak investigations and public registration found CFRs as high as 3-10% (Aaby et al. 1986a, Pfeilsticker, 1863; Picken, 1921). The decline in CFR to the present 1-2 deaths in 10,000 cases constitutes one of the most important epidemiological changes which has ever occurred. Furthermore, where immunization is introduced, the CFR seems to fall even among those unvaccinated children who still contract measles (Aaby, 1988b).

Age

As with many other infections (Mims, 1976), the measles CFR shows a U-shaped curve, being most severe among the youngest children under 2 or 3 years of age, increasing again for young adults.

Sex

When measles was severe in the industrialized world, mortality was usually highest among males (Pfeilsticker, 1863), and this has been considered the "natural" situation (Babbott and Gordon, 1954). Few studies have reported their data by sex, but studies from Bangladesh (Bhuiya et al. 1987), Tunisia (Monastiri, 1961), The Gambia (McGregor, 1964) and Mali (Fargues and Nassour, 1988) have found higher mortality among females. This has been interpreted as a result of differential treatment of the two sexes (Bhuiya et al. 1987).

Partial Immunity

Persons who presumably have some, though insufficient, immunity due to maternal antibodies, serum prophylaxis (Aaby et al. 1987b), or measles immunization have lower mortality when they develop measles (Aaby, 1989). Thus infants under six months of age have been found to have a lower CFR when they develop measles (Aaby et al. 1986d). Some hospital studies have suggested that children with measles having a record of previous vaccination against measles have lower mortality than children with no such vaccination (Gallais et al. 1981). In Africa, this association has been examined in only one community study from Guinea-Bissau, where previously vaccinated children were found to have significantly lower mortality than other measles cases (Aaby et al. 1986c).

Concurrent Infections or Chronic Conditions

High mortality in measles has been found to be associated with certain chronic conditions or pre-existing infections, e.g. leukemia (Burnet, 1968), kwashiorkor (Sinha, 1977), tuberculosis (Christensen et al. 1953) and retrovirus infection.

Other Differences

Other risk factors have been suggested but have not yet been verified or have failed to be confirmed. Thus, one study from Senegal reported higher CFRs during the rainy season, suggesting that malaria contributed to the outcome of infection (Debroise et al. 1967). However, other studies in the region have failed to find this association (Aaby et al. 1984b). Measles mortality has also been associated with some socioeconomic indicators like maternal education (Bhuiya et al. 1987), possessions (Bhuiya et al. 1987) and polygynous family structure (Aaby et al. 1984a). In Guinea-Bissau, the latter difference was due mainly to the fact that children from large polygynous families were more likely to be multiple cases (i.e. having measles together with other children). Whether maternal education and belongings are merely associated with less crowding and better vaccination status or has a positive impact on therapeutic practices has not been demonstrated.

HOST AND CARE FACTORS

In order to understand the variation in measles mortality, emphasis has been placed on malnutrition and age at infection (Walsh, 1983). Other host and care factors will be discussed only briefly.

Malnutrition

Most hospital studies have reported that malnourished children with low weight-for-age (w/a) at admission have a higher CFR than better nourished children (Aaby, 1988b). Such studies seem to be the common basis for the belief that malnutrition is the cause of high mortality from measles in developing countries. However, this belief has not found support in any of the community studies with information on the children's state of nutrition prior to their catching measles. There now exists several such studies from Bangladesh,

India, Kenya, Gambia, Nigeria and Guinea-Bissau; all reported no association between state of nutrition and severity of measles (Table 3) (Aaby, 1988b).

Only one community study has described higher measles mortality among malnourished children (Chen et al. 1980a). This study, from Bangladesh, followed 2,019 children aged 12-23 months for two years. During this period,

TABLE 3. State of Nutrition and Severity of Measles in Community Studies

Country	Age	Index	Nutritional status (% standard) Fatal cases(N)	Controls(N)	Type of Control
Bangladesh	0-9 y	w/h [a]	86%(33)	88%(33)	No M
	0-2 y	w/h	93%(4)	86%(148)	M Sur
	3-9 y	w/h	85%(5)	88%(170)	M Sur
Guinea-	0-4 y	w/a	87%(17)	90%(27)	M Sur
Bissau	0-5 y	w/a	92%(60)	92%(1,188)	All Ch
		h/a	97%(60)	97%(1,172)	All Ch
		w/h	97%(60)	98%(1,167)	All Ch
	0-1 y	w/a	88%(10)	89%(36)	M Sur
Kenya		muac	84%(31)	86%(36)	M Sur
Zaire	0-4 y	w/a	12%(6/51) <3rd; 53% <50th centile		
		w/h	6%(3/50) <3rd; 56% <50th centile		
		muac/a	12%(6/51) <3rd; 55% <50th centile		
Gambia		w/a	29%(2/7) of fatal cases were malnourished. More than 29% malnourished among controls.		
Gambia	0-2 y	w/a	State of nutrition did not affect severity		
Nigeria		w/a	No relation between malnutrition and severe measles.		
India	0-2 y	w/a	Among malnourished children, 29% had severe measles and 29% lost weight; among children of normal nutrition, 33% had severe disease and 56% lost weight.		
India	0-4 y	w/a	Severely malnourished children (<60% of w/a) had the same rate of complications and the same immunological responses.		
Philippines			Well-nourished and undernourished had similar course and outcome of their illness.		

[a] w/h indicates weight-for-height; w/a, weight-for-age; muac/a, mid-upper arm circumference-for-age; No M, no measles infection; M Sur, measles survivors; All Ch, all children in community.

From Aaby, 1988b

children with a w/a < 65% of standard had a measles mortality rate of 1.5% (11/742) compared with 0.6% (8/1,277) for children with a w/a >65% (p=0.05; x^2=3.7). However, this comparison may be partly confounded as it is based on deaths in relation to the total population rather than in relation to the number of children who contracted measles infection. Since children from large families have lower w/a and a higher risk of contracting measles (Aaby et al. 1984a), part of the reason for the higher risk of dying of measles among the malnourished children may be that they had a higher incidence rate. Furthermore, in this community, females have lower state of nutrition and a higher CFR in measles infection (Bhuiya et al. 1987). Overrepresentation of females in the <65%-group may thus explain some of the higher mortality among children with low weight. Subsequent studies from the same area found no difference in state of nutrition between children dying of measles and controls (Koster, 1988; Koster et al. 1981).

The association between state of nutrition and severity of infection in hospital studies is likely to be due to two types of confounding. Children lose much weight already during the period of incubation (Meunier, 1898). The association may therefore simply be due to children with severe measles having lost more, thus having lower w/a at admission. Furthermore, children from large families have lower state of nutrition and a higher risk of contracting measles with other children and thus they risk more severe disease. When crowding is taken into consideration, state of nutrition does not become a risk factor (Aaby et al. 1984a).

It has been suggested that while mild to moderate malnutrition seemed not to increase severity of measles infection, it could be related to more serious forms of malnutrition (Neiburg and Dibley, 1986). Kwashiorkor is probably associated with higher CFR (Sinha, 1977), although an increased severity among marasmic children has not been documented. In the one study from India which examined a large number of marasmic children, the 41 children with a w/a below 60% of standard had immune responses and severity of infection similar to those of better nourished children (Bhaskaram et al. 1986). Although severe malnutrition may be associated with higher CFRs, this would explain little of the excess measles mortality in developing countries because the prevalence of the severe forms of malnutrition is too low and do not fit the observed variation in the severity of measles infection. For example, the difference in measles mortality between Africa and Asia is incomprehensible from the nutritional point of view. In Guinea-Bissau, mean w/h of children under three years of age was 97% (Aaby et al. 1983a) and the CFR was 25% (Aaby et al. 1984a). In Bangladesh, w/h was as low as 87%, but the CFR was no more than 3% (Koster et al. 1981).

So far there is no data to indicate that other forms of nutritional practices or specific nutrients are major determinants of the severity of measles infection. In the only community study to examine the role of breastfeeding, there was no difference in CFR between breastfed and non-breastfed children of similar age groups (Aaby et al. 1981b). Vitamin A deficiency has been suggested as a cause of severe measles infection (Neiburg and Dibley, 1986), but no community study has examined whether pre-morbid vitamin A status had an effect on outcome. In one hospital study, children who received vitamin A in large doses at admission had a lower CFR (though not significant) than children who did not (Barclay et al. 1987). Should this observation be reproduced, it may reflect that severe measles had depleted vitamin A stores rather than indicating that pre-morbid vitamin A deficiency caused severe measles infection. West

Africa which has the highest CFRs in measles is not known for the severe forms of vitamin A deficiency.

Age at Infection

Since the CFR is highest among the youngest children, it has been suggested that measles mortality is highest where many children contract infection at a young age and that age of infection in the community is a major determinant of the level of severity (Walsh, 1983). Hence it has been predicted that measles mortality should be higher in urban areas, where age at infection is lower than in rural areas (Davis 1982; Foster, 1984). All community studies, however, suggest the opposite, the CFR has been higher in rural than in urban areas (Aaby, 1988b). Within a region, CFR may in fact be higher where mean age of infection in the community is highest.

It has also been hypothesized that the decline in measles CFR in the industrialized world at the beginning of this century is related to an increase in the age at infection (Reves, 1985). The birth rate in England fell dramatically from 6.7 children/woman in 1875 to 2.6 in 1925. Since children in small families tend to become infected at a later age (Aaby et al. 1984a), it has been assumed that the reduction in family size led to an increase in the age of infection in the community (Reves, 1985; Lancet, 1985). Though the risk of contracting infection within the family diminished in this period due to change in family structure, in the same period the risk of contracting infection outside the family may well have increased. Age at infection would have dropped due to more widespread urbanization, improved means of transport, higher degree of schooling and more public child care. There is no general data for all of England to show that the age at infection did in fact rise in this period (Reves 1985). The few community studies and reports of notifications do not suggest a major change (Aaby et al. 1988). We know that mean age at infection fell from 5.5 to 4.4 years between 1944 and 1968 (Anderson and May, 1982). When measles was severe in the United Kingdom, Picken (1921) noted that variation in measles mortality could not be explained by differences in age distribution of cases.

In Guinea-Bissau, we observed a decline in the CFR at the same time as the state of nutrition deteriorated and the mean age at infection decreased (Aaby et al. 1988a). Recent studies have not found that the immune response to measles was related to age (Coovadia et al. 1984). One reason for higher mortality among young children may be that they are more likely to be intensively exposed (Aaby et al. 1986d). Thus, there is little evidence indicating that age at infection in the community has an important impact on the level of mortality. The major differences between communities are due to different age-specific CFRs (Morley, 1973).

Genetics

One study has found that the HLA-Aw32 antigen was associated with profound lymphocytopenia, which is a significant index of severe clinical measles (Coovadia et al. 1981). It has also been observed that American Indians have a higher response to measles vaccination, which could suggest that they may react more strongly to the natural infection. This has been suggested as a possible explanation of the very severe virgin-soil epidemics reported among American Indians (Black et al. 1977). However, severe (virgin-soil) epidemics have occurred in all racial groups and the severe character of these epidemics may result as much from the extreme degree of clustering of cases encountered in these situations as from the genetic constitution of the virgin-soil populations (Aaby, 1985). Though there may be subgroups which are genetically susceptible to severe measles, genetic differences can hardly explain the association with overcrowding nor the rapid decrease in severity of measles within a relatively short time span in the industrialized world (Aaby et al. 1985).

Concurrent Infections and Chronic Conditions

Differences in the prevalence of complications or chronic conditions could account for some of the difference between different regions. For example, intercurrent malaria has been suggested as a major cause of the high mortality in Africa (Hendrickse, 1975), though this could hardly account for the severity of measles in Northern Europe at the turn of the century. The severity of the underlying measles infection could also influence the risk of complications. For example in Bissau, secondary cases of measles had more diarrhoea and lung complications than did index cases (Aaby et al. 1985). Therefore, it is unclear to what extent complications are the cause or the symptoms of severe measles. Furthermore, the presence or absence of potentially complicating pathogens or chronic conditions can probably not explain the internal variation in the communities with higher mortality among secondary cases and the youngest children.

TRANSMISSION FACTORS

Thus none of the possible host factors seems to contribute much to a better understanding of variations in measles mortality. There is therefore a need for a different research perspective. Whereas the common research strategies have emphasized the distinguishing features of those who became severely ill, it would seem more productive to examine the determining factors in the infection itself, or its transmission process.

Intensity of Exposure

A common feature of most of the situations where measles has a high acute mortality is that many individuals contracted infection at the same time. In Guinea-Bissau, we found in several outbreaks that age-specific mortality was considerably higher in houses with multiple cases compared with houses with only a single isolated case (see Table 4) (Aaby, 1989). This tendency has been confirmed in all other studies where it has been tested (Aaby, 1988b, 1989). Though this tendency could be a result of distinguishing features between families with respectively many and few children, we have also examined whether the difference was related to intensity of exposure. Whereas all single cases are index cases infected from someone outside the home, some of the multiple cases are secondary cases infected by an index case in the same house (Table 4). Secondary cases have presumably been more intensively exposed than index cases. The relative severity of index and secondary cases has now been examined in several studies from both developing countries and in reanalyzes of historical data from Europe (Table 5). All studies have found a significantly or nearly significantly higher mortality among secondary cases, often two-three times higher. In studies from Bissau, morbidity rates of iso-

TABLE 4. CFR in Measles Infection According to Age, Clustering and Type of Exposure - Bandium, Guinea-Bissau, 1979

Age (mos)	CFR (%) (deaths/no. ill)		
		Houses with multiple cases	
	Isolated cases	Index cases	Secondary cases
0-5	0%(0/1)		24%(4/17)
6-11	14%(1/7)	0%(0/15)	42%(11/26)
12-23	11%(2/19)	21%(3/14)	33%(14/43)
24-35	0%(0/10)	14%(2/14)	38%(14/37)
36-59	0%(0/10)	5%(2/38)	13%(5/39)
60+	33%(1/3)	6%(2/36)	0%(0/50)
Total	8%(4/50)	8%(9/117)	23%(48/212)

From Aaby, 1989

lated cases and index cases in houses with multiple cases have been very similar (see Table 4). This suggests that the contrast between single and multiple cases is not due solely to different socio-cultural and therapeutic practices or differences in genetic constitution, nor to the prevalence of complicating infections between families with many and with few susceptible children. Nevertheless, the difference between index and secondary cases could be confounded by higher general mortality in larger families whose proportion of secondary cases would also be higher. In the studies from Kenya (Aaby and Leeuwenburg, 1990) and Copenhagen (Aaby, 1988a) where this possibility has been examined, the relative risk (RR) between index and secondary cases was found to be the same in both small and large families. Maternal fatigue could also be a confounding factor if mothers gave less care for secondary cases when having to care for an index case already. However, an analysis of data from several epidemics in Bissau showed that secondary cases who had the same mother as the index cases did not have higher mortality (25%, 14/55) than secondary cases whose mother did not have to care for an index case (42%, 25/60) (Aaby et al. 1988c). Thus, differences in maternal care due to fatigue are unlikely to explain why secondary cases have higher CFR than index cases.

If no confounding factor can be found to explain the difference in mortality between index and secondary cases (Aaby, 1989), it is likely to be a biological phenomenon related to the difference in exposure. If so, a dose-response effect should be expected. This has been tested by examining whether the number of index cases had an impact on the CFR of the secondary cases. In the hospital records from Copenhagen, the CFR was significantly higher among children exposed to two or more index cases compared to a single index case (RR=1.81; 95% CI:1.05-3.11) (Aaby, 1990a). A similar tendency was found in a reanalysis of data from the Machakos area, in Kenya, where secondary cases exposed to a single index case had a CFR of 6% compared with 14% for those exposed to more than one index case (RR=2.5; 95% CI: 0.9-6.6) (Aaby and Leeuwenburg, 1990).

TABLE 5. Severity of Measles Infection According to Type of Exposure in Different Community Studies

Country (reference)	Age (yrs)	CFR% (deaths/number ill) Index	Secondary	Relative risk (95%CI)
Urban studies				
Guinea-Bissau (Aaby, 1989)	0-4	8%(10/128)	30%(48/162)	3.8(2.1-6.7)
Guinea-Bissau (Aaby et al. 1988a)	0-4	3%(1/37)	38%(10/26)	14.2(3.4-59.5)
Guinea-Bissau (Aaby et al. 1988a)	0-4	5%(3/66)	17%(14/81)	3.8(1.3-11.4)
England (Aaby et al. 1986a)	0-2	8%(4/48)	22%(8/36)	2.7(0.9-7.8)
Copenhagen (Aaby, 1988a)	0-2	11%(28/252)	27%(49/183)	2.4(1.6-3.6)
Rural studies				
Guinea-Bissau (Aaby, 1989)	0-4	7%(1/15)	38%(33/86)	5.8(1.4-24.5)
Senegal (Garenne and Aaby, 1990)	0-4	4%(8/198)	14%(37/226)	3.4(1.7-6.7)
Kenya (Aaby and Leeuwenburg, 1990)	0-4	3%(11/381)	7%(21/287)	2.5(1.3-4.9)
Bangladesh (Koster, 1988)	0-2	1%(1/134)	18%(4/22)	24.4(5.6-106.1)
Bangladesh (Bhuiya et al. 1987)	0-4	1%(38/2,551)	3%(17/630)	1.8(1.0-3.1)
Germany (Pfeilsticker, 1863)	0-14	2%(2/93)	11%(10/95)	3.7(1.1-12.1)
Ratio of severe cases				
Gambia (Lamb, 1988)	0-23	7%(3/41)	31%(4/13)	4.2(1.2-15.3)

From the same perspective, we should expect severity to increased with the intensity of contact. In a community study in Senegal, we have had an opportunity to examine the importance of variation in the closeness of contact (Garenne and Aaby, 1990). In this rural area, the Serer population lives in large compounds divided into several households. Within a household, the children may sleep in several different huts. In this setting, we assume a gradual increase in intensity of contact from a secondary case infected by another child from a different household in the same compound to a secondary case infected by someone from the same household but not sleeping in the same hut, to infection by another child sleeping in the same dwelling. As in Table 6, there

TABLE 6. Measles CFR by Age and Intensity of Exposure - Niakhar, 1983-1986

Age (mos)	Index cases	CFR%(deaths/number of cases) Secondary cases infected from: Same compound; different household	Same household; different hut	Same household; same hut
0-5	0%(0/4)	0%(0/2)	0%(0/6)	13%(2/15)
6-41	7%(8/115)	11%(10/89)	12%(18/152)	18%(31/174)
42-65	0%(0/79)	2%(1/50)	6%(4/66)	5%(4/77)
66+	0%(0/122)	0%(0/62)	0%(0/86)	2%(2/129)
Total	3%(8/320)	5%(11/203)	7%(22/310)	10%(39/395)
Odds Ratio	1.0	1.9	2.3	3.8

Based on Garenne and Aaby, 1990

was a corresponding increase in the case fatality ratio. Most studies have only analyzed the difference between index and secondary cases in the house or household. However, as the Senegalese study suggests, there may be grounds to examine the variation in intensity of contact between the infecting and the infected child.

Differences in the risk of becoming a secondary case seem to explain much of the known variation in measles mortality. Mortality is high where a high proportion of the children become secondary cases (Aaby, 1988b). For example, the contrast in measles mortality between Guinea-Bissau and Bangladesh, incomprehensible from the malnutrition theory, is logical from the exposure perspective. In Guinea-Bissau, 61% of the children under three years of age were secondary cases and the CFR was 25%, whereas in Bangladesh only 14% of the children of the same age group were secondary cases and the CFR was only 3%. Measles mortality is particularly high in West Africa, where larger families and high proportion of polygyny give children a much higher risk of becoming secondary cases (Aaby, 1988b). Consistent with the difference between West and East Africa, polygyny is more frequent in West Africa where married men had an average of 1.6-1.7 wives compared with only 1.2 in East Africa (Goody, 1973).

The apparent contradiction between a high mean age at infection and high CFR may also be comprehensible from the exposure perspective. In rural areas, the long interval between epidemics implies that more children from the same family are susceptibles and become ill in the same epidemic. The risk of being intensively exposed is thus higher in rural areas. A high mean age of infection indicates relative isolation of the community and a higher risk of having several susceptible children in a given family at the same time (Aaby et al. 1988a).

The lower mortality among unvaccinated children once immunization has been introduced can be explained along the same lines. Immunization reduces the risk of several children contracting infection during the same outbreak in the same family unit. The risk of becoming a secondary case is therefore also reduced (Aaby et al. 1988a).

Amplification of Severity/Mildness in the Community

Most studies on severity and transmission of infection have only emphasized the difference between index and secondary cases (Aaby, 1988b). However, since severe cases excrete more measles virus (Scheifele and Forbes, 1972) and presumably other pathogens as well, the severity of the secondary cases could depend on the severity of the index case. This has been tested among the children in the infectious disease hospital in Copenhagen, where both the index and the secondary case were hospitalized and clinical severity of both cases assessed. Compared with secondary cases infected by an index case without pneumonia, secondary cases infected by a severe index cases (pneumonia/ death) had significantly more pneumonia and a significantly higher CFR. The relative risk of dying was 4.6 times higher (95%, CI:2.1-10.1) for secondary cases infected by a severe index case compared with someone infected by an index case without pneumonia (Aaby, 1990). A similar observation was made in Kenya, where secondary cases infected by an index case who died had significantly higher mortality than a secondary case infected by an index case who survived (RR=4.7; 95% CI:1.6-13.4) (Aaby and Leeuwenburg, 1990). Though poor treatment of both the index case and the secondary case could play some role, these observations would seem to suggest that severe cases generate more severe secondary cases. The pattern of transmission of measles in the community may generate both a positive and a negative feedback. Where most index cases are mild, they will give rise to new, relatively mild index and secondary cases, the net result being a lowered mortality. However, where index cases are severe, they will transmit an even more severe disease to new index and secondary cases, ultimately producing higher mortality. It is probably this kind of process which causes measles to become particularly severe in institutions and in virgin-soil epidemics. For example, in a measles outbreak among soldiers in the Highland Regiment in Bedford in 1914-1915, the CFR increased from 2% (2/87) in the first month to 14% (63/442) during the two subsequent months of the epidemic (Kinnear, 1923).

As a special instance of this amplification pattern, we found in Senegal that the CFR increased exponentially with each new generation of cases. In the very large compounds, there could be several waves (Table 7) (Garenne and Aaby, 1990).

There is no adequate explanation for the decline in measles CFR in the industrialized world prior to the advent of antibiotics (Lancaster, 1952; Reves, 1985). However, the change may be related to the pattern of transmission and a process of mild feedback. Community studies from the turn of the century suggest that the proportion of secondary cases in Europe was as high as that found in Africa today (Aaby et al. 1986a; Pfeilsticker, 1863). Though there are no proper epidemiological studies of the transmission of measles under present day conditions in the industrialized countries, it seems likely that reduced family size as well as changes in schooling and public child care have led to a significant reduction in the proportion of secondary cases since the beginning of this century. It may also have been important that doctors in the 20s and 30s started to prevent or modify measles with the use of convalescence serum (and later gamma globulin). This treatment would have been applied precisely to the high risk group for measles mortality, namely, small children exposed to

TABLE 7. Measles CFR by Age and Generation of Cases (Wave) - Niakhar, 1983-1986

Age (mos)	Index cases	CFR% (deaths/number of cases) Secondary cases according to generation of cases 2nd wave	3rd wave	4th wave	5th wave
0-5	0%(0/4)	0%(0/13)	22%(2/9)	0%(0/1)	
6-41	7%(8/115)	12%(36/290)	12%(11/90)	29%(7/24)	45%(5/11)
42-65	0%(0/79)	4%(5/142)	5%(2/38)	0%(0/9)	50%(2/4)
66+	0%(0/122)	1%(1/177)	1%(1/81)	0%(0/15)	0%(0/4)
Total	3%(8/320)	7%(42/622)	7%(16/218)	14%(7/49)	37%(7/19)
Odds Ratio	1.0	2.3	3.6	5.5	16.1

From Garenne and Aaby, 1990

a sibling at home. The impact on mortality of this type of prophylaxis has not been properly assessed. In virgin soil epidemics among indigenous peoples, gamma globulin has had a marked effect on the CFR (Aaby et al. 1987b).

Though this reduction in frequency of secondary cases would explain a large part of the drop in mortality, it cannot explain the whole change, as there are still secondary cases in the industrialized world, and mortality is virtually nil. However, when the proportion of secondary cases in the community is reduced, an increasing number of index cases will be mild due to infection from mild index cases rather than severe secondary cases. Thus, secondary cases presumably became milder and the general severity of measles in the community may have been gradually reduced through a process of positive feedback.

In order to examine this process further, we have used the records from Copenhagen's infectious disease hospital. It will be seen in Table 8 that children who contracted measles in a day-care institution had significantly lower mortality than index cases who contracted measles from another child in the same apartment building (Aaby, 1988a). Presumably, exposure has been less intense in the kindergarten because children with measles have been kept at home, whereas playmates may have continued to have contact even if one child were sick. Since children infected in kindergartens also had much lower mortality than secondary cases, the process of establishing public or private day care institutions may have contributed to the drop in measles mortality by moving small children out of the home, away from infection by a school age sibling. Public child care may thus have reduced mortality by reducing the proportion of secondary cases, but it may also have led to the transmission of milder forms of measles. It will be seen in Table 9, that there was a marked variation in measles mortality from one year to another in the period from

TABLE 8. CFR in Measles According to Source of Infection and Age - Copenhagen, 1915-1925

Age (yrs)	CFR% (deaths/no. of cases) Source of infection		
	Day care	Neighbors, playmates	Secondary cases
0	13%(5/40)	40%(4/10)	29%(16/55)
1	5%(3/64)	22%(5/23)	29%(24/84)
2	6%(2/31)	22%(4/19)	20%(9/44)
0-2	7%(10/135)	25%(13/52)	27%(49/183)

Based on Aaby, 1988a

TABLE 9. CFR According to Age, Year and Frequency of Children Infected in Day Care Institutions. - Copenhagen, 1915-1925

Year	CFR% (no. of cases)			Percent infected in day care children <3 yrs
	All ages	< 3 yrs	Children < 3yrs, not in day care	
1915	10%(114)	16%(49)	20%(35)	29%(49)
1916	20%(295)	29%(123)	31%(116)	6%(123)
1917	10%(100)	21%(42)	20%(38)	10%(42)
1918	8%(153)	18%(66)	19%(58)	12%(66)
1919	14%(172)	23%(84)	25%(72)	14%(84)
1920	18%(244)	23%(125)	24%(115)	8%(125)
1921	8%(72)	10%(30)	11%(28)	7%(30)
1922	8%(167)	14%(69)	16%(57)	17%(69)
1923	16%(304)	29%(96)	30%(90)	6%(96)
1924	9%(275)	14%(108)	16%(83)	23%(108)
1925	10%(312)	12%(160)	14%(125)	22%(160)
Total	13%(2,208)	20%(952)	22%(817)	14%(952)

1915 to 1925. There was also a large variation in the proportion of children infected in a day care institution. Not surprisingly the case fatality rate was lower in the years where many children had contracted measles in a day care institution, since these children had lower CFR. However, data from Table 9 also indicate that the CFR for children below three years of age not infected in a kindergarten was lower in those years where a large proportion of all children had contracted infection in a day care institution (logistic regression, p=0.001, x^2=10.88, 1 df). These observations would seem to support the suggested pattern, more mild cases of measles will lead to less severe infection in the community in general.

Cross-sex Transmission

As an unexpected extension of the transmission perspective, we found in several studies in Guinea-Bissau that secondary cases infected by someone of the opposite sex had higher CFR compared with infection from someone of the same sex (Table 10) (Aaby et al. 1986b). This tendency did not depend on the classification of who infected whom, for it could be shown that mortality for children aged 6-59 months was higher in houses where one boy and one girl had measles together (26%) compared with houses with two boys or two girls (11%) (RR=2.65; 95% CI:1.20-5.84).

Something similar seems to have occurred in Copenhagen at the beginning of this century (Aaby, 1991). Mortality was significantly higher (p<0.05, Mantel-Haenszel, x^2=4.36) in families with one boy and one girl having measles during the same outbreak than in families with two boys or two girls (RR=1.85; 95% CI:1.04-3.30). Several other attempts have been made to examine this tendency. In a small outbreak in Keneba, in The Gambia, where the children were under close medical observation and there was no mortality, individuals infected by someone of the opposite sex were more likely to get pulmonary complications than those who contracted infection from someone of their own sex (RR=2.82, 95% CI:0.9-9.7) (Aaby and Lamb, unpublished observations). Since the severity of the index case could influence the result, we also examined whether sex-opposition or sex-sameness affected the increase in severity from index to secondary case; children infected by someone of the opposite sex were significantly more likely to have an increase in severity relative to the index case than children infected by someone of their own sex (p=0.026). In an area of Senegal, Niokholonko, under demographic surveillance since 1970, 196 deaths have been registered during three outbreaks of measles. It was found that children whose maternal sibling closest in age was of the opposite sex had a higher risk of dying of measles than those whose closest sibling was of the same sex (OR=1.71; 95% CI:1.16-2.53). Furthermore, in families with only two maternal siblings under 10 years of age, the relative risk of dying of measles during an outbreak was 1.60 times higher (95% CI: 1.08-2.37) in families with one boy and one girl than in families with two boys or two girls (Pison and Aaby, 1990). Several other studies from Senegal, Greenland, Germany and Kenya have suggested similar tendencies (author's unpublished observations). Published case reports of fatal cases of measles with information on the sex of both the secondary case and the index case indicate the same pattern of higher mortality when infection is contracted from someone of the opposite sex (Aaby et al. 1986b).

The increased severity associated with intensive exposure and cross-transmission of infection may help explain differences in mortality by sex. In European studies, there are several indications that girls have caught measles

TABLE 10. CFR Among Secondary Cases of Measles by Age and by Sex of Infecting Child - Guinea-Bissau, 1979-1983

Age group	CFR %(deaths/cases) Same-sex transmission M to M	F to F	Opposite-sex transmission M to F	F to M
6-35 mos	26%(9/35)	16%(5/31)	49%(22/45)	36%(16/45)
36-59 mos	10%(1/10)	6%(1/16)	20%(3/15)	18%(3/17)
Total	22%(10/45)	13%(6/47)	42%(25/60)	31%(19/62)

Based on Aaby et al. 1986b

infection at a younger age than boys (Aaby et al. 1984d). Girls are therefore more likely to have caught infection outside the home. The implication of this is that boys have a higher risk of getting infected at home as a secondary case. This may explain the higher mortality among boys in the industrialized countries (Aaby, 1983b). Higher measles mortality for girls has been reported from several communities, mainly Muslim (Bhuiya et al. 1987; Fargues and Nassour, 1988; McGregor, 1964; Monastiri, 1961). In these cases, for cultured reasons, girls are more likely to be bound to the home in these societies. Boys presumably have a greater chance of getting out and, therefore, of contracting infection outside the home. As a consequence, girls suffer a dual disadvantage: a higher risk of becoming a secondary case as well as a higher risk of being infected by someone of the opposite sex.

DELAYED IMPACT OF MEASLES AND INTENSIVE EXPOSURE

The importance of the transmission approach may go even further. Most studies have only dealt with acute measles mortality (within one month of the rash). Recent studies, however, suggest that measles may have a profound effect on morbidity and mortality after the acute infection (Aaby et al. 1987a). Though there are no studies of the determinants of this delayed impact, it seems likely that excess morbidity and mortality is related to intensive exposure during the acute infection. For example, among the children hospitalized in Copenhagen, secondary cases (3/152) had a higher risk of dying after the first 30 days of measles infection than did index cases (3/472) (RR=3.11; 95% CI: 0.7-14.0). In rural Senegal, the secondary cases had a significantly higher risk of dying in the year following measles infection than did index cases (OR=3.5) (Table 11) (Garenne and Aaby, 1990).

In Bissau, we have also found that children exposed at home to measles during the first six months of life had a mortality 3 times higher than community controls between six months and five years of age (34% vs 11%) (Aaby et al. 1990a). In a Cox regression analysis taking the known background factors into consideration, the mortality hazards ratio was 5.7 times higher (95% CI:2.7-12.0) for the exposed than for the control children. The difference in mortality was

TABLE 11. Post-Measles Mortality Rate (6-52 Weeks) by Age and Intensity of Exposure - Niakhar, 1983-1986

Age (mos)	Post measles mortality (deaths/measles survivors) Index cases in compound	Secondary cases in compound
0-5	0%(0/4)	5%(1/21)
6-41	2%(2/107)	7%(25/356)
42-65	1%(1/79)	2%(3/184)
66+	0%(0/122)	1%(2/275)
Total	1%(3/312)	4%(31/836)
Odds Ratio	1.0	3.5

Based on Garenne and Aaby, 1990

significant in each of the age intervals 6-11, 12-23 and 24-35 months. This delayed excess mortality was found both for children who had measles, though being below six months of age, and exposed children who did not have clinical measles. The exposed children were particularly likely to die of diarrhoea. These tendencies have not been examined elsewhere, but have been systematic in several outbreaks in Guinea-Bissau (Aaby et al. 1990a). The observation of delayed mortality among both previous measles cases and children with exposure early in life would suggest that measles may be indirectly responsible for much more of childhood mortality than is usually assumed. The mechanisms behind this kind of delayed excess mortality seem to need further study. Some form of persistent infection and immunosuppression may be involved. The effect of exposure may go even further. In two outbreaks in Guinea-Bissau, it has been found that children of mothers exposed to measles during pregnancy had significantly higher perinatal mortality (Table 12) (Aaby et al. 1988b) as well as increased post-perinatal mortality (Aaby et al. 1990b).

While these tendencies have not been studied elsewhere, there are some indications that exposure at an early age may have consequences for health in later life. The delayed fatal form of measles known as subacute sclerosing panencephalitis (SSPE) occur mostly among children who had measles early in life. This suggests that the children are more likely to have been intensively exposed as secondary cases (Aaby et al. 1984d). A study from Denmark found that adults with no history of measles in childhood had four times higher risk of cancers and immunoreactive diseases than did controls who reported having had measles in childhood (Rønne, 1985). Since all suspects had antibodies to measles, the most likely explanation for not reporting a measles infection would be exposure to measles while still partly protected by maternal antibodies or immunoglobulin. Early exposure therefore seemed to be connected with excess morbidity in later life.

TABLE 12. Perinatal Mortality Among Children of Mothers Exposed to Measles during Pregnancy - Bandim, Guinea-Bissau, 1979

Type of mortality	Perinatal mortality risk (deaths/at risk)		
	Exposed	Controls	OR(95%CI)[a]
Stillbirths	6.5%(7/107)	1.4%(5/346)	4.8(1.7-13.8)
Died 1st week	9.0%(9/100)	2.6%(9/341)	3.6(2.3-5.6)
Perinatal	15.0%(16/107)	4.0%(14/346)	4.2(2.1-8.5)

[a] Odds ratio (95% confidence interval)

Based on Aaby et al. 1988

TESTING HYPOTHESES

Consistency Between Data and Interpretation

The best hypothesis is capable of accounting for the known variation in data and can solve the contradictions inherent in other interpretations. Measured against these criteria, there is little to indicate that host factors such as malnutrition and young age at infection play a major role for the severity of measles infection. The host factor hypotheses are not able to explain the major contrasts in measles mortality and some of their predictions are in glaring contrast to the known epidemiology of severe measles. On the other hand, transmission factors seem capable of explaining much more of the variation in mortality; differences in mortality between historical periods, according to crowding, between regions, between rural and urban areas, between different ages and between the sexes correspond to differences in the intensity and type of exposure. Patterns contradictory to the host factor hypotheses, such as the difference between Africa and Asia and between rural and urban areas, are compatible with the transmission factor approach.

Testing the Implications

In health related matters, the ultimate test of a hypothesis is the investigation of its preventive implications. The different hypotheses of the determinants of severe measles lead to very different assessments of the value of specific disease intervention (vaccination, immunoprophylaxis). If measles primarily kill weak and malnourished children, measles immunization would have a limited effect on survival (Hendrickse, 1975; Mosley, 1985; The Kasongo Project Team, 1981). However, if the major determinant of severe measles is intensity of exposure, there is no reason that it should be the particularly weak children who die of measles (Aaby et al. 1984c). They would not be more likely to die of other infections. Immunization against measles should therefore have significant impact on survival. All the available community studies from Nigeria, Guinea-Bissau, Senegal, Zaire, Bangladesh and Haiti, summarized in

TABLE 13. Efficacy of Measles Vaccine Against Death - Different Studies

Country	Age at Vaccination (mos)	Period of follow-up (mos)	Mortality (%) for UV[a]	P[b]	V[c]	Vaccine efficacy against death (95% CI)
Nigeria[d]		18	-	12%(25)	0%(23)	100%
Zaire[e]	7-9	30	7.0-9.5%	-	3.8%	46-60%
Guinea-Bissau[f]	6-35	12	14.3%(70) -	-	1.9%(361)	87%(70-94)
Guinea-Bissau[g]	7-24	24	-	13.2%(53)	4.8%(124)	63%(2-86)
Guinea-Bissau[h]	9-23	24	-	-	-	66%(32-83)
Senegal[i]	6-35	30	33.6%	-	23.2%(602)	31%
Haiti[j]	6-13	9-39	6.6%(1,056)		1.3%(235)	85%(36-96)
Bangladesh[k]	9	9-60	-	-	-	36%(21-48)
Bangladesh[l]		9-60	- (8,135)		- (8,135)	46%(35-54)

a UV = unvaccinated

b P = placebo

c V = vaccinated

d A small placebo study carried out in the beginning of the 1960s (Hartfield and Morley, 1963).

e The study in Kasongo, Zaire, was carried out in the mid-1970s. Vaccination was introduced in one area and mortality compared with a neighbouring, comparable area and with data from the same two areas prior to the introduction of measles vaccination (The Kasongo Project team, 1981).

f The study was carried out in 1980 in the capital of Guinea-Bissau. Mortality has been compared for children who attended a child examination and were vaccinated against measles and children did not attend mostly because they were temporarily absent. In the year prior to the introduction of vaccination, no differences in mortality between children who attended and those who did not were observed (Aaby et al. 1984c).

g The study carried out in 1984-1986 represents a natural experiment. When blood samples taken in connection with measles vaccination were analyzed with a delay of two years, it turned out that during a short period none of the children had seroconverted and can therefore be considered to have received a placebo (Aaby et al. 1989).

h The study carried out between 1984 and 1987 compared mortality for children who received vaccination and those who did not in two districts in the capital of Guinea-Bissau (Aaby et al. 1990c).

i Two systematic measles vaccination campaigns were carried out in one rural area in Senegal in 1965 and 1967. Mortality has been compared with one area where measles vaccination was not available (Garenne and Cantrelle, 1986).

j The study compares mortality for children who had participated in a serological vaccination study in 1982. The estimate of vaccine efficacy against death takes account of background factors with a significant impact on mortality (socio-economic status, literacy, knowledge and use of oral rehydration and birth interval) (Holt et al. 1990).

k The study from the Matlab area in 1982-1984 compares vaccination status for 536 children who died and 1072 controls (Clemens et al. 1988).

l This study of 8135 vaccinated children and controls from the Matlab area covers 1982-1985. The study partially overlaps with the previous study(Koenig et al. 1991).

Table 13, indicate reductions in child mortality after the age of vaccination of at least 30%, and seven of the studies found a reduction of at least 45-50%. None of these studies have failed to find a marked effect on mortality. Most of the studies have potential methodological problems because they are not randomized, double blind trials. However, in the study which compared mortality among children who seroconverted with those children who failed to seroconvert due to having received an ineffective vaccine (placebo), mortality was 3 times higher for the non-seroconverters than for the seroconverters (Aaby et al. 1989). In all studies, the reduction in mortality was greater than expected on the basis of the relative importance of acute measles mortality for overall child mortality. For example, in Bangladesh, the reduction in mortality between 10 and 60 months of age was 36% but measles accounted only for 4% of deaths (Clemens et al. 1988). Hence, it seems that immunization is highly effective in preventing both acute and delayed mortality following measles infection.

MECHANISMS OF SEVERE DISEASE

The three transmission factors presented here have so far turned out to be systematic. There are therefore good reasons to discuss the mechanisms explaining their impact on severity of infections.

There seems to be two different mechanisms which could explain the greater severity associated with intensive exposure: a) a higher risk of intercurrent infections, and/or b) absorption of a higher dose of measles virus due to the intensive contact with the index case, secondary cases presumably have a higher risk of getting complicating infections than the index case itself. It also seems reasonable to assume that secondary cases receive and absorb a larger dose of measles virus than the index case who had a brief contact with a child from a different family; for example, it has been shown that more immunoglobulin was needed to neutralize infection under conditions of intensive exposure (Karelitz and Karelitz, 1938).

Secondary cases should have been exposed to the same number of complicating infections whether they were susceptibles or had partial immunity due to maternal antibodies, serum prophylaxis or vaccination. If the risk of complicating infections were the major determinant of the severity of measles infection, susceptibles and children with partial immunity should have similar CFRs. However, all studies of children with partial immunity indicate that they have lower mortality than susceptible children of comparable ages (Aaby et al. 1985, 1986c, 1986d; Herrman, 1917). Therefore, the risk of contracting intercurrent infection is unlikely to be the single major mechanism explaining the higher mortality of secondary cases.

The risk of contracting infection is clearly related to the dose of infection (Sabin, 1983). However, it has generally been considered that the dose of absorbed virus cannot be important for the outcome of infection; any difference in initial dose is assumed to be levelled out within 12-24 hours due to rapid multiplication of the virus. However, several types of epidemiological evidence seem to contradict this line of reasoning.

Since a high dose leads to a shorter period of incubation in animal studies (Aaby et al. 1985), we have examined the period between rashes for index and secondary cases as an approximation for the incubation period among children hospitalized in Copenhagen in the period 1915-1925. Compared with the survivors, the fatal secondary cases clearly had a much shorter interval between rash in the index case and own onset of eruption (9.3 days (SD+/-2.0)) compared with secondary cases who survived (11.9 days (+/-2.6)) ($p<0.05$, Wilcoxon two sample test). Thus, fatal cases apparently have a shorter period of incubation than survivors (Aaby, 1990). The difference may be even stronger because the prodromal period has been found to be prolonged in severe cases (Aaby et al. 1986a). Using the interval between rashes as an approximation for the period of incubation will therefore lead to an overestimation of the length of the period in severe cases. Other studies have likewise suggested that severe cases have a shorter period of incubation than milder cases of measles (Aaby, 1989).

The dose of absorbed virus presumably depends both on the intensity of exposure and the degree of susceptibility. No study seems to have measured the variation in intensity of exposure and dose of virus transmitted. However, there are studies which can be said to have measured susceptibility. Where the host has partial immunity it seems reasonable to assume that the dose of absorbed virus is reduced. We should therefore expect to find prolongation of the period of incubation, shorter prodromes, and lower risk of complications and mortality. Indeed, all studies seem to indicate this (Aaby, 1989).

The available data on the period of incubation and effect of partial immunity suggest that an imbalance between dose of infection and measles immune capacity is an important mechanism in the pathogenesis of severe measles. In developing countries, most children are probably at risk of complicating infections. Therefore, the more severe complications, e.g. longer diarrhoea (Koster et al. 1981), seen in fatal cases may depend more on the extent of immunosuppression induced by measles than on variation in the risk of intercurrent infections. In contrast to the complication hypothesis, the dose hypothesis is compatible with all of the epidemiological tendencies noted here: a) increased severity of secondary cases, b) shorter period of incubation in severe cases, and c) reduced severity of measles infection in children with partial immunity.

The amplification pattern as described here is presumably related to the dose of infection model (Garenne and Aaby, 1990) since severe cases apparently excrete more measles virus than milder cases (Scheifele and Forbes, 1972). However, variation in the likelihood of mild and severe cases transmitting complications may also play a role for the resulting pattern.

There is no simple explanation of the higher mortality associated with cross-sex transmission of infection. It is not obvious that children of opposite sex have more intensive contact. In fact, available studies indicate that contact seems more likely between individuals of the same sex (Aaby et al. 1986b). It seems probable therefore, that a biological mechanism is involved. Since virus takes material from the cell it has grown in, it might be that male and female-grown virus have different potential for infecting cells of the opposite sex (Aaby et al. 1986b).

There is clearly a need for further studies of the mechanisms suggested here, since they do not correspond to the usual assumptions about causes of severe infection.

CONCLUSION: HOST OR TRANSMISSION FACTORS

In much of medical research, the underlying perspective has been to search for the distinguishing features of the individual who becames severely ill, those having these features being weaker than others. We assume that there is something wrong with the child who dies. For example, "the children whose death might be prevented by measles vaccine are at risk of dying not because of the severity of measles per se, but because they are on the "road to death", and their nutritional status is so poor that they are more likely to die of any infectious disease. Thus, preventing a death among these children may not necessarily save a life, but only change the cause of death" (Mosley, 1985).

For measles, this paradigm has not been able to provide an acceptable interpretation of the known data on severity of infection and impact of prevention. Rather the problem lies in the infection itself, i.e. in observing how the severe infections were contracted. Among such transmission factors 1) intensity of exposure, 2) amplification of severity/mildness in the community, and 3) cross-sex transmission seem to be important. Though further studies of the underlying mechanisms are clearly needed, it seems likely that the dose of absorbed virus plays an important role in explaining both the impact of intensity of exposure and the amplification of severity or mildness. The delayed impact on morbidity and mortality following measles infection is not sufficiently recognized. It seems likely that delayed morbidity and mortality is related to the intensity of exposure during acute infection. In contrast to host factor approaches, the transmission perspective emphasizes disease specific interventions. All available studies strongly confirm the importance of such specific interventions against measles infection for child survival.

Comparable studies do not exist for other infections. However, secondary cases of chickenpox (Ross, 1962) and polio (Siegel and Greenberg, 1953) have been found to be more severe. Furthermore, data from Guinea-Bissau indicate that male-female twins have a higher risk of post-neonatal mortality compared to same sex twin pairs, and it is a risk factor for post-neonatal child mortality if the closest older sibling is of the opposite sex (Aaby and Mølbak, 1990). These observations make it likely that transmission factors play a role in diseases other than measles. Likewise, some of the childhood infections may have delayed consequences for morbidity and mortality which have not yet been discovered. Given that the contrasting hypotheses have very different implications for our understanding of disease patterns and preventive priorities, future studies should examine the relative importance of host and transmission factors.

REFERENCES

Aaby P (1985) Epidemics among Amerindians and Inuits. A preliminary interpretation. In: Brøsted J, Dahl J, Gray A, Gulløv HC, Henriksen G, Jørgensen JB, Kleivan I (eds): Native Power. Bergen: Universitetsforlaget, pp. 329-339.

Aaby P (1988a) Severe measles in Copenhagen, 1915-1925. Rev Infect Dis 10:452-456.

Aaby P (1988b) Malnutrition and overcrowding-exposure in severe measles infection. A review of community studies. Rev Infect Dis 10:478-491.

Aaby P (1989) Malnourished or overinfected. An analysis of the determinants of acute measles mortality. Dan Med Bull 36:93-113.

Aaby P (1990) Patterns of exposure and severity of measles infection. Copenhagen 1915-1925 (submitted).

Aaby P (1991) Severity of measles and cross-sex transmission of infection in Copenhagen, 1915-1925 (submitted).

Aaby P, Clements J (1989) Measles immunization research. A review. Bull WHO 67:443-448.

Aaby P, Leeuwenburg J (1990) Patterns of transmission and severity of measles infection. A reanalysis of data from the Machakos area, Kenya. J Infect Dis 161:171-174.
Aaby P and Mølbak K (1990) Sex opposition as a risk factor for child mortality. Br Med J 301:143-145.
Aaby P and Lamb WH (1991) Sex and transmission of measles in a Gambian village (submitted).
Aaby P, Bukh J, Lisse IM, Smits AF (1981a) Measles vaccination and child mortality. Lancet 2:93.
Aaby P, Bukh J, Lisse IM, Smits AJ, Smedman L, Jepsson O, Lindeberg A (1981b) Breastfeeding and measles mortality in Guinea-Bissau. Lancet 2:1231.
Aaby P, Bukh J, Lisse IM, Smits AJ (1983a) Measles mortality, state of nutrition, and family structure: A community study from Guinea-Bissau. J Infect Dis 147:693-701.
Aaby P, Bukh J, Lisse IM, Smits AJ (1983b) Les hommes sont-ils plus faibles ou leurs soeurs parlent-elles trop? Essai sur la transmission des maladies infectieuses. Anthropologie et Societe 7:2:47-59.
Aaby P, Bukh J, Lisse IM, Smits AJ (1984a) Overcrowding and intensive exposure as determinants of measles mortality. Am J Epidemiol 120:49-63.
Aaby P, Bukh J, Lisse IM, Smits AJ, Gomes J, Fernandes MA, Indi F, Soares M (1984b) Determinants of measles mortality in a rural area of Guinea-Bissau: Crowding, age and malnutrition. J Trop Pediatr 30:164-168.
Aaby P, Bukh J, Lisse IM, Smits AJ (1984c) Measles vaccination and reduction in child mortality: a community study from Guinea-Bissau. J Infect 8:13-21.
Aaby P, Bukh J, Lisse IM, Smits AJ (1984d) Risk factors in subacute sclerosing panencephalitis (SSPE): Age- and sex-dependent host reactions or intensive exposure. Rev Infect Dis 6:239-250.
Aaby P, Coovadia H, Bukh J, Lisse IM, Smits AJ, Wesley A, Kiepiela P (1985) Severe measles: A reappraisal of the role of nutrition, overcrowding and virus dose. Med Hypoth 18:93-112.
Aaby P, Bukh J, Lisse IM, Smits AJ (1986a) Severe measles in Sunderland, 1885: A European-African comparison of causes of severe infection. Int J Epidemiol 15:101-7.
Aaby P, Bukh J, Lisse IM, Smits AJ (1986b) Cross-sex transmission of infection and increased mortality due to measles. Rev Infect Dis 8:138-43.
Aaby P, Bukh J, Leerhøy J, Lisse IM, Mordhorst CH, Pedersen IR (1986c) Vaccinated children get milder measles infection: a community study from Guinea-Bissau. J Infect Dis 154:858-63.
Aaby P, Bukh J, Hoff G, Leerhøy J, Lisse IM, Mordhorst CH, Pedersen IR (1986d) High measles mortality in infancy related to intensity of exposure. J Pediatr 109:40-44.
Aaby P, Clements J, Cohen N (1987a) Key issues in measles immunization research: A review of the literature. Geneva: WHO, EPI/GAG, WP.10.
Aaby P, Bukh J, Hoff G, Lisse IM, Smits AJ (1987b) Humoral immunity in measles infection: A critical factor? Med Hypoth 23:287-302.
Aaby P, Bukh J, Lisse IM, da Silva CM (1988a) Measles mortality decline: Nutrition, age at infection or exposure? Br Med J 296:1225-1228.
Aaby P, Bukh J, Lisse IM, Seim E, da Silva MC (1988b) Increased perinatal mortality among children of mothers exposed to measles during pregnancy. Lancet 1:517-22.
Aaby P, Bukh J, Lisse IM, da Silva CM (1988c) Measles mortality: Further community studies on the role of overcrowding and intensive exposure. Rev Infect Dis 10:474-477.

Aaby P, Bukh J, Kronborg D, Lisse IM, da Silva MC (1990a) Delayed excess mortality after exposure to measles during the first six months of life. Am J Epidemiol 132:211-219.
Aaby P, Bukh J, Lisse IM, Seim E, da Silva MC (1990b) Increased post-perinatal mortality among children of mothers exposed to measles during pregnancy. Am J Epidemiol 132:531-539.
Aaby P, Knudsen K, Jensen TG, Thaarup J, Poulsen A, Sodemann M, da Silva MC, Whittle H (1990c) Measles incidence, vaccine efficacy and mortality in two urban African areas with high vaccination coverage. J Infect Dis (in press).
Aaby P, Pedersen IR, Knudsen K, da Silva MC, Mordhorst CH, Helm-Petersen NC, Hansen BS, Thaarup J, Poulsen A, Sodemann M, Jakobsen M (1989) Child mortality related to seroconversion or lack of seroconversion after measles vaccination. Pediatr Infect Dis J 8:197-200.
Anderson RM, May RM (1982) Directly transmitted infectious diseases: Control by vaccination. Science 215:1053-1060.
Babbott FL, Gordon JE (1954) Modern measles. Am J Med Sci 228:334-61.
Barclay AJG, Foster A, Sommer A (1987) Vitamin A supplements and mortality related to measles: a randomised clinical trial. Br Med J 294:294-296.
Bhaskaram P, Reddy V, Rajk S, Bhatnager RC (1984) Effect of measles on the nutritional status of preschool children. J Trop Med Hyg 87:21-25.
Bhaskaram P, Madhusudhan J, Radhakrishna RV, Reddy V (1986) Immune responses in malnourished children with measles. J Trop Pediatr 32:123-6.
Bhuiya A, Wojtyniak B, D'Souza S, Nahar L, Shaikh K (1987) Measles case fatality among under-fives: a multivariate analysis of risk factors in a rural area of Bangladesh. Soc Sci Med 1987; 24:439-43.
Black FL, Pinheiro F de P, Hierholzer WJ, Lee RV (1977) Epidemiology of infectious diseases: the example of measles. In: Health and Disease in Tribal Societies. Amsterdam: Elsevier, pp. 115-135.
Burnet M (1968) Measles as an index of immunological function. Lancet 2:610-612.
Chen LC, Chowdhury AKMA, Huffman SL (1980) Anthropometric assessment of energy-protein malnutrition and subsequent risk of mortality among pre-school aged children. Am J Clin Nutr 33:1836-45.
Christensen PE, Schmidt H, Bang HO, Andersen V, Jordal B, Jensen O (1953) An epidemic of measles in Southern Greenland, 1951. Acta Med Scand 144:313-22, 430-541.
Clemens JD, Stanton BF, Chakraborty J, Chowdhury A, Rao MR, Ali M, Zimicki S, Wojtyniak B (1988) Measles vaccination and childhood mortality in rural Bangladesh. Am J Epidemiol 128:1330-1339.
Coovadia HM, Wesley A, Hammond MG, Kiepiela P (1981) Measles, histocompatibility leukocyte antigen polymorphism, and natural selection in humans. J Infect Dis 144, 142-147.
Coovadia HM, Kiepiela P, Wesley AG (1984) Immunity and infant mortality from measles. S Afr Med J 65:918-921.
Davis R (1982) Measles in the tropics and public health practice. Trans Roy Soc Trop Med Hyg 76:268-75.
Debroise A, Sy I, Satgé P (1967) La rougeole en zone rural. L'Enfant en Milieu Tropicale 38:20-36.
Fargues P, Nassour O (1988) Douze ans de mortalite urbaine au Sahel. Paris: Presses Universitaires de France.
Foster SO (1984) Immunizable and respiratory diseases and child mortality. Pop Dev Rev 10(suppl):119-140.

Gallais H, Kadio A, Odehouri K, Cornet C (1981) La rougeole a Abidjan. A propos de 500 cas. Bull Soc Pathol Exot Filiales 74:283-292.
Garenne M, Cantrelle P (1986) Rougeole et mortalite au Senegal: etude de l'impact de la vaccination effectue a Khombole 1965-1968 sur la survie des enfants. In: P Cantrelle, S Dormont, P Farques, J Goujard, J Guignard, C Rumeau-Rouquette (eds). Estimation de la mortalite du jeune enfant (0-5 ans) pour guider les actions de sante dans les pays en developpement. Paris: INSERM, pp. 515-32.
Garenne M, Aaby P (1990) Pattern of exposure and measles mortality in Senegal. J Inf Dis 161:1088-1094.
Goody, J (1973). The Character of Kinship. Cambridge University Press, Cambridge, pp 175-190.
Hartfield J, Morley D (1963) Efficacy of measles vaccine. Journal of Hygiene (Cambridge) 61:143-147.
Hendrickse RG (1975) Problems of future measles vaccination in developing countries. Trans R Soc Trop Med Hyg 69:31-34.
Hermann C (1917) Observations on measles. Arch Pediatr 34:38-42.
Hertz JB, Sørensen TB, Vejerslev L (1977) Morbillidødsfald i Danmark 1958-1969. Ugeskr Laeger 138:589-593.
Holt EA, Boulos R, Halsey NA, Boulos IM, Boulos C (1990) Childhood survival in Haiti: Protective effect of measles vaccination. Pediatrics 85:188-194.
Hull, HF(1988). The effect of crowding on measles mortality in The Gambia, 1981. Rev Infect Dis 10:463-467.
Hull HF, Williams PJ, Oldfield F (1983) Measles mortality and vaccine efficacy in rural West Africa. Lancet 1:972-975.
Karelitz S, Karelitz RF (1938) The significance of the conditions of exposure in the study of measles prophylaxis. J Pediatr 13:195-207.
Kinnear W (1923) Epidemic of measles in the highland division at Bedford, 1914-1915. Edinb Med J 30:593-599.
Koenig MA, Khan MA, Wojtyniak B, Clemens JD, Chakraborty J, Fauveau V, Phillips JF, Akbar J, Barua US (1991) The impact of measles vaccination upon childhood mortality in Matlab, Bangladesh. Bull WHO (in press).
Koster FT (1988) A review of measles in Bangladesh with respect to mortality rates among primary versus secondary cases. Rev Infect Dis 10:471-473.
Koster FT, Curlin GC, Aziz KMA, Haque A (1989) Synergistic impact of measles and diarrhoea on nutrition and mortality in Bangladesh. Bull WHO 59:901-908.
Lamb, WH (1988)Epidemic measles in a highly immunized rural West African (Gambian) village. Rev Infect Dis.
Lancaster HO (1952) The mortality in Australia from measles, scarlatina and diphtheria. Med J Aust 2:272-276.
Lancet (1985) Infant mortality and family structure. Lancet 2:650
Maxcy KF (1948) Principles and methods of epidemiology. In: Rivers TM (ed) Viral and Rickettsial Infections of Man. Philadelphia: Lippencott.
McGregor IA (1964) Measles and child mortality in the Gambia. West Afr Med J 14: 251-257.
Meunier H (1898) Sur un symptome nouveau de la periode precontagieuse de la rougeole et sur sa valeur prophylactique. Gazette Hbdo Medicine Chir. 1057-1061.
Mims CA (1976) The pathogenesis of infectious diseases. London: Academic Press.
Monastiri H (1961) Quelques donnees statestiques relatives a la mortalite par rougeole dans la Commune de Tunis. La Tunisie Medical 39:179-187.

Morley D (1973) Paediatric priorities in the developing world. London: Butterworth.
Mosley WH (1985) Will primary health care reduce infant and child mortality? A critique of some current strategies. With special reference to Africa and Asia. In Vallin J, Lopez AD (eds), Health Policy, Social Policy and Mortality Prospects.Liege: Ordina, pp. 103-137.
Neiburg P, Dibley MJ (1985) Risk factors for fatal measles infections. Int J Epidemiol 15:309-11.
Osagie, H.F. (1986) Delayed mortality and morbidity 12 months after measles in young children in Nigeria. Institute of Child health, University of London (MSc Thesis), London.
Pfeilsticker A (1863) Beitrage zur Pathologie der Masern mit besonderer Berucksichtung der statistischen Verhaltnisse. Tubingen: L Fr Fues.
Picken RMF (1921) The epidemiology of measles in rural and residential area. Lancet 1:1349-53.
Pison G, Aaby P (1991) Increased risk of measles mortality for children with a sibling of the opposite sex among the Fula Bande and Niokholonko, Senegal (submitted).
Reder J (1918) Die Bedeutung der Masern in Sammelneiderlassungen nach den im k.k. Flüchtlingslager Gmünd gemachten Wahrnehmungen. Z Kinderheilk 18:355-370.
Reves R (1985) Declining fertility in England and Wales as a major cause of the twentieth century decline in mortality. The role of changing family size and age structure in infectious disease mortality in infancy. Am J Epidemiol 122:112-26.
Ross AH (1962) Modification of chickenpox in family contacts by administration of gamma globulin. N Engl J Med 267:369-376.
Sabin A (1983) Immunization against measles by aerosol. Rev Infect Dis 5:514-523.
Scheifele DW, Forbes CE (1972) Prolonged giant cell excretion in severe African measles. Pediatrics 50:867-873.
Shahid NS, Claquin P, Shaikh K, Zimicki S (1983) Long-term complication of measles in rural Bangladesh. J Trop Med Hyg 87:77-80.
Siegel M, Greenberg M (1953) Passive immunization in relation to multiple cases of poliomyelitis in the household. N Engl J Med 249:171-176.
Sinha DP (1977) Measles and malnutrition in a West Bengal village. Trop Geogr Med 29:125-134.
The Kasongo Project Team (1981) Influence of measles vaccination on survival pattern of 7-35-month-old children in Kasongo, Zaire. Lancet 1:764-767.
van de Walle, E (1986) Anatomie d'une epidemie de rougeole vue par la lorgnette d'une enquete a passages repetes. In: Cantrelle P, Dormont S, Farques Ph, Goujard J, Guignard J, Rumeau-Rouquette C.(eds) Estimation de la mortalite du jeune enfant (0-5 ans) pour guider les actions de sante dans les pays en developpement. INSERM, Paris, pp 419-28.
Walsh JA (1983) Selective primary health care: strategies for control of disease in the developing world. IV. Measles. Rev Infect Dis 5:330-40.

DISCUSSION

Kilbourne E (Mount Sinai School of Medicine, New York, NY):

I hardly know where to start. This is really a fascinating exposition, and I found myself, at the beginning, very hostile and resistant to what you were saying, but you completely demolished my mental reservations as you went along. Thinking mechanistically, as a virologist, there are a number of comments I would like to make. First of all, it's not customary to think of much variability in the incubation period of viral diseases. Also, I think our prejudices are that most respiratory viral infections are initiated by relatively few particles under conditions of aerosol transmission or perhaps, sort of a hit probability matter. In experimental animal models, which we have used, for example, for influenza, you get somewhat conflicting results in the sense that there are models, as you pointed out, where one can actually relate virus dose to the shortness of the incubation period. For the most part, those are with enormous doses of virus, but at the same time, if you have a highly virulent strain of virus, in general, you get really no significant difference between one infectious dose and 10,000. So, I think that has to be considered, as far as the dose business is concerned. On the other hand, what's more attractive to me, and I don't think this argues against anything you are saying, but perhaps give it a virologic mechanism, is the thought that RNA viruses are extremely mutable. We're beginning to appreciate this more and more. It's not restricted to diseases such as influenza. But the mutation rate, probably of measles virus, is not much different than that of influenza which is notorious for changing its guise continually. I would put it to you that it is not unlikely that what you're having in going from the primary to the secondary cases, is perhaps a selection of more virulent virus in that period. That would also be consistent with the slowness in starting the propagation of an epidemic and that you may have to sort of rejuvenate the virus by successive host passage as this happens. I have a few other questions. This is really fascinating. I think also, your argument about the partially immune subjects must be tempered a little bit, because it might be that they would have the same risk of complications from some other source. At least in the United States, I think the mortality for measles is chiefly occasioned by secondary bacterial infection such as pneumococcal or streptococcal pneumonia. One has to think also, of co-pathogens. Going along with increased virulence in a small community might also be the greater opportunity for endogenous respiratory tract pathogens to get going, like the activation of streptococci, which apparently fare better in a virus damaged trachea than they do otherwise. So, that's something else to consider. By in large, I think that these are fascinating observations. It's time the virologists began to think about them.

Aaby P:

Many thanks for the comments. Your first comment, as far as I understood you, was that the exposure effect could be a question of selection of virulent virus. This is in no way, my domain, and from my reading, it has been that measles virus is extremely stable. I have played with the idea of the selection of viral strains. If this was true, it should produce within a community a tendency of gradual increase in severity. We don't find that in a given area. You don't find any increase in severity throughout the epidemic. You find the increase in severity in closed societies, so to speak. Within a compound or within an institution you would have an increase in severity and a larger production of virus, and it is probably so that more severe cases produce larger amounts of virus. I would be pleased for any suggestion of how to actually test these things about virulence of strains. The second thing, about the partial immunity and consideration of the co-pathogens, you're certainly right. I attempted to hammer in a message and I am trying to emphasize these because I have always been told that virologists usually don't consider this. It's usually assumed that after the initial infection it doesn't matter very much whether your dose was 10- or 100- or 1,000-fold. This other situation may not have been sufficiently considered because we may deal with much larger differences in terms of infective dose. When sleeping with a sick measles child, you probably receive a million times more virus than you do from playing with a child outside the house. The other reason that I consider the dose attractive, is because it could provide some kind of explanation for long-term consequences. I showed you some data on longer term morbidity. I am not talking about the following month, I am talking about years after. How do we explain this? My simple thinking is that it may be that those have been more intensively exposed, harbor a larger dose latently, that's where I place emphasis. I admit that probably the transmission of other infections at the same time as measles may be very important for the outcome as well.

Kilbourne E:

We have slightly different prejudices on this. I am prejudice of the fact that, at least in the experimental model, you can study that the rapidity replication of the virus is such that you really have washed out any great difference of input dose after relatively few cycles of replication. The question about the stability of the virus is quite correct in terms of antigenic stability, but again, the more we isolate these in plaque forming systems and other methods of looking at these, and look at the DNA or RNA sequence, in this case, the more mutations you find. There are not readily at hand, as far as I know, any ways of measuring the virulence of the virus in *in vitro* systems, but I would like to see somebody approach this at the molecular level and look at strains isolated late, or for secondary cases opposed to those of primary cases. They might well find some differences.

Aaby P:

Yes, that would be interesting.

Lennette E (California Public Health Foundation, Berkeley, CA):

Any pediatricians in the audience like to comment?

Gershon A (Columbia University, New York, NY):

I have been fascinated by your articles and certainly by your lecture. As a pediatrician and a virologist, I would like to make a couple of comments. First of all, studies by Ross a number of years ago on transmission of chickenpox showed precisely the same thing. The secondary case in the family developed a more severe case than the index case. So, I think that your observation is very interesting, and in contrast to Dr. Kilbourne, I would tend to accept the dose phenomenon. As far as transmission of chickenpox goes, I'll be showing data later on, indicating that the skin rash is very important in whether the disease is transmitted or not. Clearly the first case in the family develops fewer skin lesions than subsequent cases in the family, after that type of exposure. The other thing that you said which interested me very much was the concept of partial immunity which is something that, again, we found in the chickenpox vaccine situation. It wasn't what we had initially predicted, because, until fairly recently, most of us have looked at immunity as more of an all-or-none phenomenon rather than in various shades of grey. Clearly, in people who are immunized against chickenpox, while most people will develop an all-or-none kind of immunity, there are others who develop a partial immunity, and when exposure takes place, instead of developing full blown disease, a modified disease develops. Finally, I have a question. You had mentioned that in the high dose situation, there is a short incubation period and a delayed immune response, and I wonder if you could just elaborate on that delayed immune response business. I think it's interesting and I hadn't really ever thought about that.

Aaby P:

Many thanks for your comments, which was more of less what I hoped from you from reading your papers and seeing the development in your writing on chickenpox. I have been looking at chickenpox to see if these patterns are valid for other infectious diseases, and I think it is very valid for chickenpox. Most people who die of chickenpox are secondary cases. Many of these are adults who are infected by their own children, as far as I understand the literature. There are also good indications in the literature that they have a shorter period of incubation. If you take the fatal cases and compare with the other ones that they have a shorter period of incubation than the ones who survive. I may note that the suggestion that exposure is important was already made for polio in the 1920's. The observation was that if you take chickenpox or measles in families with two or more cases, you will find the first case and then fourteen days later, you will find the second or the third case. Whereas, if you take families with two cases of polio, you will find them at the same time or, close to the same time. The most simple way to explain this is that they are secondary cases to a subclinical case. I think for polio you would probably also be able to find that the intensity of exposure has been important for outcome. The question about partial immunity that you raised, I think that this is very important. It was Panum who tried to establish that you don't get measles twice, and I think he was wrong. Somehow this belief seems to linger in the medical profession, everybody is brought up to think that immunity is an all or none phenomena. The implication is that people do not really look into the matter, and this may have very important implications. To give you one example, a lot of children get measles after vaccination in Africa, and it's usually assumed that it's because the cold chain didn't work. If you ask the mothers, usually they would say that they had milder measles. That's very important to them because it means that they are not discouraged from getting vaccinated. They observe something they can recognize as measles but milder, and that's very good for their kids. However, the official policy on measles from the WHO is based on a relatively simple study from Kenya where they examined seroconversion rates at 7, 8, 9, 10 or 11 months, and then you calculate the number of children you would protect by vaccination, let's say at 7, 8 , or 9 months. They found out that 8 and 9 months would give the same number of prevented cases. If you vaccinated by age 8 months, you would have more vaccine failures. More children would not seroconvert, and therefore, they would get measles later, and that would presumably work as a deterrent from getting them vaccinated. Therefore, they prefer to vaccinate at 9 months, and more children would get measles before vaccination. However, if I'm right, namely that vaccinated children get milder measles, the calculation should be different. I'm pretty sure that if they had actually done the control trial which they should have done, namely vaccinate at 7, 8 or 9 month and see what happened, they would have found that the survival rate from vaccination at 7 months was much better than vaccination at 9 months. I think this problem is a consequence of our belief in absolute immunity, because we haven't really looked into it. There are studies now, from the United States where they are trying to look into vaccinated children who get measles. It has been found that some of the students who got measles had detectable antibodies up to a level of

200 M International Unit. From the dose perspective, I think this makes sense. Resistance may be a function of how much immunity you have and how severely you are exposed. The effect of exposure to pregnant women may have to do with the same thing. If you are sufficiently intensively exposed, you may only have partial immunity. The mother may also have immunity, but is probably partially immunosuppressed during pregnancy. So, if she gets sufficiently exposed from having several children with measles, this may have consequences for the fetus. The last of what you said, about delayed immune response, I don't know if delayed immune response is the proper word, but it seems that the prodrome period is prolonged in severe cases. There is a correlation between the severity of infection and the length of the prodromes. You observe the opposite in children who have received immunoglobulin, namely that they have a much shorter period of prodrome.

Lennette E:

I have one brief question for you. Back in the late 1920's and early 1930's, a great emphasis was placed on susceptibility to infection with polio virus, by Aycock and Kraemer, also from Columbia University. Has that entirely been ruled out? Paralytic acquired poliomyelitic infection. Overt paralytic infections were somehow different from those that didn't develop paralysis. This may be long before your time, it's old literature.

Kilbourne E:

A person named Vivian Wyatt from England has made a strong case for this through the years. I've seen his evidence and it's pretty persuasive.

Aaby P:

It was Aycock who suggested that the severe polio cases could be secondary cases. In my experience in Africa, there may well be something genetic because you often find polio cases in the same families, not necessarily from the same epidemic, but siblings having polio. There is a whole misconception about polio among infectious disease doctors, that it's not very frequent in developing countries. It's assumed to be a consequence of getting an infection late in life, like what happened in the 1950's in Europe and the States. But few communities had a very high rate of polio. Before we started vaccinating in Bissau, we had 2%.of every generation getting paralytic polio. Two percent of all children would get permanent paralyses. We don't know about the mortality from polio, but paralytic polio is very important.

Lennette E:

It was primarily in children, infantile paralysis. And, of course, at that time, it was thought to be a water-borne disease because in Scandinavia it occurred only along the courses of rivers. Dr. Francis, the floor is yours.

Francis H (NIH, Bethesda, MD):

My comment is about the genetic contribution to the manifestation of diseases. It maybe considered to have more of a role than other people had considered. There has been a lot of talk, as you have stated very nicely in your lecture, about the contribution of intensity of exposure and perhaps duration. I would also like to present the consideration that the immune reaction to the infection, particularly the autoimmune phenomena, which we see a lot in different diseases in Africa might be important in the manifestation and pathogenesis of the disease. An example I could give would be onchocerciasis. A lot of the late manifestations of disease apparently are due to development of autoimmune phenomena to the different reactions that cause destruction in the eye and other tissues of the body. I think that the genetic contribution of this, which is peculiar to some diseases in Africans should be considered and worked out more thoroughly. The other thing was more of a question that had to do with the very early manifestation for mortality of disease in infants. Have you looked at what the immune status is of the mothers in their prenatal period, and what contribution that has to the mortality and morbidity of the disease after birth? Would vaccination of the mothers before pregnancy be useful in modulating the disease, over the time of the infants?

Aaby P:

The first thing, about genetics, it is in no way my domain. I am arguing a bit against genetics mainly because people tend to resort to it every time there is something they don't understand. Really, the observations we have made does not necessitate a genetic mechanism. There may be some genetic mechanisms, but the major differences that I am trying to point out are not comprehensible from a genetic point of view; the difference between an index patient and a secondary case. The difference between exposure from a severe case or a milder case, or the difference between exposure from your own or the other sex. The last question about male-female transmission, I think might be interesting to discuss, and there might be some genetic component, but at a different level. In terms of the global world distribution of severe disease, I don't think we really need genetics for an explanation. The point has been made strongest in relation to work on virgin soil epidemics. Black has argued that the Indians of the Americas have a much stronger response to measles vaccine, and it could be this stronger response which explains why they have very high mortality during the first virgin soil epidemics. It has been shown in Africa in one study that HLA types were correlated with leukocytopenia, which is a marker for severe disease. Apart from that, there are no good studies to suggest a correlation between the severity of disease and genetics in measles. There may well be, but it hasn't been sufficiently studied. As far as my interest goes, I don't think that we really need that mechanism. This is just a kind of general statement. These virgin soil epidemics have been extremely severe, and follow this pattern of the first cases not being severe and then you have a boom in mortality. You may have 50% of a community dying from measles. However, if you introduce gammaglobulin, and this has been done in some epidemics in Greenland and among the Inuits in Northern Canada, you find relatively low mortality, very little mortality among those who received the immunoglobulin. Which would indicate that it is not genetics alone. It's also an observation that the mortality may change very rapidly, say within a time span of one decade or two decades. You may have a fall in mortality that is not really compatible with the change in genetics and constitution in that population, that's the second half of my argument for not emphasizing genetics. This happens when you introduce vaccination. When you introduce measles vaccination, you also have a benefit for the unvaccinated, because you are changing the spread of the disease. You are reducing the risk of having secondary cases. By taking some of the children out of the community, you're reducing the risk of other children having measles at the same time. Therefore, you are also lowering mortality to those who did not get vaccinated. So you may have a very dramatic drop in mortality from introducing vaccination in a community. The question about the immune status of the mother, we haven't checked that. We haven't looked at the vaccination or how the antibodies of the mother correlated with the susceptibility of severe disease afterward. I think that we may be in a mix here, because it's clearly beneficial to have maternal antibodies for the acute infection. Children who have maternal antibodies have a higher rate of survival if they get measles. But, children who are exposed to measles in the home, particularly this age group, have very severe delayed consequences. The relative benefit of the acute infection and the relative delayed complications ought to be major questions. I'm pretty sure that what I have observed for measles, probably applies to other viral infectious diseases. I have one candidate and that's chickenpox.

THE INTERNATIONAL EPIDEMIOLOGY OF HIV-1 INFECTIONS

[1]Henry L. Francis and [2]Thomas C. Quinn

[1]National Institute of Allergy and Infectious Diseases
The National Institutes of Health
Bethesda, Maryland, USA
[2]The Johns Hopkins University
Baltimore, Maryland, USA

INTRODUCTION

The acquired immunodeficiency syndrome (AIDS) epidemic has spread to almost all countries in the world. This disease is responsible for and will be the cause of enormous morbidity and morality in all countries but particularly in the third world countries. Even with widespread underreporting of human immunodeficiency virus type 1 (HIV-1) infections in many developed and underdeveloped countries, 266,000 AIDS cases have been reported in 156 countries. Between 1981 and 1990, the World Health Organization (WHO) conservatively estimates that more than 500,000 cases have occurred worldwide and 300,000 have died (Chin and Mann, 1989; Mann and Chin, 1988a). When HIV-1 was isolated as the cause of AIDS (Barre-Sinoussi, 1983; Gallo et al. 1984), epidemiologists noted that the large number of HIV-1 infected patients with the clinical syndrome of AIDS was only a small segment of the total number of persons infected with HIV-1. These asymptomatic patients are the most prevalent group of infected persons and are the vector by which the disease continues to spread even in the face of control and prevention programs. The WHO recently estimated that by the year 2000 there will be 14 million asymptomatic HIV-1 infected persons (Chin and Mann, 1989; Piot et al. 1990; World Health Organization, 1989b), and at least 6,000,000 of these infected persons will die of AIDS within the 10 years.

PATTERNS OF SPREAD

Although the immunology and pathogenesis of HIV-1 infection appear to be similar throughout the world, there are geographic variations of clinical expression of the disease, as well as in the distribution of different human retroviruses such as HIV-1, HIV-2 and human T-cell lymphotropic virus-1 (HTLV-I) infections (Piot et al. 1988; Quinn et al. 1986, 1989). Epidemiologic patterns of spread that were distinguishable 5 years ago are less clear in current studies (Mann and Chin, 1988). Epidemiologic studies on disease transmission however have demonstrated that HIV-1 continues to be transmitted by three major

Medical Virology 10, Edited by L.M. de la Maza and
E.M. Peterson, Plenum Press, New York, 1991

modes: sexual transmission which may be homosexual, bisexual or heterosexual; parenteral transmission including transfusion of infected blood products or injection with blood contaminated needles or syringes; and perinatal transmission (Curran et al. 1988; Friedland and Klein, 1987). The relative frequency of these three types of transmission and the rate of HIV-1 introduction and dissemination determine the characteristics of the HIV-1 epidemic in different regions. The duration of the HIV-1 epidemic has been long enough so that all three transmission types have significantly contributed to the spread of AIDS and have blurred the distinctions defined earlier by WHO which are described below for historical perspective.

The first pattern observed is HIV-1 transmission among homosexual/bisexual men and drug users. This pattern is commonly seen in developed countries such as Western Europe, Australia, New Zealand, North America, some urban areas of Latin America and more recently Asia. HIV-1 was introduced first in homosexual men in the mid 1970's and in some areas over 50% of homosexual men are currently infected (Centers for Disease Control, 1987). Intravenous drug use accounts for the next largest portion of HIV-1 infection. Transmission of HIV-1 from IV drug users to their bisexual and heterosexual partners is emerging as the predominant route by which HIV-1 is entering new populations. Female sexual partners of HIV-1 infected drug users or female drug users are one of largest groups responsible for perinatal transmission of HIV-1 in the United States.

The second pattern is heterosexual transmission of HIV-1 among men and women of an endemic area. This pattern predominated in Africa, some areas of Latin America and the Caribbean and more recently is increasing in areas where drug addicts are transmitting HIV-1 to their sexual partners. The virus may have been introduced into these populations in the early or late 1970's, and recent serologic surveys demonstrate that 10 to 20% of 20 to 40 year old individuals in some areas are infected (Friedland and Klein, 1987; N'Galy and Ryder, 1988; Piot et al. 1988; Quinn et al. 1986a, 1989b; Ronald et al. 1988; Rwandan HIV Seroprevalence Study Group, 1989; Ryder and Piot, 1988). Perinatal transmission has also become a major problem in these areas where 5 to 15% of pregnant women are HIV-1 seropositive (Ryder and Hassig, 1988; Ryder et al. 1989). The ratio of infected males to females in these areas is variable though 1:1 ratios have been commonly reported. The etiology of the rapid spread of HIV-1 among heterosexuals is complex and is modulated by prevalence of HIV-1 among prostitutes, prevalence of sexually transmitted diseases, numbers of IV drug users and prevalence of HIV-1 or HIV-2 infections as well as other factors (De Cock et al. 1989; Poulsen et al. 1988; Ryder 1988). The multifactorial nature of the risk factors for HIV-1 transmission in this population represents a most difficult problem area for control of the epidemic.

The third pattern is found in countries where HIV-1 has been newly introduced and the pattern of transmission is either mixed or not clearly defined. This pattern is evident in areas of Eastern Europe, North Africa, the Middle East, Asia and most areas of the Pacific. More recent surveillance data suggests that the mixed nature of the HIV-1 epidemic will become the prevalent pattern in industrialized and third world countries. HIV-1 was probably introduced in these areas during the early to mid 1980's, and the extent of these epidemic areas such as India and Thailand, is just beginning to be appreciated. Indigenous homosexual, heterosexual and IV drug use transmission have been documented in many countries. The number of AIDS cases were until recently small, but the rapid spread of HIV-1 infection is being documented in some of these countries particularly among female prostitutes and IV drug users (De Cock et al. 1989; Mann et al. 1988; Poulsen et al. 1988; Tanphaichitra et al. 1988).

In selected countries of West Africa, infection with a second human retrovirus referred to as HIV-2 has become epidemic (Clavel et al. 1986, 1987a, 1987b; Kanki et al. 1987). High prevalence rates in high risk individuals, such as hospitalized patients, female prostitutes and STD clinic patients have demonstrated that HIV-2 appears to spread according to the same modes of transmission as HIV-1 but seems to have a different pathogenicity. Systematic surveys for HIV-2 outside of West Africa have documented isolated cases in parts of central Africa, North America and Brazil (Centers for Disease Control, 1988a, 1989d; Cortes et al. 1989, Horsburgh and Holmberg, 1988; Kloser, 1989; Neuman et al. 1989; Ruef et al. 1989; Veronesi et al. 1987). In all cases, the infected individuals had lived in some part of the HIV-2 endemic areas of West Africa and had returned home with the infection. Though the modes of transmission of HIV-2 are similar to HIV-1, it has been reported to spread in a population more slowly than HIV-1 (De Cock et al. 1989) and may not be well transmitted perinatally (Poulsen et al. 1988).

Despite the above geographical distinctions, the patterns of the HIV-1 epidemic are continuously changing. In the United States and South America, the pattern of spread has shifted from homosexuals to an increasing heterosexual spread as originated from sexual contacts of drug users. Ironically, this same pattern is seen in the poor urban centers of the United States. The implications of this blurring of transmission patterns has broad meaning to current efforts to prevent and control this epidemic. As the HIV-1 epidemic continues to grow at a frightening pace, solutions to control and prevention of this epidemic will have to change with the evolving nature of global AIDS epidemic.

AIDS SURVEILLANCE

Using the WHO/CDC case definition for AIDS, 266,098 cases of AIDS have been reported from 156 countries as of July 1, 1990 (World Health Organization, 1990) (Table 1). In the region of the Americas, 162,885 cases or 61% of the total have been reported from 44 countries. In the United States alone, 133,889 cases or 86% of the cases in the Americas have been reported. For the rest of the Americas, exponential increases in AIDS cases are being reported with 23,022 cases reported in Central and South America, and 5,974 cases reported from the countries in the Caribbean basin. In relation to its population base, the Caribbean subregion has reported a disproportionate number of AIDS cases. Excluding the United States, the English speaking Caribbean countries with only 2% of the population and Latin Caribbean countries with 6% of the population have reported 10% and 21%, respectively, of all cases for the Caribbean and Latin America (Quinn et al. 1989b). In contrast to other areas of the Americas, the male-to-female ratio reported AIDS cases in the Caribbean is 2.4:1, reflecting an increased number of female cases due to heterosexual transmission of HIV-1 (Narain et al. 1989). In countries with increasing number of female AIDS cases in a declining male-to-female ratio, the proportion of AIDS cases that are less than 13 years of age also increases proportionately due to perinatal transmission (Pape et al. 1989; Quinn et al. 1989a). For example, in the Caribbean 11% of reported AIDS cases were under 5 years of age reflecting increased mother-to-infant transmission.

TABLE 1. Global AIDS Cases Reported to The World Health Organization, July 1990

Continent	No. of Countries Reporting 1 or more cases	No. of Cases
Africa	51	65,149
Americas	44	162,885
Asia	25	655
Europe	29	35,353
Oceania	7	2,056
TOTAL	**156**	**266,098**

In Europe, over 35,000 cases have been reported from 29 European countries. AIDS in Europe is similar to the United States in relationship to the proportion of cases among high risk groups. Among adults, the male:female ratio is 7.4:1, but the sex ratio was closer to one among IV drug users, those with heterosexual contact to other high risk individuals and transfusion recipients (World Health Organization, 1989a). In 1988, there was a 60.7% increase in AIDS cases among homosexuals and bisexuals and a 130% increase among IV drug users. In some countries such as Italy and Spain over 60% of AIDS cases have occurred among IV drug users.

With a late introduction of HIV-1 into Asian and Pacific countries, the number of AIDS cases still remain low in those areas with 2,056 cases reported from seven Pacific countries and 29 Asian countries. Serologic studies in some of these countries have demonstrated high rates of HIV-1 infection among hemophiliacs due to importation of factor 8 and factor 9 concentrates from the United States and Europe prior to the institution of blood screening in 1985. In Southeast Asia the predominant HIV-1 transmission is now among IV drug users and their heterosexual partners. In India, seroprevalence studies among prostitutes have noted increasing rates of HIV-1 infection, even though the number of AIDS cases remain low (Simoes et al. 1987).

Even though 65,149 cases have been reported from 51 African countries, this is an underestimate of the true number of cases. As documented by serologic studies, HIV-1 infection has spread rapidly particularly among urban centers of Central Africa with peak infection rates among women age 20-39 and men age 30-49 (N'Galy et al. 1988; Piot et al. 1988; Quinn et al. 1986; Ronald et al. 1988; Rwandan HIV Seroprevalence Study Group, 1989; Ryder et al. 1988). The male-to-female ratio is 1:1.5, and because of the large number of infected women (up to 10% infection rates), pediatric cases comprise 15 to 20% of the total AIDS cases. Selected seroprevalent studies among diverse urban population groups clearly demonstrate the relative high frequency of HIV-1 infection among blood donors (up to 20%), pregnant women (5-10%), hospitalized patients (25-50%), men attending STD clinics (30-45%) and female prostitutes (40-90%) (De Cock et al. 1989; Mann et al. 1988; Melbeye et al. 1986; N'Galy et al. 1988; Piot et al. 1987, 1988a, 1988b; Quinn et al. 1986; Rwandan HIV Seroprevalence Study Group, 1989; Ryder et al. 1988a, 1988b, 1989). Thus, of all the regions, HIV-1 infection has probably had its greatest impact in the countries of central Africa and now poses the greatest challenge to control of this epidemic.

CURRENT STATUS OF AIDS IN THE UNITED STATES

Of the 139,765 cases reported in the United States as of July 1, 1990, 137,385 (98.3%) occurred among adults and 2,380 (1.7%) occurred among children less than 13 years of age. Of the adults, 124,385 (90.5%) were men and 13,000 (9.5%) were women (Table 2). Mean age at time of diagnosis was 37.0 years. The overall mortality rate of 55% at the time of reporting and greater than 90% five years from time of diagnosis (Curran et al. 1988). AIDS related mortality disproportionately affected persons in the 20 to 49 year old age group leading to substantial decreases in life expectancy. Deaths occurring in 1987 among persons with AIDS that were reported to CDC represented 9% of all deaths among persons 25 to 34 years of age and 7% of all deaths among persons 35 to 44 years of age. In 1987 AIDS was the 7th leading cause of premature mortality before age 65 in the United States (Centers for Disease Control, 1989a, 1990).

AIDS cases were disproportionately high in blacks and Hispanics in the United States reflecting higher reported rates of AIDS among black and Hispanic IV drug users, their sex partner and their infants (Centers for Disease

TABLE 2. Percent of Cumulative Adult AIDS Cases by Risk Category, July 1990

Category	Men	Women	Total
Number	124,385	13,000	137,385
Homosexual/Bisexual	66%	-	60%
IV Drug User	18%	51%	21%
Homosexual/IVDU	8%	-	7%
Hemophiliac	1%	0%	1%
Heterosexual	2%	32%	5%
Blood Transfusion	3%	10%	2%
Other/Undetermined	3%	7%	3%

Control, 1989b, 1990; Cortes et al. 1989). Blacks accounted for 27% of the adults and 53% of the pediatric cases, and Hispanics represented 15% of the adults and 24% of pediatric cases. In contrast, blacks and Hispanics account for only 11.6% and 6.5%, respectively of the United States population. The relative risk of AIDS for blacks and Hispanics is nearly 10 times as high in the Northeastern United States as in other regions of the country, partially because of the regional concentration of IV drug use related AIDS. In 1988, the annual incident rate of AIDS cases associated with IV drug use was 11.5 times higher among blacks and 8.8 times higher among Hispanics than among whites (Stall et al. 1988).

Of male cases, 66% had a history of homosexual/bisexual activity with IV drug use, 18% acknowledged IV drug use without homosexual/bisexual activity, and 8% acknowledged both homosexual activity and IV drug use. Another 2% had a history of blood transfusion, 1% had hemophilia or other coagulation disorders, 1% had sex partners at increased risk for HIV-1 or who were known to be infected with HIV-1, 1% were born in countries with predominantly heterosexual transmission of HIV-1, and 2% had undetermined means of exposure. This distribution has remained relatively stable, except for decrease in the proportion of men with a history of homosexual/bisexual activity without IV drug use from 70% of those reported in 1987 to 63% of those reported in 1988. There has been an increase in the proportion of men with histories of IV drug use and no homosexual/bisexual activity from 14% in 1987 to 20% in 1988. These data suggest that the rate of AIDS among homosexual men is slowing, partially due to adherence to preventive measures, such as safe sex guidelines (Centers for Disease Control, 1987,1989a; Stall et al. 1988). In contrast, AIDS is continuing to rise among IV drug users and their heterosexual partners (Table 3), groups which need more intensive educational efforts (Brickner et al. 1989a; Centers for Disease Control, 1989a; Des-Jarlais et al. 1988, 1989).

The proportion of women with AIDS increased from 8% in 1987 to 11.2% in 1989. The most commonly affected group was between 30-39 years old, and 51.8% were black, 27.1% white, 20% Hispanic and 1.1% other racial groups. As in men, the cumulative incidence of AIDS between 1981 and 1988 was 13.6 times higher among black woman and 1.2 times higher among Hispanic women than among white women. Among women with AIDS, 52% had a history of IV drug use, 31% had sex partners at increased risk for or known to be infected with HIV-1, 11% had histories of blood transfusion, and 7% had undetermined means of exposure. The proportion of women with AIDS who had sex partners at increased risk for HIV-1 rose from 15% before 1984 to 26% in 1988 (Simoes et al. 1987; Stall et al. 1988).

Between 1988 and 1989, AIDS cases due to perinatal transmission increased 17% (Table 3) (Centers for Disease Control, 1990). This was the largest increase among all HIV-1 risk groups reported in the United States. Of the 2,380 children, 83% were born to mothers who had AIDS or who were at risk of acquiring AIDS; 10% had received infected blood transfusions; and 5% were hemophiliacs who had received HIV-1 infected factor 8 or factor 9 concentrates. The remaining 3% included patients for whom risk factor information was incomplete.

When the HIV-1 infected cases are stratified by race, very different risk factors are found to be prevalent in different populations. Forty-three percent of the white pediatric HIV-1 infections are due to receipt of blood products or blood transfusion, whereas only 15% of black children and 12% of Hispanic children have this particular risk factor. In the black and Hispanic population,

TABLE 3. Changes in AIDS Cases According to Risk Group Between 1988 and 1989

Risk Groups	1988	1989	Increase
Homosexual/Bisexual	18,130	19,652	8%
IV Drug Users	7,580	7,970	5%
Homosexual/IVDU	2,129	2,138	0%
Hemophiliac	339	321	-5%
Heterosexual	1,603	1,954	22%
Perinatal	468	547	17%
Transfusion Recipient	935	808	-14%
Other/Undetermined[a]	1,012	1,848	-
Total	32,196	35,238	9%

[a] Cases in this category still under investigation

From Centers for Disease Control, 1990

being born to IV drug using mothers accounts for more than half of the HIV-1 cases. Interestingly, 16% of the black pediatric cases are from countries that have the predominantly heterosexual spread of HIV-1 infection (Pattern II countries).

MODES OF TRANSMISSION OF HIV-1

Sexual Transmission

It is evident from serologic surveys and AIDS surveillance worldwide that sexual transmission probably accounts for more than 75% of HIV-1 infection worldwide. In developed countries such as the United States homosexual transmission has been the predominate sexual mode of spread, and has been strongly associated with the number of sexual partners and the frequency by which they practice receptive anal intercourse (Curran et al. 1988; Darrow et al. 1987; Goedert et al. 1984; Kingsley et al. 1987; Mos et al. 1987; Van Greinsven et al. 1987; Winkelstein et al. 1987). Other practices that lead to rectal trauma such as receptive "fisting" and douching appear to increase the risk of infection. Since 1984, more than 50 HIV-1 serologic studies of homosexual and bisexual men in the United States have shown prevalence rates ranging from 10% to as high as 70% with most between 20% to 50% (Centers for Disease Control, 1987c, 1990). Fortunately, recent seroincidence studies of HIV-1 infection among homosexual male cohorts have demonstrated a decline in the rate of HIV-1 infection in many of these individuals, hopefully due to utilization of safer sex practices and a lower frequency of high risk behaviors (Centers for Disease Control, 1987; Stall et al. 1988).

With increasing rates of HIV-1 infection among IV drug users and bisexuals, an increasing number of women are being infected with HIV-1 due to heterosexual exposure (Centers for Disease Control, 1989c). Among heterosexuals, female prostitutes, patients attending STD clinics and heterosexual partners of HIV-1 infected individuals appear to be at high risk for HIV-1 infection (Chamberland and Dondero, 1987; Guinan and Hardy 1987). In the United States, the proportion of reported cases associated with heterosexual contact have increased from 1.1% in 1982 to 4.4% in mid-1989. Approximately 70% of the index partners for these cases were IV drug users, and 18% of the index partners for female cases were bisexual men. Interestingly, the male-to-female ratio of heterosexual contact cases in the United States is 1:1.4, similar to the male-to-female ratio in African countries where heterosexual transmission is the predominate mode of spread (Johnson and Laga, 1988; Piot et al. 1988a, 1988b; Quinn et al. 1986). This predominance of women in the heterosexual contact category in the United States is probably due to the larger pool of infected men classified as bisexual men or IV drug users, but the efficiency of male-to-female vs female-to-male transmission may also be relevant.

The efficiencies of bi-directional transmission have not been well documented, but it is evident that most heterosexual transmission occurs during vaginal intercourse and that receptive anal intercourse will also increase the risk of infection in women (Anderson et al. 1986; Holmberg et al. 1989). In one study, the relative risk of HIV-1 infection in spouses who engage in anal intercourse was 2.3 compared to those who only had vaginal sex (Anderson, 1988). In nine studies reporting on anal intercourse, four found a significantly increased risk of transmission in couples who reported anal as well as vaginal intercourse, but the other five studies showed no association with anal intercourse (Johnson and Laga, 1988). In several studies, investigators have also reported the number of genital-genital exposures for HIV-1 infected persons and their spouses or their monogamous sexual partners (Anderson, 1988; Kreiss et al. 1985; May, 1988; Padian et al. 1987; Peterman et al. 1988; Redfield et al. 1985). It is impossible to specify which sexual exposure resulted in HIV-1 transmission, but is possible to define maximum and minimum rates for single exposure (Anderson et al. 1988,1989; Holmberg et al. 1989). One can assume for the lower limit of such transmission that only one exposure resulted in infection of the susceptible partner, and for the upper limit that all exposures in couples in which both partners are HIV-1 infected involved HIV-1 transmission. The per-contact infectivity for male-to-female transmission of HIV-1 has generally been calculated at less than 0.2%, that is, the binomial probability that 15% of persons exposed will be infected after greater than 100 unprotected sexual contacts. However, some persons become HIV-1 infected after a single or few sexual exposures, whereas others remain uninfected despite hundreds of unprotected sexual contact. This finding suggests that the likelihood of HIV-1 infection may be substantially affected by intrinsic properties of the HIV-1 infected partner, the virus itself, or the non-infected partner (Holmberg et al. 1989).

Variable rates of heterosexual transmission among sex partners of HIV-1 infected individuals have been documented in studies in the United States. Reported rates of infection range from 9% to 20% for female partners of infected male hemophiliacs (Allain, 1986; Goedert et al. 1988a; Jason et al. 1986; Kreiss et al. 1985); 26% for female sexual partners of bisexual men (Fischl et al. 1987; Padian et al. 1987); 19.7% and 14.8%, respectively, for female and male sexual contacts of transfusion recipients (Peterman et al. 1988); 47.8% and 50%, respectively, for female and male sexual contacts of IV drug users (Fischl et al. 1987; Padian et al. 1987; Redfield et al. 1985). Even in African studies such as one which examined 250 married couples in which at least spouse was seropositive, 20% were concordantly seropositive and in 80% only one partner was infected despite a mean period of eight years of marriage and unprotected sexual intercourse (Nzila et al. 1988). However, this finding is in contrast to another African study of 124 HIV-1 seropositive couples compared to 150 seronegative couples in which the authors demonstrated that seropositivity in the husbands was significantly associated with sexual contact with prostitutes and with a history of STD's within the past two years (Carael et al. 1988). Thus, it is apparent that although many couples included in studies of heterosexual HIV-1 transmission have had unprotected sex over prolonged periods of time; an average of 50 to 60% of partners have been infected in most studies. This suggest that in addition to behavioral factors, biological factors also contribute to the efficiency of HIV-1 transmission.

Some infected individuals may be more efficient transmitters than others, and infectiousness may vary over the course of infection. In a prospective study of the partners of infected hemophiliacs, Goedert et al. (1988a, 1988b) reported that the best predictor of HIV-1 transmission is the absolute number of T-helper lymphocytes in the hemophiliac. Their preliminary data suggest that as these patients become more immunosuppressed with fewer than 200 CD4 positive cells/ml, and viral titers increased with increasing p24 antigenemia, their sex partners were more likely to become infected after controlling for the duration of infection and frequency of exposures. This is consistent with the finding that the ability to isolate HIV-1 significantly increases as the number of CD4 positive cells decline and the clinical course progresses (Coombs et al. 1987). These studies in hemophiliac heterosexual partners have also been confirmed in other groups. In a study of 77 heterosexual partners (62 women, 15 men) of 72 HIV-1 infected persons (39 African, 38 European) low CD4 positive cell concentrations in the index cases were associated with significantly increased risk of HIV-1 infection in their partners (Laga et al. 1988). Similarly, in two studies, one in West Germany (Stazewski et al. 1988) and one in Miami, of heterosexual partners of HIV-1 infected persons (Fischl et al. 1987), the likelihood of infection increased with severity of disease, low CD4 positive lymphocyte counts and p24 antigenemia.

Studies in the United States and in Africa have also documented a strong association of HIV-1 transmission with a history of STDs such as genital ulcers, and lack of circumcision (Bongaarts et al. 1989; Cameron et al. 1989; Pepin et al. 1989). In one of the first studies of female prostitutes in Africa, HIV-1 infection was strongly associated with a history of genital ulceration and gonorrhea (Kreiss et al. 1986). In a subsequent study of 115 heterosexual men in Nairobi who presented with genital ulcers to an STD clinic, HIV-1 infection was positively associated with lack of circumcision and again with a past history of genital ulceration (Greenblatt et al. 1988). In a larger study of 340 men attending the Nairobi STD clinic of whom 11.2% were HIV-1 seropositive, HIV-1 infection was independently associated with the presence of genital ulceration and lack of circumcision (Cannon et al. 1988). In order to confirm this association, a prospective study of HIV-1 seronegative men attending the Nairobi STD Clinic and who had a recent history of female prostitute contact was initiated (Cameron et al. 1989). In men who were circumcised and who lack genital ulcers at time of presentation, the seroconversion rate was 2.5%. This seroconversion rate increased to 13.4% in 111 uncircumcised men with genital ulcers to 29% for 27 uncircumcised men without genital ulcers and to a high of 52.6% for 61 uncircumcised men with genital ulcers. The association of HIV-1 seroconversion with the lack of circumcision could not be related to sexual behaviors of the circumcised versus uncircumcised men, nor could it be related to increased frequency of genital ulcers in the uncircumcised men. A recent ethnographic review on circumcision practices in 409 African ethnic groups demonstrated a direct correlation between lack of circumcision and increased HIV-1 seroprevalence rates in capital cities of Africa ($R=0.9$; $p > 0.001$) (Bongaarts et al. 1989). It is possible that balanitis, maceration of the skin of the glands penis, trauma to the intact foreskin, or simply the microenvironment of the tissue under the foreskin may allow for greater survival and penetration of HIV-1.

Even in the United States among patients attending STD clinics, a strong association of HIV-1 has been documented with a history of genital ulcerations caused by syphilis or herpes simplex virus type 2 (HSV-2). In combined studies of over 8,000 patients attending STD clinics in Baltimore between 1987 and 1988, 5% were HIV-1 seropositive. Of those heterosexuals who denied other traditional risk factors, 2.3% were seropositive and HIV-1 infection was significantly associated in those individuals with a history of syphilis, a positive serologic test for syphilis, and serologic evidence for HSV-2 (Cannon et al. 1988; Quinn et al. 1990). Although sexually transmitted infections may be covariates of another primary risk factor, a strong association between HIV-1 infection and syphilis and HSV-2 infection has also been documented among homosexual men even after controlling for sexual behavior, such as the number of life-time sexual partners or receptive anal intercourse (Holmberg et al. 1988; Stamm et al. 1988). In these studies HIV-1 seroconversion was also strongly associated with HSV-2 seroconversion within the year prior to HIV-1 seroconversion (Holmberg et al. 1988). Consequently, genital ulcers and other STD's may allow for penetration of HIV-1 into a susceptible host by causing epithelial disruption or by increasing the susceptibility of the individual by increasing the population of CD4 positive lymphocytes, target cells for HIV-1 at the site of infection within the genital tract (Coombs et al. 1989). In a study by Coombs et al. (1989), CD4 positive lymphocytes and HIV-1 p17 antigen were demonstrated in genital fluid in nine of 14 seropositive women. The findings that HIV-1 can be isolated from genital ulcers and from vaginal fluid is significant since it implies that HIV-1 transmission can occur at any time and that infected lymphocytes may increase in the genital tract in association with STD's, inflammation of the cervix and menstruation. Thus, the probability that any single episode of genital-genital or anogenital sexual intercourse will result in transmission of HIV-1 may be determined by multiple biologic factors of the infectious person, the virus itself or the exposed susceptible person (Hearst and Hulley, 1988; Holmberg et al. 1989). Some of these factors are known or suspected and they may explain the observed differences in the sexual transmission of HIV-1 in different parts of the world, notably in Africa, where genitoulcerative disease is more common and strongly influences the epidemiology of HIV-1 (Johnson and Laga, 1988; Piot et al. 1988a).

Perinatal Transmission

Perinatal transmission of HIV-1 may occur *in utero,* perinatally at the time of delivery or postnatally through breast-feeding. Risk factors associated with perinatal transmission remain unknown and prospective studies are currently underway to assess the efficiency and determinants of perinatal transmission. Detection of HIV-1 in fetal tissues and within the placenta support the hypothesis that infection can occur in utero (Jovaisas et al. 1985; Lapointe et al. 1985 Sprecher et al. 1986). Similarly, HIV-1 transmission at birth through exposure to infected maternal blood or vaginal secretions also seems likely, but it is difficult to differentiate from *in utero* transmission (Blanche et al. 1989; Ryder and Hassig, 1988a, 1988b; The European Cooperative Study, 1988). A series of case reports of women who were apparently infected with HIV-1 by blood transfusions given during the immediate postpartum period and who subsequently infected their infants have also suggested that breast-feeding is a route of HIV-1 transmission (Rogers, 1989; Thiry et al. 1985).

Studies on the rate of perinatal HIV-1 transmission have been complicated by the lack of reliable diagnostic procedures to detect HIV-1 infection in newborn infants. Because infants born to HIV-1 infected mothers have detectable maternally derived antibodies to HIV-1 it has been necessary to follow seropositive infants over time to detect the loss of maternal antibodies and the development of serologic or virologic markers of infection in the infant. Prospective studies in the United States, Europe and Africa have suggested a 25 to 40% rate of perinatal transmission (Blanche et al. 1989; Rogers et al. 1985; Ryder et al. 1988a, 1989; The European Cooperative Study, 1988). In one study in Kinshasa, Zaire where nearly 6% of pregnant women are infected, the rate of perinatal transmission after two years of followup was estimated to be 39% (Ryder et al. 1989). In a French collaborative study at least 27% of 117 infants followed to 18 months of age were infected (Blanche et al. 1989). In the study in Kinshasa, Zaire, the rate of perinatal transmission correlated directly with the finding of less than 200 CD4 positive lymphocytes/mm^3 in the mother (Ryder et al. 1989). In that study, perinatal transmission was higher in women of a very low social economic status and with symptomatic HIV-1 infection. Infants born to symptomatic HIV-1 seropositive women were also shown to have an increased risk of death within the first year of life. By one year of life, 21% of the children born to HIV-1 seropositive mothers were dead compared to 3.8% infants born to seronegative mothers. Of the surviving infants born to HIV-seropositive mothers, 7.9% developed AIDS within the first year. Infants born to HIV-1 seropositive mothers also had a lower mean birth-weight and a lower mean gestational age then infants in the seronegative control group. At one year the mortality rate of infants born to seropositive mothers was 284/1,000 live born children compared to 36/1,000 for live born children of seronegative mothers. Overall, HIV-1 infection may have already increased infant mortality rate by as much as 15% in Kinshasa, Zaire (Ryder et al. 1989).

The contribution of breast-feeding in perinatal transmission of HIV-1 is unclear, yet HIV-1 has been isolated from breast-milk and reports of postnatal transmission, most probably through breast-milk have been reported (Colebunders et al. 1988; Lepage et al. 1987; Thiry et al. 1985; Ziegler et al. 1985). In the French collaborative study, five of six breast-fed infants born to seropositive mothers became infected compared to 25 of 99 children of seropositive mothers who did not breast-feed (Blanche et al. 1989). This finding plus the anecdotal findings of postnatal transmission in women transfused with HIV-1 infected blood after birth suggest that breast-feeding represents a small incremental risk of mother-to-infant transmission. In the developed world where alternative nutritional support is available, breast-feeding by an HIV-1 infected mother is probably not recommended. However, in developing countries where safe and effective use of alternatives for breast-feeding are not available, breast-feeding by the biological mother should continue to be the feeding method of choice, irrespective of the mother's HIV-1 status (Colebunders et al. 1988; Mortimer and Cooke, 1988; Piot et al. 1988a; Ryder and Hassig, 1988).

The recent availability of the polymerase chain reaction (PCR), a new technique that amplifies proviral sequences of HIV-1 within DNA to detect HIV-1 in infant's lymphocytes should provide further insight into the incidence of perinatal transmission among infected women (Ou et al. 1988). In a recent study by Rogers et al. (1989), PCR was positive in five of seven infants in whom AIDS later developed. The test was positive in one of eight newborns who later developed nonspecific signs and symptoms suggestive of HIV-1 infection and no proviral sequences were detected in neonatal samples from nine infants born to HIV-1 infected women, but who remained well up to 16 months after birth. Thus, PCR was highly specific with positive results only in children who develop symptomatic HIV-1 infection, but larger studies are required to determine the overall sensitivity of the test and in determining the efficiency of perinatal transmission.

Parenteral Transmission

Sharing needles or other drug related paraphernalia results in HIV-1 transmission between IV drug users. The incidence of HIV-1 infection is currently increasing at the fastest rate among IV drug users compared to other high risk populations and the incidence varies markedly by geographical region (Centers for Disease Control 1989a, 1989b). Data from more than 18,000 drug users tested in over 90 surveys in the United States have consistently shown high prevalence rates in Northeastern United States with rates ranging from 50% to 60% (Centers for Disease Control 1987a). Since patients undergoing treatment are believed to represent only 15% of the estimated 1.1 million IV drug users in the United States, the exact number of habitual IV drug users or intermittent IV drug users infected with HIV-1 is not known. In a selected study in New York City between 1984 and 1987, the seroprevalence rate among IV drug users stabilized between 55% and 60% (Des-Jarlais et al. 1989). This relatively stable rate is attributed to new infection, new seronegative persons beginning drug injection, seropositive persons leaving drug injection and increasing conscious risk reduction. Non-white IV drug users have a higher rate of HIV-1 infection than white IV drug users with histories of similar frequencies of IV drug use and needle-sharing (Centers for Disease Control, 1989a; Schoenbaum et al. 1989). This difference may reflect a different needle use practice in various risk groups.

Parenteral exposure to whole blood, blood components and blood products has accounted for HIV-1 infection in transfusion recipients and hemophiliacs. Prior to the implementation of screening of all blood donors in 1985 (Centers for Disease Control, 1985), nearly 70% of individuals with hemophilia A and 35% of those with hemophilia B have already become infected with HIV-1. While screening for HIV-1 among blood donors is highly effective in preventing HIV-1 transmission, it does not appear to be 100% efficient. In a recent study by Ward et al. (1988) the authors demonstrated that HIV-1 infection can still occur following blood transfusion, despite wide-scale screening of all HIV-1 donations. They identified 13 patients who were seropositive for HIV-1 and who had received blood from seven donors screened as negative for HIV-1 antibody at the time of donation. All seven of these donors were later found to be infected with HIV-1 suggesting that they had been infected only recently and were seronegative at the time of blood donation. Currently, from data from over 17 million United States Red Cross blood donations, the odds of contracting HIV-1 infection are estimated to be 1:153,000/unit transfused/patient (Cumming et al. 1989). A patient who received an average transfusion of 5.4 units had odds of 1:28,000. This risk has been decreasing by more than 30% per year. The risk of undetected infectious units can possibly be further reduced by transfusing fewer units, recruiting more women and fewer men as new donors, and encouraging more frequent donation from donors who have been tested repeatedly. In testing over 15 million' blood donors between 1985 and May 1988, the seroprevalence rate for HIV-1 was 0.018% in the United States, declining from 0.035 in 1985 to 0.01 in 1988 (Centers for Disease Control, 1990).

For some developing countries, HIV-1 blood screening has not yet been introduced due to economic and technical constraints. However, technological advances using recombinant antigens have resulted in the development of highly sensitive and specific rapid diagnostic assays (Quinn et al. 1988; Spielberg et al. 1989). It is hopeful that the use of these and other inexpensive assays may allow for the immediate implementation of serologic screening for HIV-1 in all areas of the world.

Parenteral exposure to blood has also resulted in a small but definite occupational risk of HIV-1 infection for health care workers (HCWs). Of the 169 HCWs with AIDS, risk factors assessment in 28 is still considered incomplete due to death or refusal to be interviewed, 97 are still being investigated and in 44, full investigations failed to reveal any risk factors (Kelen, in press). Eighteen HCWs reported exposure to blood or other body fluids from patients during the last 10 years. Only one health care worker in this category was documented to have seroconverted to HIV-1 and developed AIDS after a needle stick exposure to infected blood. In prospective seroprevalence studiees using recombinant antigens have resulted in the development of highly sensitive and specific rapid diagnostic assays (Quinn et al. 1988; Spielberg et al. 1989). It is hopeful that the use of these and other inexpensive assays may allow for the immediate implementation of serologic screening for HIV-1 in all areas of the world.

Parenteral exposure to blood has also resulted in a small but definite occupational risk of HIV-1 infection for HCWs. Of the 169 HCWs with AIDS, risk fact sharp instruments, seroconversion has been documented in only three. Thus the risk of virus infection following needlestick transmission is calculated at 0.35% (Kelen, in press; Marcus et al. 1988). This low but very real risk emphasizes the importance of universal precautions since other studies have clearly shown increasing rates of HIV-1 infection among patients in whom HIV-1 related risk factors where unknown or unrecognized (Centers for Disease Control, 1987b; Kelen et al. 1989).

Transmission By Other Routes

After monitoring the AIDS epidemic for over 8 years, there is still is no evidence for casual or household transmission of HIV-1. In several household studies performed in the United States, Europe and Africa, the rate of HIV-1 seropositive did not differ significantly between nonspousal household contacts of AIDS cases compared with household contacts of uninfected controls

(Allain, 1986; Castro et al. 1988; Goedert et al. 1988; Mann et al. 1986). Virtually all seropositive children who have been studied have identifiable risk factors for HIV-1 infection, and the low seroprevalence rates among children 5 to 15 and low rates of HIV-1 infection rural areas as opposed to urban areas in Africa strongly argue against arthropod transmission of HIV-1 (Piot et al. 1988a; Quinn et al. 1986; Rwandan HIV Seroprevalence Study Group, 1989). Even laboratory studies of insects fed HIV-1-contaminated blood, interrupted and then allowed to feed on uncontaminated blood failed to show HIV-1 transmission (Lyons et al. 1986; Office of Technology Assessment, 1987).

FUTURE PROJECTIONS

Several methods for projecting the future course of HIV-1 infections have been proposed over the past several years. These methodologies may assist national and international organizations in preparing for future public health needs and planning health program goals. Global projections using the Delphi method (World Health Organization, 1989c) suggest that 14,000,000 persons will have HIV-1 infection by the year 2000. However, 8 million of those cases are potentially preventable. At recent workshop on the epidemiology of the AIDS epidemic in the United States. (Centers for Disease Control, 1990b) investigators reviewed data that demonstrated the different prevalence and incidence rates of HIV infection by risk factor groups. Overall, one million individuals were estimated to be infected with projections of 365,000 AIDS cases by 1992 in the United States. Incidence data regarding new HIV-1 infection suggested 2,000 infants are infected annually and that 40,000 adults may be newly infected each year. These detailed data of the AIDS epidemic in the United States will permit prevention and treatment programs to target specific groups of high risk individuals who would be considered preventable by the WHO Delphi Study (World Health Organization, 1989c).

The mathematical models used to project the future course of HIV-1 infection all suggest a continued spread of the HIV epidemic. The most important characteristic of the projections however is that many of the projected cases are preventable and that the epidemiologic data is available to maximize use of prevention and control programs in target populations.

REFERENCES

Allain JP (1988) Prevalence of HTLV-III/LAV antibodies in patients with hemophilia and in their sexual partners in France (letter). N Engl J Med 315:517-518.

Anderson RM (1988) HIV infection in heterosexuals. Nature 331:655-656.

Anderson RM, Medley GF, May RM, Johnson AM (1986) A preliminary study of the transmission of dynamics of the human immunodeficiency virus (HIV), the causative agent of AIDS. J Math Appl Med Biol 3:229-263.

Barre-Sinoussi F, Chermann JC, Rey F, Nugeyre MT, Chamaret S, Gruest J, Dauguet C, Axler-Blin C, Brun-Vezinet M, Rouzioux C, Rozenbaum W, Montagnier L (1983) Isolation of a T-lymphotropic retrovirus from a patient at risk for acquired immunodeficiency syndrome (AIDS). Science 220:868-871.

Blanche S, Rouzioux C, Moscato M-LG, Veber F, Mayaux, M-J, Jocomet C, Tricoire J, Deville A, Vial M, Firton G, de Crepy A, Douard D, Robin M, Courpotin C, Ciraru-Vigneron N, le Deist F, Grisscelli C (1989) HIV infection in newborns. French Collaborative Study Group. A prospective study of infants born to women seropositive for human immunodeficiency virus type 1. N Engl J Med 320:1643-1648.

Bongaarts J, Reining P, Way P, Conant F (1989) The relationship between male circumcision and HIV infection in African populations. AIDS 3:373-377.

Brickner PW, Torres RA, Barnes M, Newman RG, Des Jarlais DC, Whalen DP, Rogers DF (1989) Recommendations for control and prevention of human immunodeficiency virus infection in intravenous drug users. Ann Intern Med 110-833-837.

Cameron DW, Simonsen JN, D'Costa LJ, Ronald AR, Maitha GM, Gakinya MN, Cheang M, Ndinya-Achola JO, Piot P, Brunham RC, Plumner FA (1989) Female to male transmission of human immunodeficiency virus type 1: risk factors for seroconversion in men. Lancet 2:403-407.

Cannon RO, Hook EW, Glasser D, Nahmias AJ, Lee FK, Glasser D, Quinn (1988) Association of herpes simplex virus type 2 with HIV infection in heterosexual patients attending sexually transmitted disease clinics. Fourth International Conference on AIDS. Stockholm, Sweden. Abstract #4558.

Carael M, Van De Perre PH, Lepage PH, Allen S, Nsengumuremyi F, Van Goethem C, Ntaborutaba M, Nzarramba D, Chumeck N (1988) Human immunodeficiency virus transmission among heterosexual couples in Central Africa. AIDS 2:201-205.

Castro KG, Lieb S, Jaffe HW, Narkunas JP, Calisher CH, Bush TJ, Witte JJ (1988) Transmission of HIV in Belle Glade, Florida: Lessons for other communities in the United States. Science 239:193-197.

Centers for Disease Control (1985) Provisional public health service interagency recommendations for screening and donated blood and plasma for antibody to the virus causing acquired immunodeficiency syndrome. MMWR 34:1-5.

Centers for Disease Control (1987a) Human immunodeficiency virus infection in the United States: a review of current knowledge. MMWR 36:1-48.

Centers for Disease Control (1987b) Recommendations for prevention of HIV transmission in health care settings. MMWR 36:2S-10S.

Centers for Disease Control (1987c) Years of potential life lost before age 65-United States, 1987. MMWR 38:27-29.

Centers for Disease Control (1988) AIDS due to HIV-2 infection - New Jersey. MMWR 37:33-35.

Centers for Disease Control (1989a) Update: Acquired immunodeficiency syndrome associated with intravenous-drug use-United States, 1988. MMWR 38:165-70.

Centers for Disease Control (1989b) Update: Acquired immunodeficiency syndrome - United States, 1981-1988. MMWR 38:229-236.

Centers for Disease Control (1989c) Update: Heterosexual transmission of AIDS and HIV infection - United States. MMWR 38:423-434.

Centers for Disease Control (1989d) Update: HIV-2 infection-United States. MMWR 38:572-574, 579-580.

Centers for Disease Control (1989e) Guidelines for prevention of transmission of human immunodeficiency virus and hepatitis B virus to health-care and public safety workers. MMWR 38:S6.

Centers for Disease Control (1990a) AIDS and the human immunodeficiency virus infection in the United States: 1989 Update. MMWR 39:81-86.

Centers for Disease Control (1990b). Estimates of HIV prevalence and Projected AIDS Cases: Summary of a Workshop, October 31-November 1, 1989. MMWR 39:110-119.

Chamberland ME, Dondero TJ (1987) Heterosexually acquired infection with human immunodeficiency virus (HIV) (editorial). Ann Intern Med 107:763-768.

Chin J, Mann J (1989) Global surveillance and forecasting of AIDS. Bull World Health Organ. 67:1-7.

Clavel F (1987) HIV-2, the West African AIDS virus. AIDS 1:135-140.

Clavel F, Guetard D, Brun-Vezinet F, Chamaret S, Rey MA, Santos-Ferreira MO, Laurent AG, Dauguet C, Katlama C (1986) Isolation of a new human retrovirus from West African patients with AIDS. Science 233:343-346.

Clavel F, Mansinho K, Chamaret S, Guetard D, Fauier V, Nina J, Santos-Ferreira MO, Champalimaud JL, Montagnier L (1987) Human immunodeficiency virus type 2 infection associated with AIDS in West Africa. N Engl J Med 319:1180-1185.

Colebunders RL, Kapita B, Nekwei W, Bahwe Y, Baende F, Ryder R (1988) Breast-feeding and transmission of HIV. Fourth International Conference on AIDS. Stockholm, Sweden.

Coombs RW, Collier A, Nikora B, Chase M, Gjerset G, Corey L (1987) Relationship between recovery of HIV from plasma and stage of disease (Abstract WP 5.4). In Abstracts of the III International Conference on AIDS. Washington, D.C.

Coombs RW, Kreiss J, Nikora B (1988) Isolation of HIV from genital ulcers in Nairobi prostitutes (Abstract N. 1244) 28th Interscience Conference on Antimicrobial Agents and Chemotherapy.

Cortes E, Detels R, Aboulafia D, Li XL, Moudgil T, Alam M, Bonecker C, Oyafuso L, Tondo M, Boite C, Hammershlak N, Capitani C, Salmon D, Ho D (1989) HIV-1, HIV-2 and HTLV-1 infection in high risk groups in Brazil. N Engl J Med 320:953-958.

Cumming PD, Wallace EL, Schorr JB, Dodd RY (1989) Exposure to patients to human immunodeficiency virus through the transfusion of blood components that test antibody negative. N Engl J Med 321:941-946.

Curran JW, Jaffe HW, Hardy AM, Morgan WM, Selik RM, Dondero TJ (1988) Epidemiology of HIV infection and AIDS in the United States. Science 239:610-616.

Darrow WW, Echenberg DF, Jaffe HW, O'Malley PM, Byers RH, Getchell JP, Curran JW (1987) Risk factors for human immunodeficiency virus (HIV) infections in homosexual men. Am J Public Health 77:479.

De Cock KM, Porter A, Odehouri K, Barrere B, Moreau J, Diaby L, Kouadio JC, Heyward WL (1989) Rapid Emergence of AIDS in Abidjan, Ivory Coast. Lancet 2:408-410.

Des Jarlais DC, Friedman SR, Stoneburner RL (1988) HIV infection and intravenous drug use: critical issues in transmission dynamics, infection, outcome and prevention. Rev Infect Dis 10:151-158.

Des Jarlais DC, Friedman SR, Novick DM, Sotheran JL, Thomas P, Yankovitz SR, Mildvan D, Weber J, Kreck MJ, Maslansky R (1989) HIV-1 infection among intravenous drug users in Manhattan, New York. JAMA 261:1008-1012.

Fischl MA, Dickinson GM, Scott GM, Klimas N, Fletcher MA, Parks W (1987) Evaluation of heterosexual partners, children, and household contacts of adults with AIDS. JAMA 257:640-644.
Friedland GH, Klein RS (1987) Transmission of the human immunodeficiency virus. N Engl J Med 317:1125-1134.
Gallo RC, Salahuddin SZ, Popovic M, Shearer GM, Kaplan M, Haynes BF, Palker, TJ, Redfield, R, Oleske, J, Safai B, White, G, Foster P, Markham P (1984) Frequent detection and isolation of cytopathic retroviruses (HTLV-III) from patients with AIDS and at risk for AIDS. Science 224:500-503.
Goedert JJ, Sarngadharan MG, Biggar RJ, Winn DM, Grossman RJ, Green MK, Bodner AJ, Mann DL, Strong DM (1984) Determinants of retrovirus (HTLV-III) antibody and immunodeficiency conditions in homosexual men. Lancet 2:711.
Goedert JJ, Eyster ME, Biggar RJ, Blattner WA (1988a) Heterosexual transmission of human immunodeficiency virus: association with severe depletion of T-helper lymphoctye in men with hemophilia AIDS. Rev Hum Retroviruses 3:355-361.
Goedert JJ, Eyster ME, Ragni MV, Biggar RJ, Gail MH (1988b) Rate of heterosexual HIV transmission and associated risk with HIV-antigen. Fourth International Conference on AIDS. Stockholm, Sweden.
Greenblatt RM, Lukehart SA, Plummer FA, Quinn TC, Critchlow CW, Ashley RL, D'Costa LJ, Ndinya-Achola JO, Corey L, Ronald AR, Holmes KK (1988) Genital ulceration as a risk factor for human immunodeficiency virus infection. AIDS 2:47-50.
Guinan ME, Hardy (1987) Epidemiology of AIDS in women in the United States. JAMA 257:2039-2042.
Hearst N, Hulley SB (1988) Preventing the heterosexual spread of AIDS. Are we giving our patients the best advice? JAMA 259:2428-2432.
Holmberg SD, Stewart JA, Gerber AR, Byers RH, Lee FK, O'Malley PM, Nahimias AJ (1988a) Prior herpes simplex virus type 2 infection as a risk factor for HIV infection. JAMA 259:1048-1050.
Holmberg SD, Horsburg CR, Ward JW, Jaffe HW (1988b) Biologic factors in the sexual transmission of human immunodeficiency viruses. J Infect Dis 160:116-125.
Horsburgh CR, Jr., Holmberg SD (1988) The global distribution of human immunodeficiency virus type 2 (HIV-2) infection. Transfusion 28:192-195.
Jason JM, McDougal JS, Dixon G, Lawrence DN, Kennedy MS, Hilgartner M, Aledort L, Evatt BL (1986) HTLV-III/LAV antibody and immune status of household contacts and sexual partners of persons with hemophilia. JAMA 255:212-215.
Johnson AM, Laga M (1988) Heterosexual transmission of HIV. AIDS 2:S49-S56.
Jovaisas E, Koch MA, Schafer A, Stauber M, Lowenthal D (1985) LAV/HTLV-III in a 20-week fetus. Lancet 2:1129.
Kanki PJ, M'Boup S, Ricard D, Barin F, Denis F, Boye C, Sangare L, Traver K, Albaum M, Marlink R (1987) Human T-lymphotropic virus type 4 and the human immunodeficiency virus in West Africa. Science 236:827-831.
Kelen GD, DiGiovanna T, Bisson L, Kalainov D, Sivertson KT, Quinn TC (1989) Human immunodeficiency virus infection in emergency patients: Epidemiology, clinical presentations, and risk to health care workers: The Johns Hopkins experience. JAMA 262:516-522.

Kingsley LA, Detels R, Kaslow R (1987) Risk factors for seroconversion to human immunodeficiency virus among male homosexuals. Lancet 1:345-349.

Kloser PC, Mangia AJ, Leonard J (1989) HIV-2 associated AIDS in the United States. The first case. Arch Intern Med 149:18765-1877.

Kreiss JK, Kitchen LW, Prince HE, Kasper CK, Essex M (1985) Antibody to human T-lymphotropic virus type III in wives of hemophiliacs: evidence for heterosexual transmission. Ann Intern Med 102:623-626.

Kreiss JK, Koech D, Plummer FA, Holmes KK, Lightfoote M, Piot P, Ronald AR, Ndinya-Achola JO, D'Costa LJ, Roberts P (1986) AIDS virus infection in Nairobi prostitutes. Spread of the epidemic to East Africa. N Engl J Med 314:414-418.

Laga M, Taelman H, Bonneux L, Cornet P, Vercauteren G, Piot P (1988) Risk factors for HIV infection heterosexual partners of HIV infected Africans and Europeans (abstract 4004). In: Program and Abstracts of the Fourth International Conference on AIDS. Stockholm: Swedish Ministry of Health.

Lapointe N, Michaud J, Pekovic D, Chausseau JP, Dupuy J-M (1985) Transplacental transmission of HTLV-III virus. N Engl J Med 312:1325.

Lepage P, Van de Perre P, Carael M, Nsengumuremi F, Nkurunziza J, Butzler JP, Sprecher S (1987) Postnatal transmission of HIV from mother to child. Lancet 2:400.

Lyons SF, Jupp PJ, Schoub BD (1986) Survival of HIV in the common bedbug. Lancet 2:45.

Mann JM, Quinn TC, Francis H, Nilambi N, Bosenge N, Bila K, McCormick JB, Ruti K, Asila PK, Curran JW (1986) Prevalence of HTLV-III/LAV in household contacts of patients with confirmed AIDS and controls in Kinshasa, Zaire. JAMA 256:721-724.

Mann JM, Chin (1988a) AIDS: A global perspective. N Engl J Med 319:302-304.

Mann JM, Chin J, Piot P, Quinn TC (1988b) The international epidemiology of AIDS. Sci Am 10:82-89.

Marcus R, CDC Coperative Needlestick Surveillance Group (1988) Surveillance of health care workers exposed to blood from patients infected with the human immunodeficiency virus. N Engl J Med 319:1118-1123.

May RM (1988) HIV infection in heterosexuals. Nature 331:655-656.

Melbeye M, Njelesani EK, Bayley A, Mukelabai K, Manuwele JK, Bowa FJ, Clayden SA, Levin A, Blattner WA, Weiss RA, Tedder R, Biggar RJ (1986) Evidence for heterosexual transmission and clinical manifestations of human immunodeficiency virus infection and related conditions in Lusaka, Zambia. Lancet 2:1113-1116.

Mortimer PP, Cooke EM (1988) HIV infection, breastfeeding, and human milk banking. Lancet 11:452-453.

Moss AR, Osmond D, Bacchetti P, Chermann JC, Barre-Sinoussi F, Carlson J (1987) Risk factors for AIDS and HIV seropositivity in homosexual men. Am J Epidemiol 125:1035.

N'Galy B, Ryder RW (1988) Epidemiology of HIV infection in Africa. J Acq Imm Def Synd 1:551-558.

Narain JP, Hull B, Hospedales CJ, Mahabir S, Bassett DC (1989) Epidemiology of AIDS and HIV infection in the Caribbean. In: AIDS: Profile of an Epidemic. Pan American Health Organization, Sci. Publ. No. 514, p61-72.

Neumann PW, O'Shaughnessy MV, Lepine D, DSouza I, Major C (1989) Laboratory diagnosis of the first cases of HIV-2 infection in Canada. Can Med Assoc J 140:125-128.

Nzila N, Ryder R, Colebunders R, Ndilu M, LeBughe M, Kamenga M, Kashamuka M, Brown C, Francis H (1988) Married couples in Zaire with discordant HIV serology. IV International Conference on AIDS, Stockholm (abstract 4059).

Office of Technology Assessment, United States Congress (1987) Do insects transmit AIDS? Washington, D.C., U.S. Government Printing Office.

Ou CY, Kwok S, Mitchell SW, Mack DH, Sninski JJ, Krebs JW, Feorino P, Warfield D, Schokhatman G (1988) DNA amplification for direct detection for HIV-1 DNA of peripheral blood mononuclear cells. Science 239:295-297.

Padian N, Marquis L, Francis DP, Anderson RE, Retlerford GW, O'Malley PM, Winkelstein W (1987) Male-to-female transmission of human immunodeficiency virus. JAMA 258:788-790.

Pape JW, Johnson WD (1989) HIV-1 infection and AIDS in Haiti. In: The Epidemiology of AIDS. Kaslow RA, Francis DP, eds, Oxford Univ Press, New York, pp. 194-221.

Pepin J, Plummer FA, Brunham RC, Pio P, Cameron DW, Ronald AR (1989) The interaction of HIV infection and other sexually transmitted diseases: an opportunity for intervention. AIDS 3:3-9.

Peterman T, Curran JW (1986a) Sexual transmisson of human immunodeficiency virus. JAMA 256:2222-2226.

Peterman TA, Stoneburner RL, Allen JR, Jaffe HW, Curran JW (1986b). Risk of HIV transmission from heterosexual adults with transfusion-associated infection. JAMA 259:44-48.

Piot P, Plummer FA, Rey MA, Ngugi EN, Rouzioux C, Ndinya-Achola JO, Veracauteren G, DCosta LJ, Laga M, Nsanze H (1987) Retrospective seroepidemiology of AIDS virus infection in Nairobi populations. J Infect Dis 155:1108-1112.

Piot P, Plummer FS, Mhalu JL, Chin, Mann JM (1988a) AIDS: An international perspective. Science 239:573-579.

Piot P, Kreiss JK, Ndinya-Achola JO, Ngugi E, Plummer FA (1988b) Heterosexual transmission of HIV. AIDS 2:1-10.

Piot P, Laga M, Ryder R, Perriens J, Temmerman M, Heyward W, Curran J (1990) The global epidemiology of HIV infection: continuity, heterogeneity, and change. J Acq Immun Def Synd 3:403-412.

Poulsen AG, Kvinesdal B, Aaby P, Molbak K, Frederiksen K, Dias F, Lauritzen E (1989) Prevalence of and mortality from human immunodeficiency virus type 2 in Bissau, West Africa. Lancet 1:827-830.

Quinn TC, Mann JM, Curran JW, Piot P (1986) AIDS in Africa: an epidemiologic paradigm. Science 234:955-963.

Quinn TC, Riggin CH, Kline R, Francis H, Mulanga K, Sension MG, Fauci AS (1988a). Rapid latex agglutination assay using recombinant envelope polypeptide for the detection of antibody to the human immunodeficiency virus. JAMA 260:510-513.

Quinn TC, Glasser D, Cannon RO Matusak DL, Duning RW, Kline RL, Campbell CH, Israel E, Fauci AS, Hook EW (1988b) Human immunodeficiency virus infection among patients attending clinics for sexually transmitted diseases. N Engl J Med 318:197-203.

Quinn TC, Zacarias FRK, St John RK (1989a) AIDS in the Americas. An emerging public health crisis. N Engl J Med 320:1005-1007.

Quinn TC, Zacarias FRK, St John RK (1989b) HIV and HTLV-I infections in the Americas: A regional perspective. Medicine 68:189-209.

Redfield RR, Markham PD, Salahuddin SZ, Saragadharan MG, Bodner AJ, Folks TM, Ballog WR, Wright G, Gallo RC (1985) Frequent transmission of HTLV-III among spouses of patients with AIDS-related complex and AIDS. JAMA 253:1571-1573.

Rogers MF (1987) Breast-feeding and HIV infection (Letter) Lancet 1987; 11:1278. N Engl J Med 32:1649-1654.

Rogers MF, Ou CY, Rayfield M, Thomas PA, Schoenbaum EE, Abrams E, Krasinski K, Selwyn PA, Moore J, Kaul A, Grimm KT, Bamji M, Schochetman G, the New York City Collaborative Study of Maternal HIV Transmission and Montefiore Medical Center HIV Perinatal Transmission Group (1989) Use of polymerase chain reaction for early detection of the proviral sequence of human immunodeficiency virus in infants born to seropositive mothers. N Engl J Med 320:1649-1654.

Ronald AR, Ndinya-Achola JO, Plummer FA, Simonsen JN, Cameron DW, Ngugi EN, Pamba H (1988) A review of HIV-1 in Africa. Bull NY Acad Med 64:480-490.

Ruef C, Dickey P, Schable CA, Griffith B, Williams AE, D'Aquila RT (1989) A second case of of the acquired immunodeficiency syndrome due to human immunodeficiency virus type 2 in the United States: the clinical implications. Am J Med 86:709-712.

Rwandan HIV Seroprevalence Study Group (1989) Nationwide community-based serological survey of HIV-1 and other human retrovirus infections in a central African country. Lancet 1:941-943.

Ryder RW, Piot P (1988a) Epidemiology of HIV-1 infection in Africa, In: Piot P, Mann JM, eds. AIDS and HIV infection in the tropics. London: Bailliere-Tyndall, pp. 113-30.

Ryder RW, Hassig SE (1988b) The epidemiology of perinatal transmission of HIV. AIDS 2 (suppl 1):S83-S89.

Ryder RW, Nsa W, Hassig SE, Behets F, Rayfield M, Ekungola B, Nelson AM, Mulenda U, Francis H, Kashamuka M, Davachi F, Rogers M, Nzila N, Greenberg A, Mann J, Quinn TC, Piot P, Curran JW (1989) Perinatal transmission of the human immunodeficiency virus types 1 to infants of seropositive women in Zaire. N Engl J Med 320:1637-1642.

Schoenbaum EE, Hartel D, Selwyn PA, Klein RS, Davenny K, Rogers M, Feiner C, Friedland G (1989) Risk factors for human immunodeficiency virus infection in intravenous drug users. N Engl J Med 321:874-879.

Simoes EA, Babu PG, John TJ, Nirmala S, Solomon S, Lakshainarayana CA, Quinn TC (1987) Evidence for HTLV-III infection in prostitutes in Tamil Nadu (India). Indian J Med Res 85:335-338.

Simonsen JN, Cameron W, Gakinya MN, Ndinya-Achola JO, D'Costa LJ, Darasira D, Cheang M, Ronald AR, Piot P, Plummer FA (1988) Human immunodefiency virus infection among men with sexually transmitted diseases. Experience from a center in Africa. N Engl J Med 319:274-278.

Spielberg F, Kabeya CM, Ryder RW Kifuani NK, Harris J, Bender TR, Heyward WL, Quinn TC (1989) Field testing and comparative evaluation of rapid, visually read screening assays for antibody to human immunodeficiency virus. Lancet 1:580-583.

Sprecher S, Soumenkoff G, Puissant F, Degueldre M (1986) Vertical transmission of IV in a 15-week fetus. Lancet 2:288.

Quinn TC, Cannon RO, Glasser, D, Groseclose SL, Brathwaite WS, Fauci AS, Hook EW (1990) The asssociation of syphilis with risk of human immunodeficiency virus infection in patients attending sexually transmitted diseases clinics. Arch Intern Med. 150:1297-1302.

Stall RD, Coates TJ, Hoff C (1988) Behavioral risk reduction for HIV infection among gay and bisexual men: a review of results in the United States. J Am Psychol. 143:878-85.

Stamm WE, Handsfield HH, Rompalo AM, Ashley RL, Roberts PL, Corey L (1988) Association of genital ulcer disease with risk of HIV infection in homosexual men. JAMA 260:1429-1433.

Stazewski S, Rehmet S, Hofmeister WD, Helm EB, Stille W, Werner A, Doerr HW (1988) Analysis of transmission rates in heterosexual transmitted HIV infection (abstract 4068), In: Program and abstracts of the Fourth International Conference on AIDS. Stockholm: Swedish Ministry of Health and Social Affairs.

Tanphaichitra D, Armstrong D, Gold J, Chien N (1988) HIV testing in Bangkok Thailand (letter) AIDS 2:228.

The European Collaborative Study (1988) Mother to child transmission of HIV infection. Lancet 2:1039-1042.

Thiry L, Sprecher-Goldberger S, Jockheer T, Levy J, Van-de-Perre P, Henrivaux P, Cogniaux-Le Clerc J, Clumeck N (1985) Isolation of AIDS virus from cell free breast milk of three healthy virus carriers. Lancet 2:891-892.

Van Griensven GJD, Tielman RAP, Goudsmit J (1987) Risk factors and prevalence of HIV antibodies in homosexual men in the Netherlands. Am J Epidemiol 125:1048.

Van de Perre P, De Clercq A, Cogniaux H, Leclerc J, Nzaramba D, Butzler JP, Sprecher-Goldberger S (1988) Detection of HIV p17-antigen in lymphocytes but not epithelial cells from cervicovaginal secretions of women seropositive for HIV: implications for heterosexual transmission of the virus. Genitourinary Med 64:30-33.

Veronesi R, Mazzsa CC, Santos-Fereira MO, Lourenco MH (1987). HIV-2 in Brazil (Letter). Lancet 2:405.

Ward JW, Holmberg SD, Allen JR, Cohen DL, Critchley SE, Ravenholt O, Davis JR, Quinn MG, Jaffe HW (1988) Transmission of human immunodeficiency virus (HIV) by blood transfusion screened as negative for HIV antibody. N Engl J Med 318:473-478.

Winkelstein W, Lyman DM, Padian N, Grant R, Samuel M, Wiley JA, Anderson RE, Lang W, Riggs J, Levy JA (1987) Sexual practices and risk of infection by the human immunodeficiency virus: the San Francisco Men's Health Study. JAMA 257:321-325.

World Health Organization (1989a) AIDS surveillance update in the WHO European Region. Wkly Epidem Rec 64:221-228.

World Health Organization (1989b). Acquired immunodeficiency syndrome (AIDS); Global projections of HIV/AIDS. Wkly Epidem Rec 64:229-231.

World Health Organization and the Centers for Disease Control (1990) World Health Organization Global Statistics. AIDS 4:171-176.

Ziegler JB, Cooper DA, Johnson RO, Gold J (1985) Postnatal transmission of AIDS-associated retrovirus from mother to infant. Lancet 1:896-898.

DISCUSSION

Levy J (University of California Irvine Medical Center, Orange, CA):

You mentioned the possibility of detecting HIV up to, I think, three months prior to the seropositive reaction, using PCR. I'm wondering, is there any clinical significance to that sort of early detection, and if so, are there situations in which screening, using PCR, might be appropriate?

Francis H:

Let me paraphrase your question to make sure I understand it. As far as perinatal transmission of the disease, is there some use in doing early diagnosis in women?

Levy J:

No. You said that PCR could detect HIV infection up to three months prior to a seropositive response. I'm wondering if that early detection has any clinical significance, and if so, does that suggest that there are situations where screening with PCR would be appropriate?

Francis H:

Yes, I do think there is a role. The places I think that it's most significant are for people who are at high risk for acquiring disease. For example, in Africa, we were concerned about the high incidence of HIV in prostitutes. Is it worthwhile detecting women before they seroconvert? Probably, because the prostitutes may be infectious before they seroconvert. The other group of people in which we see a great deal of utility for using PCR are the false-positives. If you've worked with African sera or South American sera, you know that these patients have a tremendous burden of disease. There are numerous antibodies and protein in their sera which will interfere with our assays and may cause a lot of false positives and false negatives. That's changed since we've used recombinant antigens and antibodies. If you do Western blots, some patients have a combination of bands like a p24 and a gp120 and, we're not really sure what that indicates about the patient's serostatus. The PCR in that case might be useful. The reason PCR will be useful in African women who have multiple pregnancies and people who have multiple transfusions, is that these two groups are more likely to have a higher incidence of false positives by both the ELISA and Western blot. The other situation which is probably more germane to the Americans is where many Americans are p24 positive for unclear reasons. It has been seen in U.S. personnel abroad. PCR in that case has been very useful for defining whether they are HIV infected or not.

Minocha H (Kansas State University, Manhattan, KS):

Does that mean that in the United Stated there could be in blood banks, certain groups of blood for transfusion, maybe it's very minor, but maybe a 0.001 or less that is still positive to HIV.

Francis H:

Theoretically, that is possible. We are talking maybe one or two cases out of 100,000. There may never be a situation which completely clears the blood supply of all infectious agents. It has been a concern for all AIDS researchers. Manufacturers of HIV diagnostic kits are trying to improve the sensitivities and specificities of these assays for screening. I'm not sure there will every be a really good answer for this issue, except to develop some better screening strategies in the donor interviews. That's probably the approach that many health professionals may take, but as you know, problems of invasion of privacy complicate that kind of an approach.

Aaby P (University of Copenhagen, Denmark):

I have been working for 12 years in Guinea-Bissau, and you have presented a very interesting hypothesis on why the prevalence was so high in

Guinea-Bissau. For once, I'll try to act as a judge on the beauty of women in Guinea-Bissau. They are beautiful, but by no means, any more beautiful than women in any other country in West Africa. In the capital where we were working we found 10% positives in the normal adult population. This is not a high risk population in any way, no prostitutes, just the standard population above fifty. The more probable explanation, I think, is partially the war which has been going on which has been going on, which means that you have a very large Portuguese army in this area. You have prostitution in connection with the war and that's probably why you are also finding that Europeans who are infected with HIV-2 who are mainly Portuguese people. Secondly, it might be that it is actually Guinea-Bissau where this virus originated, because this must be the place where we had the first important spread. You're finding an AIDS distribution which is totally different from what you are finding in Central Africa. In the sense that you are not finding that the women are infected before the men, and you are not finding very many people infected before it's gone too far. So, we have the increase between 25 and 40, and about 40 we have a prevalence of 20-25%. So, of all adults, above 40, we have something like 20% who are infected which suggests that the virus may have been around for a long time in this area. If you go to the neighboring countries like Gambia or Senegal, you will find the same pattern for HIV-2 prevalence as you find in Central Africa, sexually transmitted diseases which have recently been introduced. You find it mainly among single women and men frequenting prostitutes.

Francis H:

There are a couple points that are quite important. The prevalence rates such as those I have presented for surveys are formed on whatever population is available. You can get artifactually high or low prevalences, depending on whom you study. The second point is that I had not heard that the people were thinking of Guinea-Bissau as the origin of HIV-2. The people that I work with don't particularly subscribe to that belief. However, the data on HIV-2 prevalence that you present is very interesting. I think the overall critical issue is the destruction of the social structure in Western and Central Africa which is resulting in changes in sexual behavior. This development, which I did not describe in detail, facilitated the massive spread of the disease in many countries. A case in point is Uganda where the estimates of HIV infection in the general population ranges between 25-30%, depending on where you look. The HIV-2 prevalence which you have reported for Guinea-Bissau indicates to me that the area is at risk for destruction of its social structure.

Lennette E:

This is a propos to the question that that Dr. Levy just asked, which also appeared in your slide on prolonged latency, about the paper in the New England Journal of Medicine on PCR. As I recall, in that paper there were about 15 patients who were seronegative for HIV. They were positive by PCR. To me, and several other people at this particular meeting, it just seemed too beautiful to be true. We raised the question of specificity, and also contamination in the laboratory. We were assured by all the experts on PCR that there was no problem in a well-run laboratory. If you took due care, nothing would happen. These were good presentations. However, towards the late afternoon of that meeting, people admitted that you could have problems with PCR. This was also shown well last week or two weeks ago, in a letter to the Editor of Lancet. Richard Cone, who will be here tomorrow talking about herpesvirus-6, illustrates some of the problems that can occur - how he takes care of getting DNA off of doorknobs and sink faucets and whatever. The other point is, the data will have to be accurate if you're going to interpret them, whatever connection. One of the contenders that HIV has nothing to do with AIDS, a very eminent respected scientist, has presented epidemiologic data which is flawed. You showed an example of that, because this person contends that there are so many cases of seroprevalence, actually it was in Zaire. He extrapolated from the urban areas to the rural to get a total for the country. You showed very clearly that there is a difference between urban and rural areas, so called jungle. So, you have to be careful how these data are put together.

Francis H:

That's very true. You have made a good point about PCR, I'm sure most of you know the problems of PCR. When you start amplifying things about one or ten million times, the chance for error is phenomenal. As I mentioned earlier when we were talking about serologic assays, there are many problems in working with sera from developing countries. There are several types of HTLV viruses associated with disease. When we first started amplifying sequences in African patients, all samples were positive. Factors in the primers we were using to identify our unique sequence had problems. Also, it turned out that the envelope primers which we used were generated from a New York City strain (NYCIIIB). This strain has many differences from the virus seen in Africa. For this reason, many false-positives and false-negatives were found, because the primers were sticking in the wrong spot or not sticking at all. The PCR, I think, is an extremely valuable tool, but not only in the detection of DNA. I suspect that one of its greatest features will be in the detection of messenger RNA and other nucleic acids. You see a lot of that research on that subject. For example, PCR detection of HIV mRNAs may be used to study women who transmit HIV to their children. This parameter may detect active disease and help distinguish between mothers who do not transmit HIV perinatally. PCR might have a role in that, both in the qualitative and quantitative sense. But, these are ideas that have not been confirmed.

Lennette E:

Some of that problem may be resolved by a Q-β replicase assay, where you enhance the signal rather than the target.

Francis H:

Yes, I saw the lecture title and I'm looking forward to hearing about that.

HUMAN HERPESVIRUS 6: AN UPDATE

Richard Cone

University of Washington
Children's Hospital
Seattle, Washington, USA

INTRODUCTION

The existence of human herpesvirus 6 (HHV-6) was first reported in 1986 (Salahuddin et al. 1986), after which many laboratories independently isolated other strains of HHV-6 (Agut et al. 1988; Downing et al. 1987; Lopez et al. 1988; Pietroboni et al. 1988a; Tedder et al. 1987; Yamanishi et al. 1988). Most isolates were initially derived by cultivation of peripheral blood mononuclear cells (PBMCs) from patients with immune dysfunction, such as lymphoproliferative disorders (Salahuddin et al. 1986), acquired immunodeficiency virus (AIDS) (Agut et al. 1988; Becker et al. 1988; Downing et al. 1987; Lopez et al. 1988; Pietroboni et al. 1988a; Tedder et al. 1987), and organ transplants (Asano et al. 1989c; Okuno et al. 1990). HHV-6 has also been isolated from saliva of normal adults (Harnett et al. 1990; Levy et al. 1990) and children with roseola (Asano et al. 1989a; Yoshiyama et al. 1990).

At this point the only clear disease association with HHV-6 infection is roseola. More than 90% of the human population becomes infected in early childhood (Brown et al. 1988b; Knowles and Gardner 1988; Levy et al. 1990; Okuno et al. 1989; Yanagi et al. 1990; Yoshikawa et al. 1990), with persistent and possibly latent infection occurring in almost all infected people. Increased replication appears to emerge in those who become immunocompromised. It is not surprising that circumstantial evidence has associated this ubiquitous agent with a variety of ailments, but the role of HHV-6 in human biology remains largely undefined.

PHYSICAL CHARACTERISTICS

Morphology

All herpesviruses share features of their ultrastructural morphology, a commonality that allowed the initial identification of HHV-6 as a herpesvirus. The virion core is an electron dense accumulation of double stranded DNA and nucleoproteins. The 100 nm diameter capsid is composed of 162 capsomeres forming the icosahedral shell (Biberfeld et al. 1987; Salahuddin et al. 1986). The capsids acquire a bilayered membrane as they bud out of the nucleus

Medical Virology 10, Edited by L.M. de la Maza and E.M. Peterson, Plenum Press, New York, 1991

and into the cytoplasm through the nuclear envelope. The membrane-bounded capsids acquire an electron-dense layer, called the tegument, between the inner and outer membrane layers, which probably completes the viral maturation. Cytoplasmic maturation has also been observed with cytomegalovirus (CMV) (Papadimitriou et al. 1984), the herpesvirus which is most closely related to HHV-6.

Genomic Organization and Taxonomy

The entire HHV-6 genome has been subcloned and sequencing of the genome is nearly complete (Robert Honess and Phil Pellet, personal communications), representing the fastest progress of this kind ever achieved for a herpesvirus. The overall genomic organization as well as homology between specific genes confirm that HHV-6 most closely resembles CMV at the genetic level (Efstathiou et al. 1988; Lawrence et al. 1990; Littler et al. 1990).

The genomic length of HHV-6 strain U1102, which is the most thoroughly studied, is 161.5 kb (Honess, personal communication). Two forms of the viral genome may exist: a shorter version exemplified by the U1102 strain and a longer version containing 170 kb, as seen with the Z-29 strain (Lopez et al. 1988). The difference in lengths has been attributed to variation in size of the terminal repeat elements (Lopez and Honess, 1990). The termini contain direct repeats and some coding regions that apparently have no significant homology for any of the other herpesviruses. The large central region of the genome contains genes that are unique to HHV-6 and others that have a strong homology to CMV. Eighteen open reading frames (ORFs) were identified in 22 kb of HHV-6 sequence from the central region of the genome (Lawrence et al. 1990). Fourteen of the 18 ORFs have significant homology with their CMV counterparts, whereas 6 ORFs from this region have weak homology with Epstein-Barr virus (EBV). The HHV-6 major capsid protein (MCP) shares 43.8% amino acid identity with CMV and less than 30% identity with EBV, herpes simplex virus (HSV)-1, and varicella zoster virus (VZV) (Littler et al. 1990).

Distinct isolates of HHV-6 appear to be largely homologous: subgenomic clones from one strain hybridize with other strains that have been tried (Downing et al. 1987; Josephs et al. 1986; Tedder et al. 1987). However, sequence variation among strains has been documented by restriction fragment length polymorphism (RFLP) analysis using the HHV-6 subclone pZVH-14 (GS strain) as a probe (Kikuta et al. 1989). A detailed study of RFLPs has shown that most restriction sites remain constant, but that individual isolates vary at some sites (Helene Collandre, personal communication). This situation may prove to be analogous to that for HSV and CMV, wherein transmission of specific strains can be traced by obtaining RFLP patterns from the isolates (Chandler and McDougall 1986, Chantratita and Yoosook 1990).

The lymphotropic nature of HHV-6 suggests that it could be classified as a γ herpesvirus in the α-β-γ classification scheme which is based on several biological characteristics (Roizman et al. 1981). However, the common *in vitro* growth characteristics and the sequence homologies between HHV-6 and CMV have led to classifying HHV-6 as a β herpesvirus (CMV is also a β herpesvirus) (Lawrence et al. 1990). This apparent inconsistency has added to mounting confusion about the α-β-γ classification system of herpesviruses, reinforcing the notion that herpesviruses may be better classified on the basis of phylogenetic characteristics, such as DNA and amino acid homologies, than on the basis of biological behavior.

Cell and Tissue Tropism

The initial report of HHV-6 referred to the virus as human B-lymphotropic virus (HBLV) because it was thought to preferentially infect B cells (Salahuddin et al. 1986). Subsequent studies from other laboratories showed that HHV-6 preferentially grows in T4 lymphocytes *in vitro* (Becker et al. 1988; Black et al. 1989; Downing et al. 1987), although it has also been found in a variety of other cell types both *in vitro* and *in vivo* (Table 1). The original discoverers of HHV-6 proceeded to do a more complete survey of cell types that supported viral growth, and concurred that infection was most productive in T cells (Ablashi et al. 1987; Lusso et al. 1987). The designation "lymphotropic human herpesvirus" was proposed (Downing et al. 1987), but the term human herpesvirus 6 (Lusso et al. 1987), which does not imply any particular cell tropism, is now accepted in keeping with recommendations for the classification of herpesviruses (Roizman et al. 1981).

Most investigators now agree that *in vitro* infection of cord blood lymphocytes (CBL) and PBMCs is most productive in T lymphocytes bearing the CD4 antigen (Agut et al. 1988; Black et al. 1989; Lopez et al. 1988;Lusso et al. 1988, 1989, 1990; Takahashi et al. 1989). A careful analysis of the *in vivo* tropism of HHV-6 was done by separating T4, T8 and monocyte subsets of PBMCs from patients with roseola (Takahashi et al. 1989). When each subset was co-cultivated with CBLs, the CD4 bearing cells were predominantly responsible for viral propagation. Subsets lacking the CD3 (mature T cell) marker were not sources of infection, nor were they susceptible to HHV-6 infection *in vitro,* leading to the conclusion that mature T4 positive cells harbor the virus. Other investigators have found that various HHV-6 isolates can be cultivated in some, but not all, T cell lines (Asada et al. 1989; Black et al. 1989; Lusso et al. 1988; Tedder et al. 1987). Differences between strains may account for some of the heterogeneity in cell types that can be infected *in vitro,* although variations in culture technique could also play a role. The GS strain (Salahuddin et al. 1986) has been found to grow in a wider variety of cell types than the other HHV-6 isolates, including B lymphocytic, monocytic, megakaryocytic and neuronal cell lines (Table 1) (Ablashi et al. 1987; Lusso et al. 1987, 1988; Salahuddin et al. 1986).

This virus has frequently been cultivated from saliva (Harnett et al. 1990; Pietroboni et al. 1988a), and *in situ* hybridization has shown infection in serous and mucous cells of parotid and submandibular salivary glands (Fox et al. 1990). Viral antigens have been found in the renal tubular epithelium of kidneys that were rejected after transplantation, suggesting that these epithelial cells can support HHV-6 replication (Okuno et al. 1990). Similarly, HHV-6 was isolated from a kidney biopsy even though the patient's PBMCs were HHV-6 culture negative (Asano et al. 1989c). These findings suggest that, like CMV, HHV-6 may have a wider host cell range *in vivo* than *in vitro.*

There are no published reports of isolating HHV-6 from animals, although some data suggest that infection with an HHV-6 like virus may occur. HHV-6 has been cultivated in PBMCs from chimpanzees, where it has a predilection for T4 cells (Lusso et al. 1990). Also, antibodies that specifically recognize HHV-6 antigens have been found in 8 of 10 species of wild monkeys (Higashi et al. 1989). This study found high prevalences in African green monkeys (100%) and chimpanzees (90%), although these two species lacked antibodies according to a different report (Lusso et al. 1990).

TABLE 1. Cell Types That Have Been Reported to Support HHV-6 Growth

HHV-6 STRAIN	BLOOD[a]				CELL LINES[b]					
	PBMC	CBL	T cells	B cells	HSB2	T cell	B cell	Mono	Mega	Glia
GS[c]	Y	Y	Y	Y/N[d]	Y	Y	Y	Y	wk/N	Y
LHV[e]		Y	Y		Y	Y	Y	N		Y
U1102[f]		Y	Y	N	Y	Y	Y	Y		
Z-29[g]	Y	Y	Y		N	Y	N	N	N	N
SIE[h]	Y		Y							
HST[i]	Y	Y	Y		Y	Y				

a BLOOD = freshly isolated peripheral blood cells of the type indicated (PBMC = adult peripheral blood mononuclear cells; CBL = cord blood lymphocytes).

b CELL LINES indicates continuous *in vitro* cultures (T cell and B cell = various lines from their respective origins; Mono = mononuclear cell lines; Mega = megakaryocytic cell lines; and Glia = glial cells lines).

c Two conclusions in the same position indicate the presence of contradictory reports (wk = low level of growth; Y = supports HHV-6 growth; N = does not support growth).

d GS (Ablashi et al. 1989; Lusso et al. 1988; Salahuddin et al. 1986)

e LHV (Tedder et al. 1987)

f U1102 (Downing et al. 1987)

g Z-29 (Black et al. 1989; Lopez et al. 1988)

h SIE (Agut et al. 1988)

i HST (Asada et al. 1989)

LABORATORY DETECTION

Culture

The discovery of HHV-6 was made possible by breakthroughs in lymphocyte culturing techniques which progressed rapidly in response to the AIDS crisis (Gallo et al. 1984), making the culture of lymphocytes a routine procedure by the mid-1980s. The supplementation of lymphocyte cultures with IL-2 (Mier and Gallo, 1980) was particularly significant for the development of HHV-6 cultivation. It is interesting to note that the discovery of HHV-6 occurred in the same laboratory that introduced lymphokine supplementation as a culture technique (Salahuddin et al. 1986).

Isolation is often achieved by cocultivation of a patient's PBMC with "normal" PBMC. CBLs may be more susceptible to infection than PBMCs (Lopez et al. 1988), although the opposite conclusion has also been published (Levy et al. 1990). Most isolates can be propagated in CBLs, which have a distinct advantage over PBMCs since cord blood is not likely to be infected with HHV-6. The nearly universal prevalence of HHV-6 infection in people over 2 years of age correlates well with the finding that most samples of PBMCs from normal donors harbor HHV-6 DNA. Therefore, some isolates of HHV-6 may have been derived from the "normal" PBMC donor rather than the patient. This fact led one group to republish their isolation results using CBLs instead of PBMCs (Harnett et al. 1990).

HHV-6 culture is still undergoing optimization. The basic technique involves cocultivation of the subject's PBMCs with phytohemagglutinin (PHA, 0.0005%) stimulated CBLs in RPMI with 10% fetal calf serum and IL-2 (Black et al. 1989). Infection generally occurs within 7-10 days post-inoculation as measured by anti-complement immunofluorescence (ACIF) (Lopez and Honess, 1990) or the immunofluorescent antibody technique (Ablashi et al. 1988b). Infected cultures typically display a subpopulation of enlarged cells, syncytia, and cytoplasmic and nuclear inclusions, although these morphological changes may be absent from some infected cultures.

The amount of IL-2 supplementation varies widely in different reports, from 0.1 U/ml (Asada et al. 1989) to 30 U/ml (Black et al. 1989). Some reports have proposed that 5-10 U/ml of recombinant IL-2 can suppress HHV-6 growth, whereas lower concentrations do not (Roffman and Frenkel, 1990; Wrzos and Gibbons, 1990). A state of apparent dormancy can be induced at these recombinant IL-2 concentrations, and this effect may provide a model of *in vitro* latency. CBL susceptibility to HHV-6 infection varies widely among different cell donors (Black et al. 1989), and the variation cannot be attributed to differences in the level of passive maternal antibodies in the cord blood. Several more refinements in HHV-6 culture techniques were recently reported. A 100 fold enhancement of infection was achieved by centrifugation of the inoculum with the susceptible cells (Pietroboni et al. 1989). HHV-6 infection of PBMCs was also enhanced beyond the stimulatory effect of PHA by the addition of anti-CD3 antibodies (Kikuta et al. 1990a). Interferon produced by the cultured PBMCs inhibited HHV-6 replication (Kikuta et al. 1990b). CBLs do not produce as much interferon as PBMCs under these circumstances, which could account in part for the observations that CBLs are the more productive cells. These data also suggest that interferon may play a role in the host defense against HHV-6 infection.

Serology

The first report of HHV-6 included serologic data using an indirect immunofluorescence assay (IFA) for antibody detection (Salahuddin et al. 1986). HHV-6 infected cells prepared *in vitro* are dried onto a glass slide. Incubation of serum layered onto the prepared slides results in binding of serum antibodies to the cellular antigens. The bound anti-viral antibodies are labelled using a second incubation with fluorescein labelled anti-IgG or IgM antibodies and visualized by fluorescent microscopy. IFA represents one of the least complicated serologic procedures, but suboptimal sensitivity results in artifactually low estimates of serologic titers because the serum must be diluted in order to decrease background fluorescence. The widespread use of a relatively insensitive IFA for HHV-6 serologic determinations contributed to the initial impression of low seroprevalence in normal adults (Table 2). Several groups (Ablashi et al. 1988a; Knowles and Gardner, 1988; Tedder et al. 1987) developed improvements in the HHV-6 IFA technique which culminated in a report of nearly universal seroprevalence when lower serum dilutions were analyzed (Knowles and Gardner, 1988). A modification of the IFA procedure is also widely used for detection of HHV-6 in cell cultures.

ACIF, a serologic technique developed at the Centers for Disease Control, provided increased sensitivity and specificity (Lopez et al. 1988). Infected cells on slides are incubated with serum containing antibody and complement

TABLE 2. HHV-6 Seroprevalences from 10 Studies of Normal Populations

SUBJECTS	SEROPREVALENCE	REFERENCE
	serum dil 1:10 (1:40)	
Adults[a]	(8%)	(Huemer et al. 1989)
Adults	(18%)	(Tedder et al. 1987)
Children (6-11 mos)	36%	(Levy et al. 1990)
Adults (41-88 yrs)	45%	(Levy et al. 1990)
Adults	63% (26%)	(Krueger et al. 1988b)
Children (0-4 yrs)	71% (46%)	(Knowles and Gardner, 1988)
All[b] (8 mos-27 yrs)	77%	(Yoshikawa et al. 1989)
All	79%	(Okuno et al. 1989)
All	95%	(Levy et al. 1990)
Adults	97%	(Saxinger et al. 1988)
Adults	97%	(Linde et al. 1988)
Adults	98%	(Knowles and Gardner, 1988)

a "Adults" reflects studies which only described the population as adults without other age specifications.

b "All" indicates children and adults

which is "fixed" to the cells when the antibodies bind. The fixed C3 component is then detected with fluoresceinated anti-C3 antibodies. Although this procedure is more complicated than IFA, the improved results have led to its use for measuring serologic titers as well as detecting HHV-6 infection in culture.

Increasing experience with HHV-6 serologies has revealed that adults undergo changes in anti-HHV-6 IgM or IgG titers in association with primary EBV or CMV infection (Andre and Matz, 1988, Chou and Scott 1990, Enders et al. 1990; Irving et al. 1990; Linde et al. 1988, 1990; Morris et al. 1988). In fact, the major reason for increases in HHV-6 IgG titers in adults is concurrent primary infection with another herpesvirus (Linde et al. 1988). Cross-reacting antibodies between CMV or EBV and HHV-6 do not explain the titer rises since the non-HHV-6 antibodies can be adsorbed out without affecting the HHV-6 serology (Irving et al. 1990; Larcher et al. 1988).

Nucleic Acid

There are three methods of DNA detection that have played an important role in detection of HHV-6: Southern blot, *in situ* hybridization, and polymerase chain reaction (PCR). Southern blotting has been used to study RFLPs from various isolates of HHV-6, as discussed above in the Genomic Organization and Taxonomy section. The Southern blot/RFLP technique has been shown to distinguish strains of HHV-6 from different people in the same population, and will prove useful for epidemiologic studies of viral transmission. *In situ* hybridization for HHV-6 has been used to detect viral DNA in lymph nodes and saliva (Buchbinder et al. 1988; Ensoli et al. 1989; Fox et al. 1990; Krueger et al. 1988a).

PCR, invented just before HHV-6 was discovered, finds many applications in clinical studies on this virus. It has already been used to identify virus in blood, saliva, brain and liver tissues (Asano et al. 1990; Buchbinder et al. 1988; Gopal et al. 1990; Kondo et al. 1990; Qavi et al. 1989; Wrzos et al. 1990). The remarkable sensitivity of PCR has enabled detection of HHV-6 DNA in anatomic locations that are routinely culture negative, especially peripheral blood. As the PCR technique matures, it will undoubtedly play a major role in clinical HHV-6 research.

CLINICAL FEATURES

Epidemiology

HHV-6 is the among the most ubiquitous viral infections of mankind. Estimates of seroprevalence in the general population have varied between 8% and 97% (Table 2) (Brown et al. 1988a; Huemer et al. 1989; Krueger et al. 1988b; Linde et al. 1988; Morris et al. 1988; Okuno et al. 1989; Saxinger et al. 1988; Tedder et al. 1987; Yoshikawa et al. 1989). A major source of variation in estimation of the seroprevalence has been the titer cutoff above which a specimen is considered positive. Although some authors have demanded a positive reaction at serum dilutions of ≥ 1:40, most have used ≥ 1:10 dilution. Other reasons for the widely varying estimates include differences in technique and the HHV-6 strain used as antigen. Most experts agree that the seroprevalence probably exceeds 90% in the 2-40 year old age group, and a global distribution seems likely (Levy et al. 1990). Several studies have demonstrated that the seroprevalence drops after age 40, reaching a low of about 30-60% (Briggs et al.

1988; Brown et al. 1988b; Enders et al. 1990; Levy et al. 1990; Yanagi et al. 1990). A profile of serologic changes with age is presented in Figure 1.

Culture studies of saliva also verify the extraordinary prevalence of HHV-6 infection. In one series of normal adults, all 9 of the subjects had viable HHV-6 in both whole saliva and the cell free fractions (Pietroboni et al. 1988a). The same group reported another series using CBLs to isolate the virus instead of PBMCs, and again found that 100% (10/10) of salivas from normal people yielded HHV-6 by culture (Harnett et al. 1990). Levy et al. (1990) cultured saliva from 45 normal individuals, simultaneously documenting the HHV-6 antibody status of the subjects. Of the 43 seropositive subjects, 41 (95%) had culture positive saliva. A subset of these subjects were sampled over time, showing that culture positivity was sometimes intermittent. The PCR has also been applied to saliva for detection of HHV-6 DNA. In saliva samples from 18 HHV-6 seropositive normal adults, 16 had PCR detectable HHV-6 DNA (R. Cone, unpublished observations). Saliva from thirty healthy adults was evaluated by PCR in another study, rendering a DNA prevalence of 63% (Gopal et al. 1990). As an aside, both of these studies identified HHV-6 seronegative adults who harbored salivary HHV-6 DNA.

Even though most HHV-6 isolates have been obtained from PBMCs, mononuclear cells from normal individuals are usually culture negative. PBMC cultures from normal people are commonly maintained without inoculation to serve as a negative control for cocultivation. These uninoculated cultures rarely exhibit HHV-6 infection (Lopez et al. 1988), in spite of the fact that many probably harbor HHV-6 DNA. It is possible that some additional stimulation, such as lymphocyte mixing in co-cultures, is required to activate viral replication. Even so, attempts to cocultivate normal adult PBMCs *in vitro* have not generally resulted in productive HHV-6 infection (Levy et al. 1990, Yoshiyama et al. 1990). Roseola studies provided further evidence that normal PBMCs are culture negative since PBMCs from the infected (roseola) children become culture negative during the convalescent phase of disease (Takahashi et al. 1988; Yoshiyama et al. 1990). The PBMC sources that do produce virus are usually from immunosuppressed individuals.

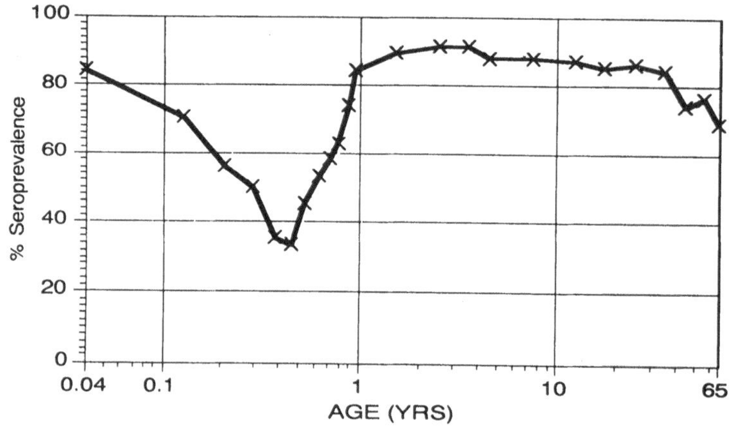

Figure 1. HHV-6 seroprevalence as a function of age. Age is expressed on a logarithmic scale to display detail for the first year. Data from seven studies have been synthesized to produce this curve (Brown et al. 1988b; Knowles and Gardner, 1988; Levy et al. 1990; Okuno et al. 1989; Yanagi et al. 1990; Yoshikawa et al. 1989, 1990).

Detection of HHV-6 DNA with the PCR indicates that HHV-6 DNA does reside in PBMCs from asymptomatic adults. Kondo et al. (1990) found that PBMCs from normal adults were uniformly negative by HHV-6 PCR, but only about 3×10^4 cells were used in each reaction. Forty-nine percent (n=45) of adult PBMC samples contained HHV-6 DNA when $1x10^5$ cells were evaluated in each reaction (Gopal et al. 1990). When $5x10^5$ PBMC cells per PCR reaction were evaluated, 94% of 17 seropositive healthy adults harbored HHV-6 DNA and 2 seronegative people did not have HHV-6 DNA in their PBMCs (R. Cone, unpublished observations). The amount of HHV-6 DNA was determined in selected samples by performing PCR on dilutions of the purified PBMC DNA. The number of HHV-6 genomes ranged from about 100 to 0.5 copies per $1x10^5$ cells, suggesting that some of the negative results in the former studies could be attributed to an inadequate sample size. Taken together, these data support the idea that HHV-6 DNA is present in the saliva and blood of almost all seropositive individuals, and in the saliva of some seronegative people. This argues strongly for a latent and/or persistent state after primary HHV-6 infection, as is found with the other herpesviruses.

Horizontal transmission via oral secretions represents the most likely route of infection since HHV-6 achieves high concentrations in saliva. Levy et al. (1990) used the time to antigen positivity in culture as a semi-quantitative estimate of infectious units in the cell-free fraction of saliva, providing a range of $<10^2$ to 10^6 infectious units/ml, with an average of about $2.6x10^4$ infectious units/ml of cell-free saliva. PCR detection of HHV-6 DNA in the cellular fraction of saliva (cell-free fraction removed) indicated that there was only about 1 HHV-6 genome in $5x10^4$ oral cells (Gopal et al. 1990). Semiquantitation of 6 unfractionated saliva samples by PCR yielded concentrations ranging from 400 to $8x10^5$ HHV-6 genomes/ml, averaging about 10^5 genomes/ml (R. Cone, unpublished observations). The cell-free nature of oral virus was supported by data showing that the concentrations of HHV-6 DNA in saliva samples varied independently of the cellular DNA concentrations. Therefore, it seems likely that HHV-6 establishes a persistent oral infection that leads to shedding of viable virus at high titer.

Several investigators have proposed that the major route of transmission is from adult to child early in life, after the infant's passive maternal HHV-6 antibodies wane (Harnett et al. 1990; Levy et al. 1990; Ueda et al. 1989; Yoshiyama et al. 1990). This is supported by a wealth of data, including serologic studies documenting decreases in maternal antibodies followed by seroconversion (Figure 1), high titers of virus in adult saliva, and evidence that the same strain of HHV-6 can be found in the mother and her infant (Koichi Yamanishi, personal communication). In a series of 12 infants with roseola, 11 had not received breast milk, making this an unlikely source of transmission (Takahashi et al. 1988). Although sexual transmission has been observed with HSV and CMV, venereal transmission of HHV-6 seems unlikely because no virus has been identified at genital sites (Cone, unpublished observations; Harnett et al. 1990). The possibility of reinfection has not been carefully investigated. The high seroprevalence at an early age has been used to argue against reinfection (Brown et al. 1988a). However, seropositivity for HSV, EBV, VZV or CMV has not prevented reinfection with different strains of the same herpesvirus, suggesting that reinfection with HHV-6 may also occur in spite of the seroepidemiology.

Disease Associations

Roseola. Roseola is the only illness yet identified with a definite etiological link to HHV-6 infection. This syndrome, variously called *roseola infantum, exanthem(a) subitum,* sixth disease, pseudorubella, *exanthem criticum* or 3-day fever, was first reported in 1910 (Zahorsky, 1910). The clinical presentation is characterized by sudden onset of a 102-105° F fever lasting for 2-5 days, followed by quick defervescence and a truncal, non-vesicular rash which persists for 1-3 days (Krugman and Katz, 1981). The syndrome is self-limited, but convulsions associated with the high fever can cause permanent neurological injury (Burnstine and Paine, 1959). Before HHV-6 was discovered, roseola had been induced in humans and monkeys using throat washings instilled intranasally and serum given by intravenous, intramuscular or subcutaneous routes (Hellstrom and Vahlquist, 1951; Kempe et al. 1950). The attack rate in the general population is at least 30% with an incubation period ranging from 5-15 days (Breese, 1941). Atypical roseola, defined by a 4-fold increase in anti-HHV-6 titers and virus isolation without the characteristic clinical presentation, has also been reported. Five patients presented with fever but no rash (Suga et al. 1989), and 2 others were documented with rash but no fever (Asano et al. 1989a). The majority of primary HHV-6 infections are apparently asymptomatic or do not come to the attention of doctors. There are also cases of illness that present with the roseola-like symptomatology that are not associated with HHV-6 seroconversion or virus isolation (Yoshiyama et al. 1990), suggesting that the clinical syndrome may sometimes result from etiologies other than HHV-6.

Several lines of evidence have implicated HHV-6 as the causative agent of roseola. HHV-6 was readily cultured using PBMCs from about 60% of affected children, with continued culture positivity for 5 to 120 days after onset (Asano et al. 1989b; Yamanishi et al. 1988; Yoshiyama et al. 1990). The rate of culture positivity dropped from 100% during the first 2 days before onset of the rash to less than 20% by days 5-7. Evidence for productive HHV-6 infection during roseola was also provided by finding cell-free virus in plasma from 28% of cases during days 1-4 (Asano et al. 1989b). Culture from PBMCs of other children and normal adults was non-productive (Kikuta et al. 1989). In addition, the convalescent phase was associated with decreasing amounts of HHV-6 in PBMCs (Kondo et al. 1990).

The typical age range of roseola attacks, 6-24 months, corresponds with the age range that includes most seroconversions to HHV-6 (Figure 1). Maternal IgG to HHV-6 crosses the placenta, providing 90-100% seropositivity in cord blood (Knowles and Gardner, 1988; Saxinger et al. 1988). The maternal IgG appears to have a protective effect, because primary HHV-6 infection is rarely seen in children less than 5 months of age. As the infant's maternal antibody titers decline, they apparently become increasingly susceptible to infection (Takahashi et al. 1988). Serologic studies of children in the acute and convalescent phases showed de novo synthesis of anti-HHV-6 IgM starting about 5 days after onset of symptoms (Fox et al. 1988; Kondo et al. 1990), followed by 4-fold rises of anti-HHV-6 IgG titers in over 90% of children with roseola (Yamanishi et al. 1988; Yoshiyama et al. 1990).

AIDS. The existing clinical evidence does not support a role for HHV-6 in HIV-1-related disease. In spite of this, there are compelling reasons to suspect HHV-6 infection as a cofactor in HIV-1 disease progression. Some investigators are hopeful that if HHV-6 has a role in HIV-1 progression, then anti-HHV-6 therapy might slow the development of AIDS (Laurence, 1990).

The high prevalence of HHV-6 assures that almost every AIDS patient is coinfected. Frequent culture positivity of PBMCs from AIDS patients indicates that HHV-6 replication may be more active in this group than in the normal population (Agut et al. 1988; Becker et al. 1989; Downing et al. 1987; Lopez et al. 1988; Tedder et al. 1987). Interest in this area is further spurred by the fact that the other herpesviruses are responsible for significant morbidity and mortality during HIV infection, pointing to HHV-6 as a likely candidate by association.

Most data supporting an HIV-1/HHV-6 link come from *in vitro* studies. Coinfection of CAT-HIV-LTR constructs with HHV-6 show augmentation of HIV-1 promoter activity (Ensoli et al. 1989). However, transactivation of the HIV-LTR is not unique, since HSV, CMV and adenovirus also perform that function (Davis et al. 1987; Horvat et al. 1989; Mosca et al. 1987; Rice and Mathews 1988). HHV-6 mediated transactivation takes on more importance when considering that HHV-6 and HIV-1 naturally infect the same CD4 positive cell types. If fact, both viruses can coinfect the same cell, as demonstrated by two color immunofluorescence (Spira et al. 1990) and *in situ* hybridization (Ensoli et al. 1989). These data support the hypothesis that HHV-6 might stimulate HIV-1 replication. Alternatively, one could argue that HHV-6 infection may have a protective role on the basis that HHV-6 interferes with *in vitro* propagation of HIV-1 (Lopez et al. 1988; Pietroboni et al. 1988b). This mechanism has been proposed to explain some negative results of oral HIV-1 cultures from HIV-1-infected patients (Pietroboni et al. 1988a).

Serologic studies on this immunocompromised population do not support a role for HHV-6 in AIDS. An initial HHV-6 seroprevalence study reported that 82% of HIV seropositives with advanced disease had HHV-6 antibodies as compared with 44% of HIV-1 seropositives with less advanced disease (Krueger et al. 1988b). In retrospect, antibody detection in this study was probably insensitive since the seroprevalence reported in the normal population was only 26%. The high HHV-6 antibody titers required for positivity might easily produce this pattern in light of new data showing that the geometric mean titer (GMT) of anti-HHV-6 antibodies increases with the stage of HIV-1 disease (Levy et al. 1990). Similar increases in GMT were also noted in patients with autoimmune diseases and "chronic fatigue", suggesting that this phenomenon is not specific to HIV-1 infected individuals. In spite of the changes in GMT, HHV-6 seroprevalence remained unchanged when HIV-1 infected and uninfected groups were compared (Brown et al. 1988a; Enders et al. 1990; Fox et al. 1988). Other studies have reported that the seroprevalence of HHV-6 decreases as HIV-1 infection progresses (Fox et al. 1988; Spira et al. 1990). Finally, a prospective study of asymptomatic, HIV-1-infected individuals showed that neither HHV-6 seropositivity or antibody titer predicted which patients would progress to AIDS (Spira et al. 1990). This data joins similar studies of CMV or EBV serologies, all of which show that herpesvirus serologies are not good predictors of HIV-1 progression (Lang et al. 1989).

In summary, *in vitro* data suggests some tantalizing interactions between HIV-1 and HHV-6, but the existing clinical data does not support the hypothesis that HHV-6 plays an important role in HIV-related disease.

Kawasaki Disease. Kawasaki disease (KD) is a febrile vasculitis occurring in infancy or early childhood. The etiologic agent remains unknown, but seasonal epidemic peaks have suggested an infectious etiology. The ubiquitous nature of HHV-6 and onset in infancy and early childhood make this a potential cause of KD, although seasonality has not been identified as a feature of HHV-6 infection. A serologic study addressing this issue was done with 22 KD patients (8 months to 5 years old) and 16 age-matched controls (Okano et al. 1989). Overall HHV-6 seropositivity was 82% in the disease group and 63% in the controls. As might be expected in this age group, HHV-6 IgM seroprevalence was high: 36% in KD versus 50% of controls. The authors concluded that "...HHV-6 infection is not the direct cause of KD...". Similarly, Marchette et al. (1990) reported that 75% of KD patients had HHV-6 IgG as compared with 70% of age-matched controls. Another series of 20 KD patients had an IgG seroprevalence of 40% (Enders et al. 1990). Paired acute and convalescent sera available on 6 KD patients showed no rises in anti-HHV-6 IgG titers. While these serologic studies do not rule out an association between HHV-6 infection and KD, no clear trend has emerged.

Lymphoproliferative Disease. Most of the data regarding HHV-6 infection with lymphoproliferation has been done using Southern blots to examine the affected tissues. DNA samples from a series of 165 lymphoid tissues were analyzed by Southern blot and probed with a subclone of HHV-6 (Josephs et al. 1988). Broadly classifying groups of patients, HHV-6 DNA was detected in 3/82 B cell lymphomas, 0/22 T cell lymphomas, 0/15 myeloid lymphomas, 0/8 Hodgkin's lymphomas, 0/38 benign lymphadenopathies and 0/14 Kaposi's sarcomas. The 3 positive cases consisted of one patient with Burkitt's lymphoma and 2 patients with Sjogren's syndrome. *In situ* hybridization on Burkitt's lymphoma tissues from 10 patients produced positive results in 7 cases, although less than 1/1000 cells hybridized with the HHV-6 probe (Ablashi et al. 1988a). This suggests that the low amount HHV-6 DNA in Burkitt's lymphoma tissues may have precluded detection by Southern blot.

In another large study (Jarrett et al. 1988) 117 tissues were examined for HHV-6 DNA by Southern blot. Two positives were detected: one in a T zone lymphoma from a patient with angioimmunoblastic lymphadenopathy, and the other in tissue from a gastric lymphoma from a patient with Sjogren's syndrome. Viral sequences were also detected in non-malignant tissues from both of these patients, suggesting that generalized reactivation of HHV-6 might explain the presence of HHV-6 DNA in the malignant tissues. The number of HHV-6 positive tissues/all tissues examined can be summarized as follows: 1/35 B cell lymphomas, 0/29 Hodgkins lymphomas, 1/18 T-cell lymphomas and 0/35 nonlymphomatous tissues with various pathologies.

Tissues from a series of patients with histiocytic necrotizing lymphadenitis were screened for HHV-6 antigen by ACIF (Eizuru et al. 1989). Although tissue sections were HHV-6 antigen positive in 17/18 cases of histiocytic necrotizing lymphadenitis, 6/8 tissues from patients without the disease were also positive. The authors concluded that these data demonstrate the extraordinary prevalence of HHV-6 infection.

Serologic IFA titers were used to investigate 3 patients with a syndrome of lymphadenopathy for 1-3 months duration and transient atypical lymphocytosis (Niederman et al. 1988). Acute and convalescent sera all showed transient HHV-6 IgM and titers of specific IgG that were considered "high" by these authors. None of these patients had serologic evidence of active EBV or CMV infection, suggesting that this mild lymphadenopathy syndrome may be associated with HHV-6 infection. Increases in HHV-6 IgG titers were also demonstrated in acute and convalescent sera taken from a patient with pancreatic cancer who had an enlarged mesenteric lymph node (Chappuis et al. 1989).

Taken together, these data do not support an etiologic role for HHV-6 in most lymphoproliferative diseases, a conclusion that has also been reached by other reviewers (Anonymous, 1990; Komaroff, 1990). Although the interesting association of HHV-6 with lymphadenopathy rests only on serologic changes, more data with culture and DNA detection may elucidate this problem. The weak association with Sjogren's syndrome in two large series also deserves further investigation.

Hepatitis. Several research groups have attempted to link clinical episodes of hepatitis with HHV-6 infection (Asano et al. 1990; Dubedat and Kappagoda, 1989; Irving and Cunningham, 1990; Steeper et al. 1990; Tajiri et al. 1990). The most straightforward account describes a 2 month old baby with neonatal hepatitis (Tajiri et al. 1990). Serologies done at that time ruled out hepatitis A virus (HAV), hepatitis B virus (HBV), CMV, toxoplasmosis, rubella, HSV and HHV-6 as etiologic agents. The baby recovered from the acute attack, although his serum alanine aminotransferase (ALT) remained elevated. Five months later (7 months old) he had an acute exacerbation with increased ALT, vomiting, and a 4 day fever followed by a rash. PBMCs taken during the exacerbation were HHV-6 culture positive and HHV-6 seroconversion was documented, confirming the diagnosis of roseola with primary HHV-6 infection. This well-documented case suggests that primary HHV-6 infection may exacerbate idiopathic hepatitis. Another case involved a 3 month old with sudden onset of fever, anorexia, vomiting, and jaundice (Asano et al. 1990). He was admitted 3 days after onset with stage III coma, and died 7 days after onset in spite of supportive therapy. HHV-6 was isolated from PBMCs (day 3), and HHV-6 DNA was found in liver and brain tissues using PCR. The authors asserted that seroconversion to HHV-6 occurred, including IgM production, although the serologies apparently followed exchange transfusion and the serologic results were not presented. The interpretation of HHV-6 DNA in brain and liver remains obscure without knowing whether the same result might occur in normals or in routine cases of roseola. Similarly, isolation of HHV-6 from PBMCs during an attack of hepatitis deserves comparison with other hepatitis cases to show that HHV-6 is not isolated when the etiology of the hepatitis is known to be another agent.

A report from Australia (Irving and Cunningham, 1990) describes HHV-6 antibody screening of 341 patients who had "histories compatible with acute CMV infection". This process identified 2 patients with HHV-6 IgG, HHV-6 IgM and self-limited hepatitis in the absence of CMV IgM, EBV IgM, HAV IgM, HBsAg or HIV IgG. One of these patients acquired hepatitis during a trip to Asia and had a vesicular rash of the lower extremities, which is uncharacteristic of HHV-6 infection. The second case involved a 25 year old man with fever, maculo-papular rash and lymphadenopathy. HHV-6 IgM was not present at 7 days post-onset, went to 1:40 at 20 days, and disappeared by day 74.

A study of 27 subjects with idiopathic infectious mononucleosis (IM)-like illness revealed 2 who had acute hepatitis. One IM-hepatitis patient was a 27 year old woman with 4 fold increases in HHV-6 IgG (IFA and ELISA) and a rise in HHV-6 IgM titers by ELISA, but not by IFA. Hepatitis C virus (HCV) was not ruled out as a cause of this illness. Additionally, one negative anti-HBc test is reported, but this could have been obtained in the core window, and no HBc-IgM or follow up HBV serologies are reported. The authors tested for HAV IgG, but there is no mention of HAV IgM testing. HHV-6 IgM serology on the other patient, a 40 year old female, was interpreted as positive but it did not exceed the lower cutoff value designated elsewhere in the paper. Assuming that the authors simply failed to mention that HAV was ruled out with an IgM test, the clinical syndromes could have been due to HCV, which was not tested. HHV-6 isolation or DNA detection would have strengthened their assertion that the etiologic agent was HHV-6. Lastly, a case report describes a 25 year old woman with a history of malignant melanoma, operated 1 year prior, who presented with hepatitis that resolved over 2 weeks (Dubedat and Kappagoda 1989). HBsAg, HAV IgM and a monospot test were negative in the acute serum. Unfortunately, only the 4-month convalescent serum was tested for HHV-6 antibodies: IgG was titered at 1:512, and IgM was 1:10. No evidence regarding the specificity or sensitivity of the HHV-6 serologies was presented, and HCV was not tested. The lack of temporal proximity between HHV-6 serologies and the illness make it difficult to relate these findings.

Summarizing the evidence, an association between primary HHV-6 infection and hepatitis is supported by two case reports (Asano et al. 1990; Tajiri et al. 1990). The data suggest that HHV-6 infection may exacerbate hepatitis of unknown etiology in very young children. In two cases of adult hepatitis (Irving and Cunningham 1990), most other common etiological agents were ruled out. While a host of identified and unidentified agents can cause mild hepatitis, and any of these might cause non-specific elevations of HHV-6 antibodies, the connection between HHV-6 and adult hepatitis deserves further investigation.

Mononucleosis. Mononucleosis represents a likely candidate for association with HHV-6 infection because the two other lymphotropic herpes viruses, CMV and EBV, have both been shown to induce mononucleosis. A retrospective study of 27 subjects with non-EBV, non-CMV "mononucleosis-like" illnesses revealed 8 (30%) with HHV-6 serologies that were interpreted as abnormal. Of the 8 patients, 3 had symptoms that were not very convincing for mononucleosis: coryza, cough, and ectopic heart beats (1), transient atypical lymphocytosis without other symptoms (1), and cervical lymphadenopathy without other symptoms (1). The remaining 5 had increases in HHV-6 IgG and were positive for HHV-6 IgM. In another report 1,135 normal adults were serologically screened, finding that 295 (26%) had HHV-6 IgG titers >=1:40 (Krueger et al. 1988b). Forty-one of the 295 subjects (14%) had symptoms of mononucleosis. Unfortunately, the authors do not mention how many of the 840 HHV-6 seronegative subjects also had symptoms of mononucleosis by the same criteria.

Another study employed HHV-6 IgG and IgM serologies to screen 341 patients with a wide variety of clinical presentations, including hepatitis, mononucleosis, fever with immunosuppression, various neurological or hematological syndromes, childhood rash and splenomegaly (Irving and Cunningham, 1990). This process identified 3/341 patients with seropositivity for HHV-6 IgM (>1:20) and 4-fold rises in HHV-6 IgG without serologic evidence of EBV or CMV. All 3 patients had fever and sore throat. In addition, one patient had hepatitis, hepatosplenomegaly, and a vesicular rash, another had hepatitis, hepatosplenomegaly, lymphadenopathy and a maculopapular rash, and the third had lymphadenopathy, atypical lymphocytosis and a maculopapular (morbilliform) rash.

A major problem with these data is that "active" HHV-6 infection is defined as the presence of HHV-6 IgM and increases in IgG. While antibody increases may be associated with a clinical syndrome resembling mononucleosis, there is little evidence that the association is causally related. In addition, only a small proportion of patients presenting with this syndrome have serologic markers suggesting so-called "active" HHV-6 infection. Further studies that identify viral replication in otherwise sterile sites may shed light on the clinical manifestations of HHV-6.

Chronic Fatigue (CFS). So many infectious diseases have been associated with CFS that it seems like a necessary rite of passage for this newly discovered virus to be blamed for that difficult syndrome (Komaroff and Goldenberg, 1989). At the present time there is no convincing evidence that CFS is caused by HHV-6. A role for HHV-6 in CFS was proposed shortly after the virus was discovered (Ablashi et al. 1988a). In one study, less than 1% of seropositive adults (seroprevalence of 24%) progressed to a vaguely defined state of chronic fatigue (Krueger et al. 1988b). Even this rate of association may be related to that fact that CFS patients have increased titers of HHV-6 antibodies (Levy et al. 1990). Another study showed increased "seroconversion" to HHV-6 (IgM and IgG) in CFS, although some of these patients simultaneously seroconverted to CMV, an event that can elevate HHV-6 antibody titers (Kirchesch et al. 1988). Other studies include data on a HHV-6/CFS link (Krueger and Sander 1989, Rodier et al. 1990; Steeper et al. 1990; Wakefield et al. 1988), but the etiology of persistent fatigue has continued to elude us.

CONCLUSION

HHV-6 represents one of the most ubiquitous infections of mankind. The serconversion rate approaches 100% in the first few years of life, and viral replication persists in over 90% of the adult population. Investigations from geographical locations around the world indicate the global nature of HHV-6 infection. The consequences of HHV-6 infection remain largely unidentified. While some lines of evidence suggest an association between HHV-6 and hepatotoxicity or lymphadenopathy, roseola is the only clinical syndrome that is clearly related to HHV-6 infection, when young children acquire a self-limited fever and rash with primary infection.

Many major questions regarding HHV-6 remain to be answered. Latency has not been clearly demonstrated for this virus. The humoral and cellular immune responses to HHV-6 infection remain largely unexplored. In addition, many herpesvirologists suspect that HHV-6 may influence other disease processes, but these have not yet been discovered.

Amazing progress has been made in understanding HHV-6 during these few years since its discovery, largely thanks to contributions from molecular biology. In particular, antibody, DNA and protein detection methods with extraordinary sensitivity and specificity have already emerged. If the other herpesviruses serve as any example, then future studies on HHV-6 will undoubtedly reveal an interesting complex of host-virus interactions.

REFERENCES

Ablashi DV, Salahuddin SZ, Josephs SF, Imam F, Lusso P, Gallo RC, Hung C, Lemp J, Markham PD (1987) HBLV (or HHV-6) in human cell lines (letter). Nature 329:6136.

Ablashi DV, Josephs SF, Buchbinder A, Hellman K, Nakamura S, Llana T, Lusso P, Kaplan M, Dahlberg J, Memon S, Imam F, Ablashi K, Markham P, Kramarsky B, Krueger G, Biberfeld P, Wong-Staal F, Salahuddin S, Gallo R

(1988a) Human B-lymphotropic virus (human herpesvirus-6). J Virol Methods 21:29-48.
Ablashi DV, Lusso P, Hung CL, Salahuddin SZ, Josephs SF, Llana T, Kramarsky B, Biberfeld P, Markham PD, Gallo RC (1988b) Utilization of human hematopoietic cell lines for the propagation and characterization of HBLV (human herpesvirus 6). Int J Cancer 42:787-791.
Ablashi DV, Lusso P, Hung CL, Salahuddin SZ, Josephs SF, Llana T, Kramarsky B, Biberfeld P, Markham PD, Gallo RC (1989) Utilization of human hematopoietic cell lines for the propagation and characterization of HBLV. Develop Biol Standard 70:139-146.
Agut H, Guetard D, Collandre H, Dauguet C, Montagnier L, Miclea JM, Baurmann H, Gessain A (1988) Concomitant infection by human herpesvirus 6, HTLV-I, and HIV-2 (letter). Lancet 1:712.
Andre M and Matz B (1988) Antibody responses to human herpesvirus 6 and other herpesviruses (letter). Lancet 1:426.
Anonymous (1990) The newest human herpesvirus comes into focus (editorial). Lancet 335:325-326.
Asada H, Yalcin S, Balachandra K, Higashi K, Yamanishi K (1989) Establishment of titration system for human herpevisurs 6 and evaluation of neutralizing antibody response to the virus. J Clin Microbiol 27:2204-2207.
Asano Y, Suga S, Yoshikawa T, Urisu A, Yazaki T (1989a) Human herpesvirus type 6 infection (exanthem subitum) without fever. J Pediatr 115:2264-2265.
Asano Y, Yoshikawa T, Suga S, Yazaki T, Hata T, Nagai T, Kajita Y, Ozaki T, Yoshida S (1989b) Viremia and neutralizing antibody response in infants with exanthem subitum. J Pediatr 114:535-539.
Asano Y, Yoshikawa T, Suga S, Yazaki T, Hirabayashi S, Ono Y, Tsuzuki K, Oshima S (1989c) Human herpesvirus 6 harbouring in kidney. Lancet 2:1391.
Asano Y, Yoshikawa T, Suga S, Yazaki T, Kondo K, Yamanishi K (1990) Fatal fulminant hepatitis in an infant with human herpesvirus-6 infection [letter]. Lancet 335:862-863.
Becker WB, Engelbrecht S, Becker ML, Piek C, Robson BA, Wood L, Jacobs P (1988) Isolation of a new human herpesvirus producing a lytic infection of helper (CD4) T-lymphocytes in peripheral blood lymphocyte cultures-Another cause of acquired immunodeficiency? S Afr Med J 74:610-614.
Becker WB, Engelbrecht S, Becker ML, Piek C, Robson BA, Wood L, Jacobs P (1989) New T-lymphotropic human herpesviruses [letter]. Lancet 1:41.
Biberfeld P, Kramarsky B, Salahuddin SZ, Gallo RC (1987) Ultrastructural characterization of a new human B lymphotropic DNA virus (human herpesvirus 6) isolated from patients with lymphoproliferative disease. J Natl Cancer Inst 79:933-941.
Black J, Sanderlin K, Goldsmith C, Gary H, Lopez C, Pellet P (1989) Growth properties of human herpesvirus-6 strain Z29. J Virol Meth 26:133-146.
Breese B (1941) Roseola infantum (exanthema subitum). NY State Med 41:1854.
Briggs M, Fox J, Tedder RS (1988) Age prevalence of antibody to human herpesvirus 6 (letter). Lancet 1:1058-1059.
Brown NA, Kovacs A, Lui CR, Hur C, Zaia JA, Mosley JW (1988a) Prevalence of antibody to human herpesvirus 6 among blood donors infected with HIV (letter). Lancet 2:1146.
Brown NA, Sumaya CV, Liu CR, Ench Y, Kovacs A, Coronesi M, Kaplan MH (1988b) Fall in human herpesvirus 6 seropositivity with age (letter). Lancet 2:396.

Buchbinder A, Josephs SF, Ablashi D, Salahuddin SZ, Klotman ME, Manak M, Krueger GR, Wong SF, Gallo RC (1988) Polymerase chain reaction amplification and *in situ* hybridization for the detection of human B-lymphotropic virus. J Virol Methods 21:191-197.

Burnstine R and Paine R (1959) Residual encephalopathy following roseola infantum. Am J Dis Child 98:144-152.

Chandler SH and McDougall JK (1986) Comparison of restriction site polymorphisms among clinical isolates and laboratory strains of human cytomegalovirus. J Gen Virol 67:2179-2192.

Chantratita W and Yoosook C (1990) Restriction endonuclease analysis of deoxyribonucleic acid (DNA) from genital herpes simplex virus type 2 isolates in Thailand. Sex Trans Dis 17:42-47.

Chappuis B, Ellinger K, Niepel F, Kirchner T, Kujath P, Fleckenstein B, Muller-Hermelink H (1989) Human herpesvirus 6 in lymph nodes (letter). Lancet 1:40-41.

Chou S and Scott K (1990) Rises in antibody to human herpesvirus 6 detected by enzyme immunoassay in transplant recipients with primary cytomegalovirus infection. J Clin Microbiol 28:851-854.

Davis M, Kenney S, Kamine J, Pagano J, Huang E (1987) Immediate-early gene region of human cytomegalovirus trans-activates the promoter of human immunodeficiency virus. Proc Natl Acad Sci USA 84:8642-8646.

Downing RG, Sewankambo N, Serwadda D, Honess R, Crawford D, Jarrett R, Griffin BE (1987) Isolation of human lymphotropic herpesviruses from Uganda. Lancet 2:390.

Dubedat S and Kappagoda N (1989) Hepatitis due to human herpesvirus-6. Lancet 2:1463-1464.

Efstathiou S, Gompels UA, Craxton MA, Honess RW, Ward K (1988) DNA homology between a novel human herpesvirus (HHV-6) and human cytomegalovirus (letter). Lancet 1:63-64.

Eizuru Y, Minematsu T, Minamishima Y, Kikuchi M, Yamanishi K, Takahashi M, Kurata T (1989) Human herpesvirus 6 in lymph nodes (letter). Lancet 1:40.

Enders G, Biber M, Meyer G, Helftenbein E (1990) Prevalence of antibodies to human herpesvirus 6 in different age groups, in children with exanthema subitum, other acute exanthematous childhood diseases, Kawasaki syndrome, and acute infections with other herpesviruses and HIV. Infection 18:12-15.

Ensoli B, Lusso P, Schachter F, Josephs SF, Rappaport J, Negro F, Gallo RC, Wong SF (1989) Human herpes virus-6 increases HIV-1 expression in co-infected T cells via nuclear factors binding to the HIV-1 enhancer. EMBO J 8:3019-3027.

Fox J, Briggs M, Tedder RS (1988) Antibody to human herpesvirus 6 in HIV-1 positive and negative homosexual men (letter). Lancet 2:396-397.

Fox J, Briggs M, Ward P, Tedder R (1990) Human herpesvirus 6 in salivary glands. Lancet 336:590-593.

Gallo R, Salahuddin S, Popovic M, Shearer G, Kaplan M, Haynes B, Palker T, Redfield R, Oleske J, Safai B, White G, Foster P, Markham P (1984) Frequent detection and isolation of cytopathic retroviruses (HTLV-III) from patients with AIDS and at risk for AIDS. Science 224:500-503.

Gopal M, Thomson B, Fox J, Tedder R, Honess R (1990) Detection by PCR of HHV-6 and EBV DNA in blood and oropharynx of healthy adults and HIV-seropositives [letter]. Lancet 335:1598-1599.

Harnett GB, Farr TJ, Pietroboni GR, Bucens MR (1990) Frequent shedding of human herpesvirus 6 in saliva. J Med Virol 30:128-130.
Hellstrom B and Vahlquist B (1951) Experimental inoculation of roseola infantum. Acta Pediatrica 40:189-197.
Higashi K, Asada H, Kurata T, Ishikawa K, Hayami M, Spriatna Y, Sutarman X, Yamanishi K (1989) Presence of antibody to human herpesvirus 6 in monkeys. J Gen Virol 70:3171-176.
Horvat RT, Wood C, Balachandran N (1989) Transactivation of human immunodeficiency virus promoter by human herpesvirus 6. J Virol 63:970-973.
Huemer H, Larcher C, Wachter H, Dierich M (1989) Prevalence of antibodies to human herpesvirus 6 in HIV1-seropositive and -seronegative intravenous drug addicts. J Infect Dis 160:549-550.
Irving W and Cunningham A (1990) Serological diagnosis of infection with human herpesvirus type 6. Br Med J 300:156-159.
Irving W, Ratnamohan V, Hueston L, Chapman J, Cunningham A (1990) Dual antibody rises to cytomegalovirus and human herpesvirus type 6: frequency of occurrence in CMV infections and evidence for genuine reactivity to both viruses. J Infect Dis 161:910-916.
Jarrett RF, Gledhill S, Qureshi F, Crae SH, Madhok R, Brown I, Evans I, Krajewski A, OBrien CJ, Cartwright RA (1988) Identification of human herpesvirus 6-specific DNA sequences in two patients with non-Hodgkin's lymphoma. Leukemia 2:496-502.
Josephs SF, Salahuddin SZ, Ablashi DV, Schachter F, Wong SF, Gallo RC (1986) Genomic analysis of the human B-lymphotropic virus (HBLV). Science 234:601-603.
Josephs SF, Buchbinder A, Streicher HZ, Ablashi DV, Salahuddin SZ, Guo HG, Wong SF, Cossman J, Raffeld M, Sundeen J (1988) Detection of human B-lymphotropic virus (human herpesvirus 6) sequences in B cell lymphoma tissues of three patients. Leukemia 2:132-135.
Kempe C, Shaw E, Lackson J, Silver H (1950) Studies on the etiology of exanthem subitum (roseola infantum). J Pediatr 27:561-568.
Kikuta H, Lu H, Matsumoto S, Josephs SF, Gallo RC (1989) Polymorphisms of Human Herpesvirus 6 DNA from five Japanese patients with exanthem subitum. J Infect Dis 160:550-551.
Kikuta H, Lu H, Tomizawa K, Matsumoto S (1990a) Enhancement of human herpesvirus 6 replication in adult human lymphocytes by monoclonal antibody to CD3. J Infect Dis 161:1085-1087.
Kikuta H, Nakane A, Lu H, Taguchi Y, Minagawa T, Matsumoto S (1990b) Interferon induction by human herpesvirus 6 in human mononuclear cells. J Infect Dis 162:35-38.
Kirchesch H, Mertens T, Burkhardt U, Kruppenbacher JP, Hoffken A, Eggers HJ (1988) Seroconversion against human herpesvirus-6 (and other herpesviruses) and clinical illness (letter). Lancet 2:273-274.
Knowles WA and Gardner SD (1988) High prevalence of antibody to human herpesvirus-6 and seroconversion associated with rash in two infants (letter). Lancet 2:912-913.
Komaroff A (1990) Human herpesvirus-6 and human disease [editorial]. Am J Clin Path 93:836-837.
Komaroff A and Goldenberg D (1989) The chronic fatigue syndrome: definition, current studies and lessons. J Rheumatol Suppl 19:23-27.

Kondo K, Hayakawa Y, Mori H, Sato S, Kondo T, Takahashi K, Minamishima Y, Takahashi M, Yamanishi K (1990) Detection by polymerase chain reaction amplification of human herpesvirus 6 DNA in peripheral blood of patients with exanthem subitum. J Clin Microbiol 28:970-974.

Krueger G and Sander C (1989) What's new in human herpesvirus-6? Clinical immunopathology of the HHV-6 infection. Pathol Res Pract 185:915-929.

Krueger GR, Ablashi DV, Salahuddin SZ, Josephs SF (1988a) Diagnosis and differential diagnosis of progressive lymphoproliferation and malignant lymphoma in persistent active herpesvirus infection. J Virol Meth 21:255-264.

Krueger GR, Koch B, Ramon A, Ablashi DV, Salahuddin SZ, Josephs SF, Streicher HZ, Gallo RC, Habermann U (1988b) Antibody prevalence to HBLV (human herpesvirus-6, HHV-6) and suggestive pathogenicity in the general population and in patients with immune deficiency syndromes. J Virol Meth 21:125-131.

Krugman S and Katz S. Exantema subitum (roseola infantum) In: Infectious Diseases of Children. St. Louis: C. V. Mosby Co., 1981: 59-62.

Lang D, Kovacs A, Zaia J, Doelkin G, Niland J, Aledort L, Azen S, Fletcher M, Gauderman J, Gjerset G, Lusher J, Operskalski E, Parker J, Pegelow C, Vyas G, Mosley J (1989) Seroepidemiologic studies of cytomegalovirus and Epstein-Barr virus infections in relation to human immunodeficiency virus type 1 infection in selected recipient populations. J Acquired Immune Deficiency Syndromes 2:540-549.

Larcher C, Huemer HP, Margreiter R, Dierich MP (1988) Serological crossreaction of human herpesvirus-6 with cytomegalovirus (letter). Lancet 2:963-964.

Laurence J (1990) Molecular interactions among herpesviruses and human immunodeficiency viruses. J Infect Dis 162:338-346.

Lawrence GL, Chee M, Craxton MA, Gompels UA, Honess RW, Barrell BG (1990) Human herpesvirus 6 is closely related to human cytomegalovirus. J Virol 64:287-299.

Levy J, Ferro F, Greenspan D, Lennette (1990) Frequent isolation of HHV-6 from saliva and high seroprevalence of the virus in the population. Lancet 335:1047-1050.

Linde A, Dahl H, Wahren B, Fridell E, Salahuddin Z, Biberfeld P (1988) IgG antibodies to human herpesvirus-6 in children and adults both in primary Epstein-Barr virus and cytomegalovirus infections. J Virol Meth 21:117-123.

Linde A, Fridell E, Dahl H, Andersson J, Biberfeld P, Wahren B (1990) Effect of primary Epstein-Barr virus infection on human herpesvirus 6, cytomegalovirus, and measles virus immunoglobulin G titers. J Clin Microbiol 28:211-125.

Littler E, Lawrence G, Liu M-Y, Barrell B, Arrand J (1990) Identification, cloning and expression of the major capsid protein gene of human herpesvirus 6. J Virol 64:714-722.

Lopez C and Honess R (1990) Human herpesvirus-6. In Fields B, Knipe D (ed.), Virology, 2nd. New York: Raven Press, 2055-2062.

Lopez C, Pellett P, Stewart J, Goldsmith C, Sanderlin K, Black J, Warfield D, Feorino P (1988) Characteristics of human herpesvirus-6. J Infect Dis 157:1271-1273.

Lusso P, Ensoli B, Markham PD, Ablashi DV, Salahuddin SZ, Tschachler E, Wong SF, Gallo RC (1989) Productive dual infection of human CD4+ T lymphocytes by HIV-1 and HHV-6. Nature 337:370-373.

Lusso P, Markham P, DeRocco S, Gallo R (1990) *In vitro* susceptibility of T lymphocytes from chimpanzees (Pantroglodytes) to human herpesvirus 6 (HHV-6): a potential animal model to study the interaction between HHV-6 and human immunodeficiency virus type 1 *in vivo*. J Virol 64:2751-2758.

Lusso P, Salahuddin SZ, Ablashi DV, Gallo RC, Di, Marzo, Veronese F, Markham PD (1987) Diverse tropism of HBLV (human herpesvirus 6) (letter). Lancet 2:743.

Lusso P, Markham PD, Tschachler E, di, Marzo, Veronese F, Salahuddin SZ, Ablashi DV, Pahwa S, Krohn K, Gallo RC (1988) *In vitro* cellular tropism of human B-lymphotropic virus (human herpesvirus-6). J Exp Med 167:1659-1670.

Marchette N, Melish M, Hicks R, Kihara S, Sam E, Ching D (1990) Epstein-Barr virus and other herpesvirus infections in Kawasaki syndrome. J Infect Dis 161:680-684.

Mier J and Gallo R (1980) Purification and some characteristics of human T-cell growth factor from phytohemagglutinin-stiumulated lymphocyte-conditioned media. Proc Nat Acad Sci USA 77:6134-6138.

Morris D, Littler E, Jordan D, Arrand J (1988) Antibody responses to human herpesvirus 6 and other herpesviruses (letter). Lancet 2:1425-1426.

Mosca J, Bednarik D, Raj N, Rosen C, Sodroski J, Haseltine W, Hayward G, Pitha P (1987) Activation of human immunodeficiency virus by herpesvirus infection: identification of a region within the long terminal repeat that responds to a trans-acting factor encoded by herpes simplex virus 1. Proc Natl Acad Sci USA 84:7408-7412.

Niederman JC, Liu CR, Kaplan MH, Brown NA (1988) Clinical and serological features of human herpesvirus-6 infection in three adults. Lancet 2:817-819.

Okano M, Luka J, Thiele GM, Sakiyama Y, Matsumoto S, Purtilo DT (1989) Human herpesvirus 6 infection and Kawasaki disease. J Clin Microbiol 27:2379-2380.

Okuno T, Takahashi K, Balachandra K, Shiraki K, Yamanishi K, Takahashi M, Baba K (1989) Seroepidemiology of human herpesvirus 6 infection in normal children and adults. J Clin Microbiol 27:651-653.

Okuno T, Higashi K, Shiraki K, Yamanishi K, Takahashi M, Kokado Y, Ishibashi M, Takahara S, Sonoda T, Tanaka K (1990) Human herpesvirus 6 infection in renal transplantation. Transplantation 49:519-522.

Papadimitriou J, Shellam G, Robertson T (1984) An ultrastructural investigation of cytomegalovirus replication in murine hepatocytes. J Gen Virol 65:1979-1990.

Pietroboni G, Harnett G, Bucens M (1989) Centrifugal enhancement of human immunodeficiency virus (HIV) and human herpesvirus 6 (HHV-6) infection *in vitro*. J Virol Meth 24:85-90.

Pietroboni GR, Harnett GB, Bucens MR, Honess RW (1988a) Isolation of human herpesvirus 6 from saliva (letter) (title corrected from "Antibodies" to "Isolation" in erratum, Lancet 1988 May 28;1(8596):1235). Lancet 1:1059.

Pietroboni GR, Harnett GB, Farr TJ, Bucens MR (1988b) Human herpes virus type 6 (HHV-6) and its *in vitro* effect on human immunodeficiency virus (HIV). J Clin Pathol 41:1310-1312.

Qavi HB, Green MT, SeGall GK, Font RL (1989) Demonstration of HIV-1 and HHV-6 in AIDS-associated retinitis. Curr Eye Res 8:379-387.

Rice A and Mathews M (1988) Trans-activation of the human immunodeficiency virus long terminal repeat sequences, expressed in an adenovirus vector by the adenovirus E1A 13S protein. Proc Natl Acad Sci, USA. 85:4200-4204.

Rodier G, Fox E, Constantine N, Abbatte E (1990) HHV-6 in Djibouti--an epidemiological survey in young adults. Trans R Soc Trop Med Hyg 84:148-150.
Roffman E and Frenkel N (1990) Interleukin-2 inhibits the replication of human herpesvirus-6 in mature thymocytes. Virology 175:591-594.
Roizman B, Charmichael L, Deinhardt F, de The G, Nahmias A, Plowright W, Rapp F, Sheldrick P, Takahashi M, Wolf K (1981) Herpesviridae. Definition, provisional nomenclature and taxonomy. The Herpesvirus Study Group, the International Committee on Taxonomy of Viruses. Intervirology 16:201-217.
Salahuddin SZ, Ablashi DV, Markham PD, Josephs SF, Sturzenegger S, Kaplan M, Halligan G, Biberfeld P, Wong SF, Kramarsky B (1986) Isolation of a new virus, HBLV, in patients with lymphoproliferative disorders. Science 234:596-600.
Saxinger C, Polesky H, Eby N, Grufferman S, Murphy R, Tegtmeir G, Parekh V, Memon S, Hung C (1988) Antibody reactivity with HBLV (HHV-6) in U.S. populations. J Virol Methods 21:199-208.
Spira T, Bozeman L, Sanderlin K, Warfield D, Feorino P, Holman R, Kaplan J, Fishbein D, Lopez C (1990) Lack of correlation between human herpesvirus-6 infection and the course of human immunodeficiency virus infection. J Infect Dis 161:567-570.
Steeper T, Horwitz C, Ablashi D, Salahuddin S, Saxinger C, Saltzman R, Schwartz B (1990) The spectrum of clinical and laboratory findings resulting from human herpesvirus-6 (HHV-6) in patients with mononucleosis-like illnesses not resulting from Epstein-Barr virus or cytomegalovirus. Am J Clin Path 93:776-783.
Suga S, Yoshikawa T, Asano Y, Yazaki T, Hirata S (1989) Human herpesvirus-6 infection (exanthem subitum) without rash. Pediatrics 83:1003-1006.
Tajiri H, Nose O, Baba K, Okada S (1990) Human herpesvirus-6 infection with liver injury in neonatal hepatitis [letter]. Lancet 335:863.
Takahashi K, Sonoda S, Kawakami K, Miyata K, Oki T, Nagata T, Okuno T, Kamanishi K (1988) Human herpesvirus 6 and exanthem subitum (letter). Lancet 1:1463.
Takahashi K, Sonoda S, Higashi K, Kondo T, Takahashi H, Takahashi M, Yamanishi K (1989) Predominant CD4 T-lymphocyte tropism of human herpesvirus 6-related virus. J Virol 63:3161-3163.
Tedder R, Briggs M, Cameron C, Honess R, Robertson D, Whittle H (1987) A novel lymphotropic herpesvirus. Lancet 2:390-392.
Ueda K, Kusuhara K, Hirose M, Okada K, Miyazaki C, Tokugawa K, Nakayama M, Yamanishi K (1989) Exanthem subitum and antibody to human herpesvirus-6. J Infect Dis 159:750-752.
Wakefield D, Lloyd A, Dwyer J, Salahuddin SZ, Ablashi DV (1988) Human herpesvirus 6 and myalgic encephalomyelitis (letter). Lancet 1:1059.
Wrzos H and Gibbons J (1990) Effect of interleukin 2 on human herpesvirus-6 latent infection *in vitro.* (abstract Abstract # Abstract Number) 15th International Herpesvirus Workshop, Georgetown University, Washington, D.C.
Wrzos H, Gibbons J, Abt PL, Gifford R, Yang HC (1990) Human herpesvirus 6 in monocytes of transplant patients. Lancet 335:486-487.
Yamanishi K, Okuno T, Shiraki K, Takahashi M, Kondo T, Asano Y, Kurata T (1988) Identification of human herpesvirus-6 as a causal agent for exanthem subitum. Lancet 1:1065-1067.
Yanagi K, Harada S, Fumihiko B, Oya A, Okabe N, Tobinai K (1990) High prevalence of antibody to human herpesvirus-6 and decrease in titer with increase in age in Japan. J Infect Dis 161:153-154.

Yoshikawa T, Suga S, Asano Y, Yazaki T, Kodama H, Ozaki T (1989) Distribution of antibodies to a causative agent of exanthem subitum (human herpesvirus-6) in healthy individuals. Pediatrics 84:675-677.
Yoshikawa T, Suga S, Asano Y, Yazaki T, Ozaki T (1990) Neutralizing antibodies to human herpesvirus-6 in healthy individuals. Ped Infect Dis J 9:589-590.
Yoshiyama H, Suzuki E, Yoshida T, Kajii T, Yamamoto N (1990) Role of human herpesvirus 6 infection in infants with examthema subitum. Pediatr Infect Dis J 9:71-74.
Zahorsky J (1910) Roseola infantilis. Pediatrics 22:60-64.

DISCUSSION

Cesario T (University of California Irvine Medical Center, Orange, CA):

This being a herpes virus family, I believe it is thymidine kinase negative so that it wouldn't respond to acyclovir or other therapeutic applications?

Cone R:

HHV-6 does respond to acyclovir, but only in very high concentrations. It's like CMV in that respect, and it does have a thymidine kinase, but it seems not to be as important or as sensitive to acyclovir. It does respond to other antiviral drugs like Foscarnet.

Cesario T:

How about relationships to lymphoma?

Cone R:

I didn't really review that data. Let me just say that the largest series has been done by Gallo's group, where they examined 220 or so lymph nodes from various lymphomas and other lymphoproliferative disorders. They identified three people who had HHV-6 in their lymph nodes. Given that these people were also immunosuppressed, I interpret that to mean no association. In fact, it's remarkably negative. However, their interpretation was that an association exists.

Cesario T:

That means that the initial report by Gallo et al. in Science was accidental probably.

Cone R:

The initial discovery of HHV-6? In terms of their finding it in lymphoproliferative disorders in AIDS patients, it was almost predictable. Any immunosuppressed population seems to reactivate HHV-6 and become more likely to be culture positive. As far as being a causal factor, no it probably is not causally related. You're right.

Finch W (University of California, Irvine, Irvine, CA):

When you found youngsters who had not acquired HHV-6, did I understand that you looked at some of the parents and they didn't have it either? Have these people been followed? Did they pick it up later?

Cone R:

That is an interesting question: If you have seronegative parents, do the youngsters get it? Those studies have not been done. It seems clear that the maternal antibody levels are not predictive of whether or not a person is going to get roseola. Whether or not there are some people who are less likely to get infected is just unclear. We've documented at least one adult who is truly negative for HHV-6 infection, I think that's remarkable given that there is constant daily exposure to very high titers of this virus, since we're all shedding it. I can't comment any further than that data, but I think those studies on patients with roseola and the maternal link would be very interesting in terms of how it's transmitted.

Doyle P (University of British Columbia, Vancouver, British Columbia, Canada):

I saw a case of a 32 year old lady that had acute febrile illness with a high fever. It then resolved with the appearance of a rash very typical of roseola. This was about the time when the first association with roseola came out, and we never had it confirmed one way or the other.

Cone R:

Well, Chris Sherlock called me about a similar case just a couple of weeks ago. As I discussed with him, the likelihood is that we're going to find DNA in the peripheral blood from that person, because we find it in adults in general.

Peterson E (University of California Irvine Medical Center, Orange, CA):

The one patient was kind of interesting where you had either culture or PCR positive for the virus, serology negative. Did that person respond normally to the other herpesviruses? Did they have titers to HSV-1, for instance, which most of us do?

Cone R:

That's a good question. In that particular person, I have not characterized the response to the other herpesviruses, but I think I will.

Peterson E:

Similarly, these supposedly seronegative, PCR negative, really negative families, do they have normal response?

Cone R:

For one thing, let me just say that this is very preliminary data. We are only starting to look at the families and I wouldn't claim that this one is a seronegative, PCR negative family.

Peterson E:

HSV-1 is just as prevalent as HHV-6, so probably somebody in that family has seen herpes virus.

Cone R:

Right. And, actually that person has been characterized for the other herpes viruses and responds normally. I believe she is CMV positive.

PAPILLOMAVIRUSES AND HUMAN CANCER

Magnus von Knebel Doeberitz, Claudia Rittmüller, Matthias Dürst, Tobias Bauknecht and Harald zur Hausen

Institut für Angewandte Tumorvirologie
Deutsches Krebsforschungszentrum
Germany

HISTORICAL ASPECTS

The infectious nature of human epidermal papillomas has been known for a long time. It was first demonstrated by Cuiffo in 1907, that the intradermal inoculation of cell free extracts of warts induced hyperplastic papillomatous lesions (Cuiffo, 1907). Using electron microscopy, it was possible to demonstrate viral particles within papillomas of the skin (Almeida et al. 1962). These particles were subsequently classified on the basis of their morphological and biochemical features as distinct genus "papillomaviruses" in the family of papovaviridae.

Papillomaviruses attracted special interest when it emerged that they were associated with malignant tumors in a variety of different species. In 1935 reports were published that demonstrated that wart-like skin lesions of cottontail rabbits evoked by an infection with the cottontail papillomavirus (CRPV) may convert to malignant invasive growing carcinomas (Rous and Beard, 1935; Severton and Berry, 1935). About forty years later it was postulated that papillomaviruses might also be involved in the etiology of human cancers (zur Hausen, 1976). In 1979 Orth and coworkers reported that lesions on sun exposed areas of the skin of patients suffering from the rare hereditary condition epidermodysplasia verruciformis (EV) (Lutzner, 1978), which were associated with specific types of human papillomaviruses, especially HPV 5 and 8, might convert into malignant squamous cell carcinomas (Orth et al. 1979). In the last decade, the close association between the infection with some other human papillomavirus types and anogenital cancers, in particular cervical carcinoma, was established (zur Hausen, 1989).

BIOLOGY AND CLASSIFICATION OF HUMAN PAPILLOMAVIRUSES

Papillomaviruses infect the basal cells of the epidermis. However, expression of viral genes appears to be strictly controlled in these cells as has been shown by *in situ* hybridization experiments (Stoler et al. 1989). Viral replica-

Medical Virology 10, Edited by L.M. de la Maza and
E.M. Peterson, Plenum Press, New York, 1991

tion is only encountered in warts, condylomas or other non-malignant HPV related lesions. Even here no viral RNAs can be detected in the basal cell layer. Viral genes are only expressed to start the viral replication cycle if the keratinocytes have reached a defined state of differentiation in the stratum spinosum, finally leading to the production of infectious virions at the surface of the epithelium (Figure 1). The viral DNA can obviously persist in a latent state in the basal cells of the epithelium for a long time. It is then presumably multiplied in phase with the cellular DNA and distributed during mitosis to both daughter cells. However, independent viral genetic activity appears to be tightly controlled in the basal cells.

Due to the special requirements pertaining to the differentiation status of the host cells, no permissive system for the replication *in vitro* of papillomaviruses has yet been established. Classification of papillomaviruses based on serological methods was hampered by the limited amount of available natural viral antigens. The taxonomy of papillomaviruses is therefore based on the heterogeneity of their nucleic acid composition and distinct types are assigned as genotypes and not as serotypes. The method still in use to differentiate between specific genotypes is a hybridization procedure in liquid phase: Under defined stringent hybridization conditions (Tm -25°C), two papillomavirus genomes are incubated. If they cross-hybridize to more than 50%, they are defined as identical genotypes, which may still be further subdivided as subtypes of one genotype on the basis of direct sequence data or of different RFLPs (Coggin and zur Hausen, 1976). However, if they cross-hybridize under these conditions to less than 50%, they are defined as different genotypes. The liquid hybridization procedure is laborious and often gives conflicting results. There-

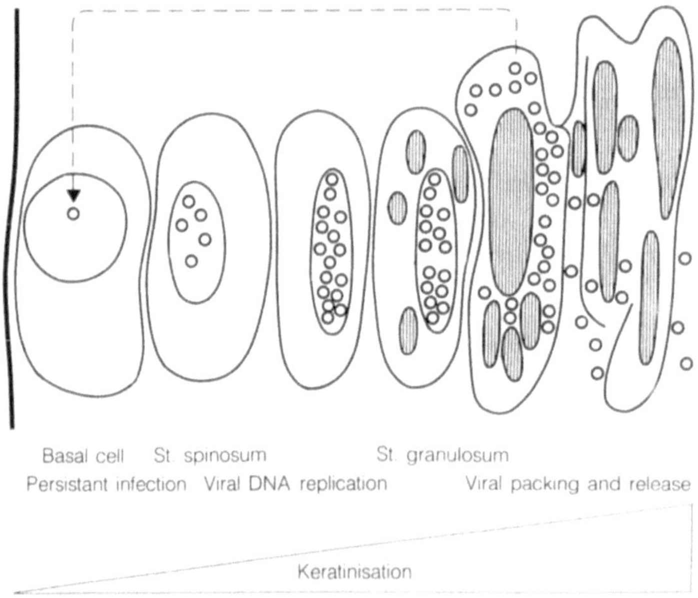

Figure 1. Schematic presentation of papillomavirus replication: Only in differentiated cells within the stratum spinosum granulosum of a lesion (wart, condyloma or low grade cervical intraepithelial neoplasia) viral genes are expressed although the virus infects the basal cells. Here the genetic activity of the virus is tightly controlled. This control mechanism is lost through cellular differentiation permitting viral replication in the superficial layers of the epithelium.

fore, alternative methods in the future for the classification of papillomaviruses are mandatory and probably will be based on direct sequence data. However, so far no other classification procedure than the liquid phase hybridization kinetic has been generally accepted.

Papillomavirus particles contain a highly twisted, supercoiled double stranded DNA genome of about 8 kb in length. In the late seventies, the viral DNA of some human papillomaviruses could be isolated by equilibrium gradient centrifugation methods from DNA preparations of warts. The DNA was then molecularly cloned and characterized (Gissmann and Schwarz, 1985). These were the first classified types of the human papillomaviruses (HPV 1, 2, 3, etc.) (Gissmann et al. 1977). Isolation of the DNA of these types was relatively easy, since they were encountered in very high copy numbers in the differentiated epithelial areas of the warts (Figure 1).

By using the same technique, it was later attempted to purify DNA molecules of papillomaviruses from DNA preparations of condylomata acuminata. Although significantly less viral genomes were found in these lesions, it was finally possible in 1980 to clone and characterize the HPV 6 genome, and a similar approach performed with DNA of laryngeal papillomas led to the isolation and cloning of the HPV 11 prototype (de Villiers et al. 1981; Gissmann and zur Hausen, 1980; Gissmann et al. 1982). Subsequent studies showed that both HPV genotypes are closely related and can be regularly detected in condylomata acuminata and juvenile laryngeal papillomas.

In cervical cancers, however, viral DNA molecules could never be purified by centrifugation techniques and virus particles have never been demonstrated in carcinoma cells. In 1982 it was possible to detect the DNA of a new papillomavirus in a cervical carcinoma biopsy specimen using a different approach. DNA extracted from the cervical carcinoma biopsy was subjected to Southern blotting and then hybridized with a labeled HPV 11 DNA probe. Non-stringent hybridization conditions (Tm -40°C) were chosen, permitting the detection of related DNA sequences in the cervical carcinoma DNA that were obviously of papilloma virus origin, but not identical with the HPV type used as probe. These sequences were subsequently cloned and characterized and the new papillomavirus DNA from the cervical carcinoma was assigned as HPV 16 genotype (Dürst et al. 1983). The same method led to the identification of the HPV 18 DNA in a variety of cell lines derived from cervical carcinomas and also tumor biopsies (Boshart et al. 1984).

Sequencing of the DNA of various genotypes revealed several open reading frames (orfs) which are encoded by only one strand (Figure 2). The orfs encoded by the papillomavirus genome are subdivided in late genes encoding viral capsid proteins (L1 and L2), and those involved in regulatory features of the viral replication cycle defined as early genes (E1-E7) (Figure 2). A third DNA segment referred to as upstream regulatory region (URR) was shown to have enhancer and promoter activity, but revealed no major orfs. It contained many motifs that interact with cellular and viral transcriptional regulatory proteins, for example, a fragment binding the viral E2 protein, thereby leading to decreased activity of the viral promoter and inhibition of gene expression of depending genes (Bernard et al. 1989; Cripe et al. 1987; Thierry and Yaniv, 1987). Furthermore, in some but not all HPV genotypes a glucocorticoid responsive element has been demonstrated in this region (Chan et al. 1989; Gloss et al. 1987).

Up to now, the DNAs of more than 62 different human papillomaviruses have been isolated (de Villiers, 1989). There is a remarkable tissue specificity of

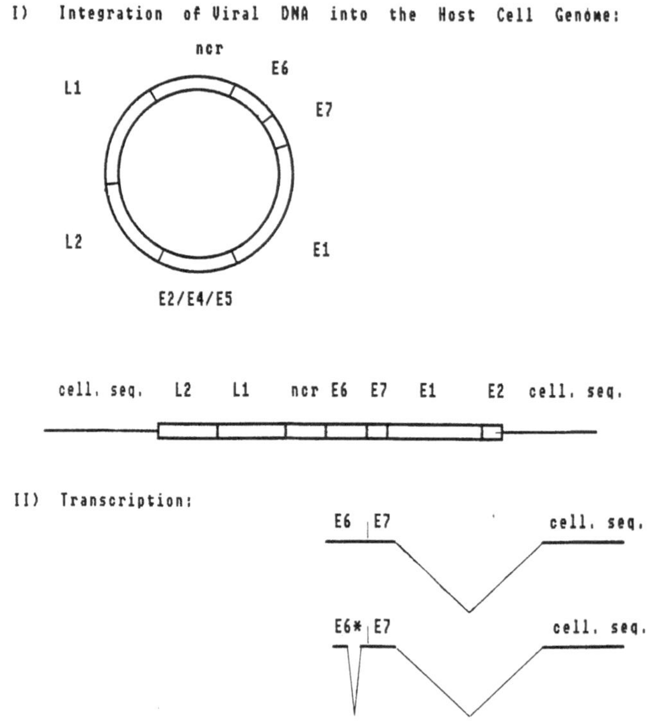

Figure 2. Human papillomavirus in cervical cancer cells. Papillomavirus genome in the episomal and the integrated state as it is typically found in cervical carcinoma cells: The orfs, referred to as E1-E7, encode early proteins involved in regulation of viral replication and packaging. The late orfs L1 and L2 encode structural proteins of the viral capsid. In cervical cancer cells, the URR and the orfs E6 and E7 are consistently preserved in the integrated state. They are regularly transcribed as polycistronic mRNAs with a cellular flanking sequence. The respective proteins can be detected within malignant cervical cancer cells. Some of the viral transcripts are spliced within the E6 orf leading to two new transcripts referred to as E6* and E6**. Whether the spliced E6 orfs are translated still remains questionable.

the different types (Table 1). Some types infect only cornified epithelium and are here associated with benign lesions. Others are encountered in the cornified epithelium, but only in immunosuppressed individuals or those suffering from EV (Lutzner, 1978). Some of these viruses are associated with malignant progression of lesions, particularly on sun exposed areas of the skin.

A different group of viruses (e.g. HPV 6 and 11) infects predominantly the mucosal epithelium, especially of the oropharyngeal and anogenital tract and are associated with benign lesions as, for example, condylomata acuminata. Others (e.g. HPV 16 and 18) are preferentially found in premalignant or malignant lesions as, for example, bowenoid papulosis, cervical, vaginal or penile intraepithelial neoplasia (CIN, VIN or PIN) or anogenital carcinomas. However, in rare instances, HPV 16 or 18 sequences are also detected in condylomata with no sign of malignancy. It is therefore not practical to classify papillomaviruses according to their association with malignant or non-malignant lesions.

TABLE 1. LIST OF HUMAN PAPILLOMAVIRUSES AND THEIR TISSUE TROPISM

Cutaneous types	1,2,3,4,7,10,28,41*,60
Cutaneous types in EV and or immunosuppressed patients	5*,8*,9,12,14*,15,17*,19,20* 21,22,23,24,25,26,27,37,46 47,48*,49,50
Mucosal types	_6_,11,13,16*,18*,_30*_,_31*_,32,_33*_, _34_,35*,_39*_,_40_,_42_,_43_,_44_,_45*_, _51*_,_52*_,53,54,55,_56*_,_57*_,_58*_,_59*_

Association of various HPV genotypes with either cornified (cutaneous) or mucosal epithelium. Types marked with * are associated with malignant conversion of the lesion. Italic typed HPVs are found in dysplastic lesions, however, some have not been isolated from malignant lesions.

EPIDEMIOLOGICAL ASPECTS OF HUMAN PAPILLOMAVIRUS INFECTIONS AND CANCER

About 90% of all cervical cancer samples tested worldwide are positive for distinct HPV genotypes, HPV 16 and 18 being the most prevalent types, since they are detected in about 70 to 80% (zur Hausen, 1989).

The almost regular detection of particular papillomavirus genotypes in cervical carcinoma biopsy specimens strongly suggested that these agents are involved in the carcinogenic process. This epidemiological evidence became questionable when it emerged that papillomavirus infections with the types detected in cervical carcinoma cells are extremely widespread also in the healthy population (Munoz et al. 1988). A detailed study by Schneider and colleagues (1987) based on Southern blot investigations revealed that in approximately 30% of pregnant women the DNA of papillomaviruses can be detected in cervical scrappings if they are repeatedly screened for the presence of HPV 6, 16 and 18, indicating that latent infections with these genotypes are very widespread among women without clinical detectable lesions. It is estimated that less than 3% of all infected individuals will develop malignant lesions. The high rate of asymptomatic HPV infections in the healthy population also emerges for HPV types associated with lesions in EV patients. This condition is extremely rare, but patients regularly develop lesions induced by the particular EV HPV genotypes (Table 1) (Lutzner, 1978). If only EV patients were infected with these virus types, the chance for one particular genotype to meet another suitable host (i.e. an EV patient) is very low. Therefore, no sufficient reservoir would be provided for virus replication. It is thus very likely that infectious particles of these particular genotypes are highly prevalent in the general population without causing disease in normal individuals.

Turning back to the association of particular HPV types and cervical cancer, it is obvious that the papillomavirus infection is not sufficient to cause malignant disease. There are several other arguments supporting this point of view. For example, there is a long latency period between infection, which probably is mediated by sexual contacts, and the development of malignant cells. Furthermore, studies on the clonality of cervical cancer cells have revealed that the malignant cells represent one single cell clone derived from only one progenitor cell, although many thousands of cells are infected by the papillomavirus.

THE MULTIPLE STEP PROCESS OF CERVICAL CARCINOGENESIS

Cervical cancer usually arises at the squamous columnar junction. Here, the part of differentiated cells covering the basal cells becomes thinner and the squamous epithelium finally ends as a single cell layer at the margin to the columnar epithelium of the endocervix. Therefore, in the junction area, papillomaviruses have easy access to their natural host cell at the bottom of the epithelium.

As other cancers, cervical carcinoma does not develop in a single step mechanism. Various cytologically and histologically defined precursor lesions proceed the malignant carcinoma. These can often be diagnosed in time and the appropriate treatment can be started, a procedure that significantly decreases the incidence of malignant cervical cancer in the Western communities (Hakama et al. 1985; Papanicolaou and Traut, 1943).

Premalignant lesions are classified according to their degree of histologically detectable dysplasia as low grade lesions (CIN I and II) or high grade lesions (CIN III). The more advanced lesions develop from the less dysplastic by the selection of new, more dysplastic cells. Several repeated selection steps finally lead to the development of an invasively growing cell clone.

TRANSFORMING FUNCTIONS OF PAPILLOMAVIRUS

What is the role of the papillomavirus during the process of cervical carcinogenesis? One line of experimental evidence for a potential causative role of the papillomaviruses as oncogenic agents is derived from the following approach: If the DNA of the papillomaviruses associated with cervical malignancies is transfected into primary human foreskin or cervical epithelial cells, cell clones with an unlimited lifespan *in vitro* arise (Dürst et al. 1988; Pirisi et al. 1987; Woodwarth et al. 1989). By definition, the virus therefore has a transforming potential. Experiments with various other HPV types have shown that this transforming potential is restricted to those genotypes associated with malignant lesions. Transfection of HPV 6 or 11 did not lead to unlimited growing keratinocytic cell clones (Pecoraro et al. 1989; Woodwarth et al. 1989). Furthermore, detailed investigations could show that the transforming potential of the human papillomaviruses are due to the E6-E7 genes (Hawley-Nelson et al. 1989; Hudson et al. 1990; Münger et al. 1989). They were sufficient to induce the unlimited lifespan *in vitro*. If these transformed cells were tested for their malignant growth potential in nude mice, they formed small epithelial cysts with clear signs of keratinocytic differentiation, but did not show any signs of invasive or metastatic growth (Dürst et al. 1989, 1990). Introduction of the genetic material of certain types of the papillomaviruses into keratinocytes therefore leads to an unlimited lifespan *in vitro*, but not to malignant growth *in vivo*. However, if an activated ras oncogene was introduced into the papillomavirus immortalized cells, they showed a significant morphological alteration and became highly malignant in nude mice (di Paolo et al. 1989; Dürst et al. 1989). Thus, the viral DNA together with a further event, for example, the introduction of a single activated oncogene, leads to malignant conversion of primary human keratinocytes.

Additional factors therefore cooperate with the papillomavirus in the carcinogenic process. These additional events can be schematically divided into those acting on the host cell without direct interaction with the virus (e.g. activation of certain oncogenes, loss of tumor suppressor gene function) or those concerning the host cell, however, interfering with viral functions, e.g. by modulating viral gene expression on various levels or the distribution of viral proteins within the cell. Only little is presently known about these host cell factors.

HUMAN PAPILLOMAVIRUS GENES IN CERVICAL CANCER CELLS

Other factors might concern the virus or its genetic information. Since molecular probes were available to study these aspects, there exists a significant amount of information. The viral genome is usually integrated into host cell chromosomes. The chromosomal location appears to be more or less randomly selected, since the viral genome is found at still other chromosomal sites in cancer cells of different clonal origin (Dürst et al. 1987; Popescu et al. 1987). Presumably, transcriptionally active regions within the host cell genome facilitate the integration process, however, there is no obvious rule where the viral sequences integrate into the cellular DNA. The viral genome in contrast has a very consistent structure in the integrated state. The viral DNA is usually disrupted within the 3' end of the E1 or the 5' end of the E2 orf. This leads to functional inactivation of the E2 orf encoding inhibitory functions which act by binding to a target sequence within the viral promoter (Bernard et al. 1989). The late orfs are uncoupled from their physiological regulation or might even be deleted. However, the URR and the two orfs E6 and E7, encoding the transforming functions necessary for the immortalization of the primary keratinocytes are consistently preserved (Baker et al. 1987; Pater and Pater, 1985; Schwarz et al. 1985; Shirasawa et al. 1987; Smotkin and Wettstein, 1986). In addition to this particular integration pattern, other modes of integration may also be encountered in the same cervical cancer cell, however, the described characteristic pattern is a consistent finding. This suggests that cells that have undergone this specific type of viral integration might be selected by a defined growth advantage compared to others that show no other features of papillomavirus integration. A further consistent finding is that in all carcinoma cells investigated with the appropriate techniques, the E6 and E7 orfs are transcribed as polycistronic mRNAs from the integrated sequences (Baker et al. 1987; Schwarz et al. 1985, Shirasawa et al. 1987; Smotkin and Wettstein, 1986; Yee et al. 1985). These transcripts consist of a viral part spanning the E6 and E7 sequences and an additional cellular part derived from the 3' cellular flanking sequences. Sequence analysis of cDNAs derived from a variety of cervical cancer cell lines have shown that these 3' cellular sequences are different in cell clones of independent origin further underlining that there is no common cellular integration site (Schneider-Gädicke and Schwarz, 1986). Furthermore, the cellular sequences contain no major orfs and it is unlikely that they encode a protein, however, the cellular flanking sequences might have a significant role for the efficiency of E6-E7 gene expression, probably by stabilizing the viral-cellular fusion transcripts (Le and Defendi, 1988). The E6 sequences are encountered in three different modes, unspliced and spliced in two different forms. Proteins derived from the spliced E6 transcripts have not been identified in the cancer cells and it has been postulated by Smotkin and coworkers (1989) that the spliced form of the E6 orf might facilitate translation of the adjacent E7 sequences without coding for a protein. Evidence has been presented that some transcripts might be initiated in the 5' cellular sequences due to use of cellular promoters. However, whether this is a general finding is not yet clarified (Inagaki et al. 1988). The E6 and E7 gene products are regularly encountered in cervical cancer cells (Androphy et al. 1987; Baker et al. 1987; Oltersdorf et al. 1987; Seedorf et al. 1987; Smotkin and Wettstein, 1986, 1987), suggesting that they have some influence on the cancerous phenotype of the respective cells. They might shift the cells in a hit and run mechanism into a transformed state without being essential for the maintenance of the the malignant phenotype. Alternatively, their continuous presence might be a prerequisite for the transformed state of cervical keratinocytes.

THE ROLE OF PAPILLOMAVIRUS E6-E7 GENES IN MAINTAINING THE MALIGNANT PHENOTYPE OF CERVICAL CANCER CELLS

To define the biological significance of the viral E6-E7 proteins for the malignant phenotype of HPV associated cervical cancer cells, we attempted to influence specifically the expression level of these proteins. First C4-1 cells (Auersperg and Hawryluk, 1962) derived from a squamous cell carcinoma and harboring one copy of the HPV 18 genome integrated on chromosome 8 (Dürst et al. 1987) were used as the experimental model.

If these cells were treated with a glucocorticoid hormone, for example, dexamethasone, transcription of the E6-E7 sequences was enhanced by a factor of 4-5. This was accompanied by a significant growth stimulation. However, other cellular gene products involved in cell growth regulation as, for example, the epidermal growth factor receptor are also expressed at increased rates in dexamethasone treated C4-1 cells (von Knebel Doeberitz et al. 1990). Therefore, this did not necessarily indicate that there is a specific correlation between growth of these cancer cells and viral E6-E7 expression. HPV unrelated effects of the hormone had to be excluded. Therefore, specific HPV 18 E6-E7 antisense RNA (Inouye, 1988; Izant and Weintraub, 1985) was expressed in these cells to interfere with the translation of the viral proteins (von Knebel Doeberitz et al. 1988). A vector system capable of expressing the E6-E7 sequences in the antisense orientation under control of the dexamethasone inducible mouse mammary tumor virus long terminal repeat (MMTV-LTR) (Fasel et al. 1982) was constructed (Figure 3) and transfected into C4-1 cells. Stable clones harboring this vector were isolated under G418 selection and raised to cell lines.

By Northern blot analysis using a strand specific hybridization probe, it could be demonstrated that upon dexamethasone treatment, these cells indeed expressed the HPV 18 antisense RNA (Figure 4). The viral protein expression was significantly affected by the antisense RNA. Treatment with dexamethasone of normal C4-1 cells led to significantly increased intracellular E7 concentration. However, in cells expressing the HPV 18 E6-E7 antisense RNA, the hormone did not alter (clones K1 and K3) or even decreased (clone K2) the E7 concentration (Figure 5) (von Knebel Doeberitz et al. 1988).

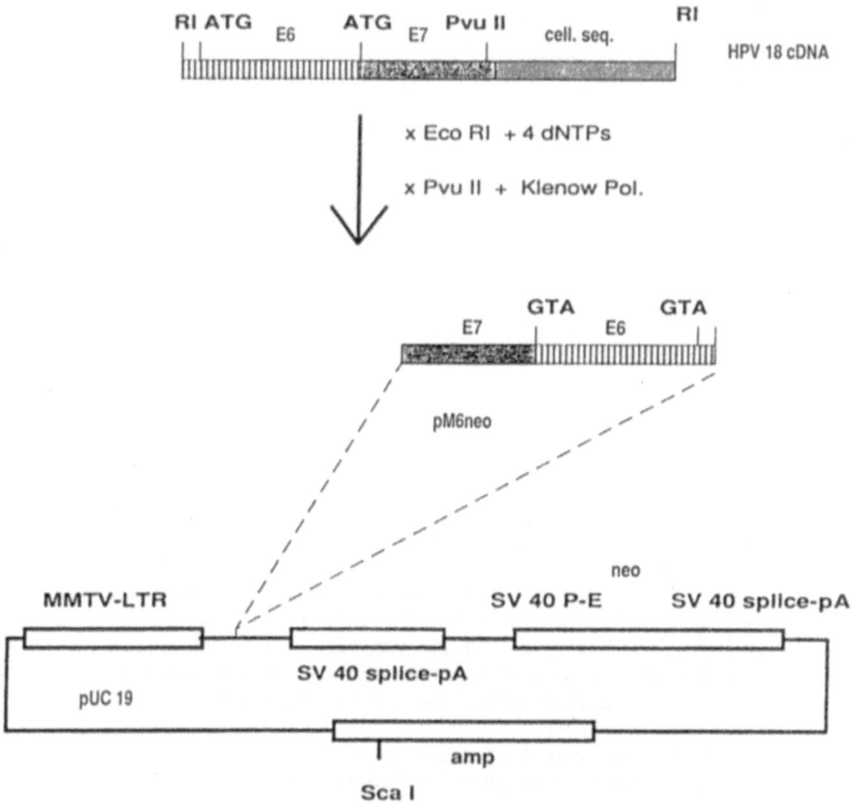

Figure 3. Construction of the dexamethasone inducible HPV 18 E6-E7 antisense RNA vector: The cloning strategy is outlined in detail elsewhere (von Knebel Doeberitz et al. 1988). Cellular sequences of the HPV 18 E6-E7 cDNA 7-23 were removed and the viral part was cloned in an inverted orientation downstream to the central Hae III fragment of the MMTV-LTR. RNA processing signals were derived from the SV 40 early polyadenylation and splice signal. Additionally, the neo dominant marker gene derived from the pSV2neo plasmid (Southern and Berg, 1982) was cloned on the vector.

The proliferative capacity of C4-1 cells reflected the expression level of the viral genes. Those cells expressing increased rates of E6-E7 gene products (C4-1 dex. and C4-1neo dex.) have an increased proliferation potential as compared to those with antisense RNA reduced HPV 18 E6-E7 expression (K1-K3 dex.) (Figure 6a).

If cells with different intracellular concentrations of the viral gene products were injected into immunoincompetent dexamethasone treated nude mice, significant differences in their tumorigenic potential were observed. C4-1 cells or G418 resistant C4-1 cells (C4-1neo) grew out rapidly to squamous cell carcinomas upon subcutaneous injection. However, if cells of the antisense clones were injected under the same conditions, no tumorigenic growth was observed (Figure 6b). The antisense clones formed small non-growing nodules that persisted for several months during the course of the experiment. Histological investigation showed that these nodules consisted preferentially of necrotic cell material. However, *in situ* hybridization with strand specific RNA probes clearly indicated expression of the antisense transcripts in the remaining live islets.

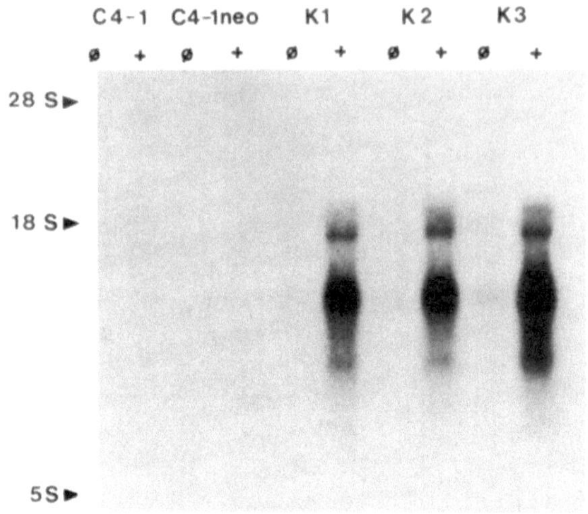

Figure 4. Northern blot analysis of HPV 18 E6-E7 antisense RNA transcripts. Ten μg of cytoplasmic RNA of cells grown either with or without dexamethasone (1μM) were transferred onto nylon membranes after agarose gel electrophoresis in a 1% denaturing formaldehyde gel. HPV 18 E6-E7 antisense RNA was detected by hybridizing the filter with a strand specific, ^{32}P-labeled, *in vitro* transcribed RNA probe. Antisense transcripts can only be detected in dexamethasone treated cells of the clones K1-K3 harboring the pM6neo vector. The parental cell line clone or a subclone (C4-1neo) harboring the dominant marker gene (neo) without HPV antisense transcription unit did not express E6-E7 antisense RNA.

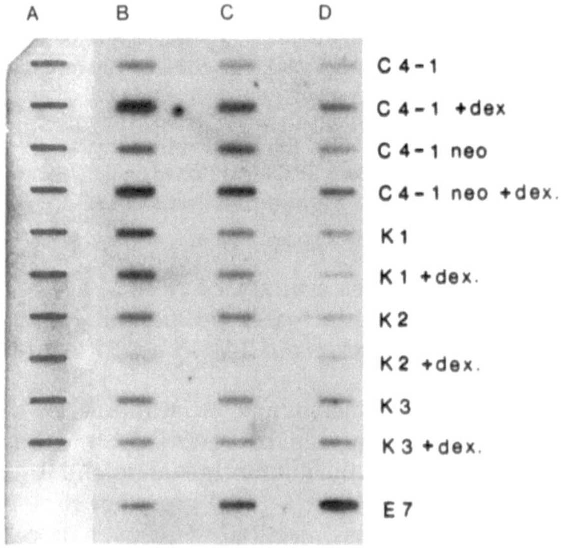

Figure 5. HPV 18 E7 protein expression in C4-1 cells with and without antisense RNA. Nitrocellulose bound protein extracts of dexamethasone treated and untreated cells were exposed to an anti-HPV 18 E7 antiserum (Seedorf et al. 1987). Lane A shows a Coomassie brilliant blue staining of a nitrocellulose filter with 20 μg of protein per slot to control and indeed comparable amounts of protein were loaded in each slot. Lanes B to D show an X-ray exposure of filters with 40 μg, 20 μg and 10 μg of soluble cytoplasmic protein incubated with the anti-E7 antiserum and ^{125}I-labeled protein A (von Knebel Doeberitz et al. 1989). As quantitative standard 2,4 and 8 ng of bacterially expressed E7-fusion protein were loaded on the same filter (von Knebel Doeberitz et al. 1988)

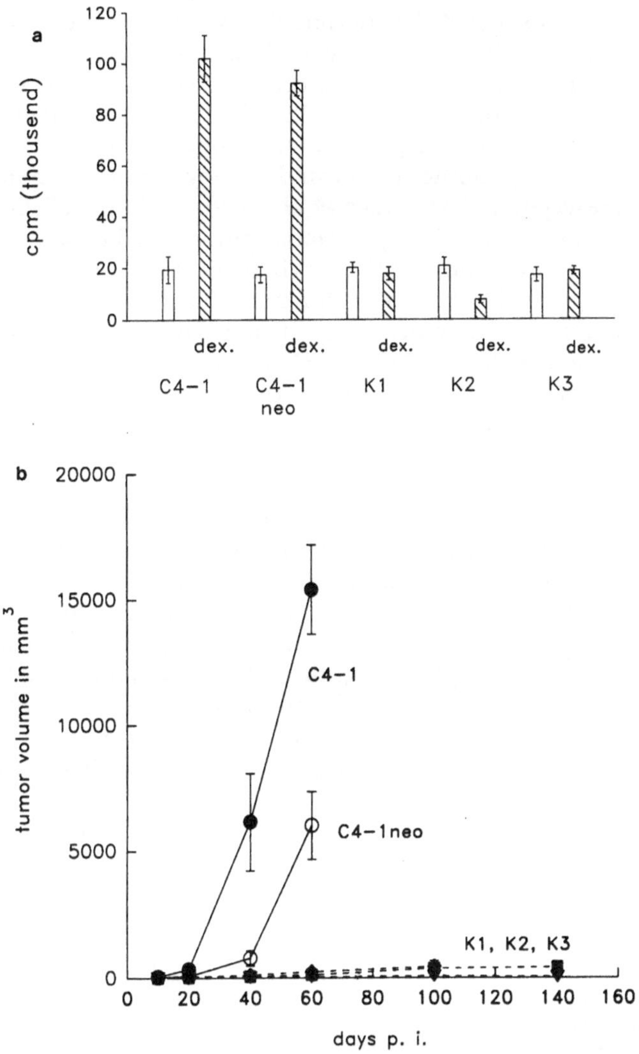

Figure 6. Growth kinetic of C4-1 cells with or without HPV 18 E6-E7 antisense RNA expression in cell culture (a) and in nude mice (b). The proliferation rate of cells *in vitro* was analyzed by determining their ^{3}H-thymidine incorporation rate (von Knebel Doeberitz et al. 1988). Dexamethasone treated cells with increased rates of viral gene expression (C4-1 dex. and C4-1neo dex.) proliferate significantly faster then those with less viral gene products (K1-K3 dex.). Bilateral subcutaneous inoculation of 10^7 cells with increased levels of viral early gene products (C4-1 and C4-1neo) into dexamethasone treated nude mice gave rise to fast growing tumors, while injection of C4-1 cells in which the HPV antisense RNA is expressed (K1-K3) only led to the formation of small persistent non-growing nodules.

The observation that dexamethasone led to enhanced viral gene expression was in good agreement with the identification of a glucocorticoid responsive element (GRE) within the viral promoter (Chan et al. 1989; Gloss et al. 1987). Presumably, upon hormone treatment, an activated glucocorticoid receptor protein binds to the consensus sequences within the promoter region and leads subsequently to enhanced transcription of the integrated viral genes.

Therefore, it was expected that dexamethasone treatment of other HPV 18 positive cervical cancer cell lines also evoked enhanced expression of the integrated viral genes. However, cell lines of different clonal origin responded differently to hormone treatment (Figure 7). C4-1 and C4-2 cells were both derived from the same tumor biopsy, are of identical clonal origin and therefore, harbor the viral genome integrated at the same chromosomal site (Auersperg and Hawryluk, 1962; James et al. 1989). In both cell lines, dexamethasone led to enhanced E6-E7 expression. However, HeLa cells that were derived from an adenocarcinoma of the cervix (Gey et al. 1952; Jones et al. 1971) and have the viral genes integrated at different chromosomal loci (Dürst et al. 1987) did not respond with significant modification of the viral gene expression upon dexamethasone treatment. In SW 756 cells derived from a squamous cell carcinoma (Freedman et al. 1982) and harboring the viral genes at still another chromosomal site, compared to HeLa or the C4 cells (Popescu et al. 1987) dexamethasone, surprisingly, induced a marked inhibition of viral gene expression (von Knebel Doeberitz et al., in press). The growth properties of all cell lines consistently correlated with the viral expression level.

SW 756 cells did not only respond with decreased growth on dexamethasone treatment, but were also morphologically modified (Figure 8). Hormone treated cells with severe inhibition of the viral expression rate did not grow to confluent cell layers. The cells detached and died after having reached a certain state of confluency. Their shape was significantly larger and some had multiple nuclei resembling syncytia. How specific are these alterations correlated to the inhibition of viral gene expression? Is this effect of the hormone on the SW 756 cells independent of the rate of viral gene expression or is it causally related to viral functions.

To answer this question, a dexamethasone inducible E6-E7 expression vector was introduced into SW 756 cells. This plasmid consists of the 7-23 HPV 18

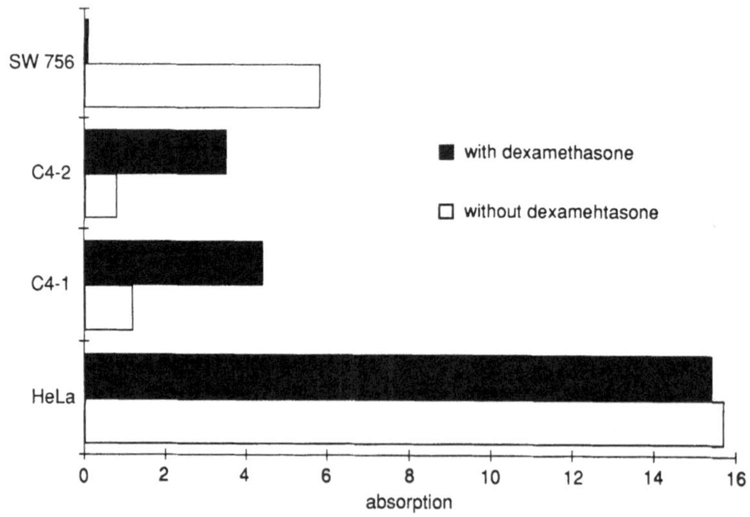

Figure 7. RNA slot blot analysis of the steady state level of HPV 18 E6-E7 transcripts in hormone treated or untreated HeLa, C4-1, C4-2 and SW 756 cells. Cytoplasmic RNA of hormone treated or untreated cells was subjected to an RNA slot blot analysis as described (Gasser et al. 1982). The filter was hybridized with a ^{32}P-labeled HPV 18 E6-E7 probe and the X-ray signals on autoradiograms were scanned. Each value was determined at least in three independent experiments.

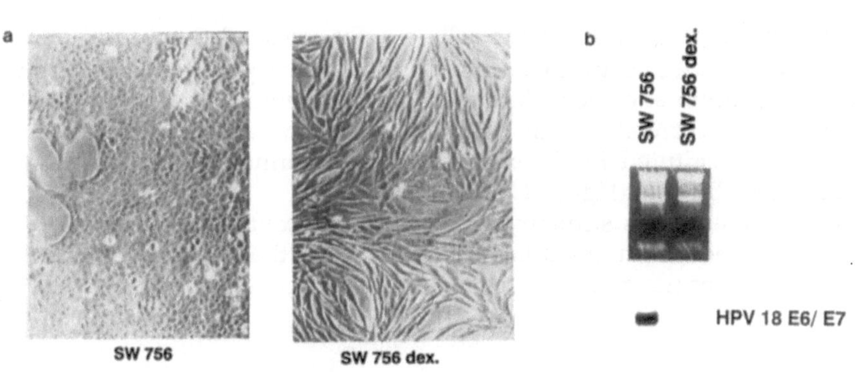

Figure 8. Morphology of SW 756 cells treated or untreated for one week with dexamethasone (1 μM). A Northern blot analysis showed the steady state level of the viral E6-E7 transcripts in the respective cells.

E6-E7 cDNA (Schneider-Gädicke and Schwarz, 1986) cloned 3' to the dexamethasone inducible MMTV-LTR. By cotransfection with the pSV2neo plasmid transferring resistance to G418, several stable cell clones could be isolated. A Northern blot analysis revealed that some SW 756 cell clones (P1 and P5) showed complete reconstitution of the viral gene expression, although they were treated with dexamethasone (Figure 9). Some others, for example, clones P2 and P3 showed only partial reconstitution, while still others like P4, were not reconstituted at all in their viral expression rate.

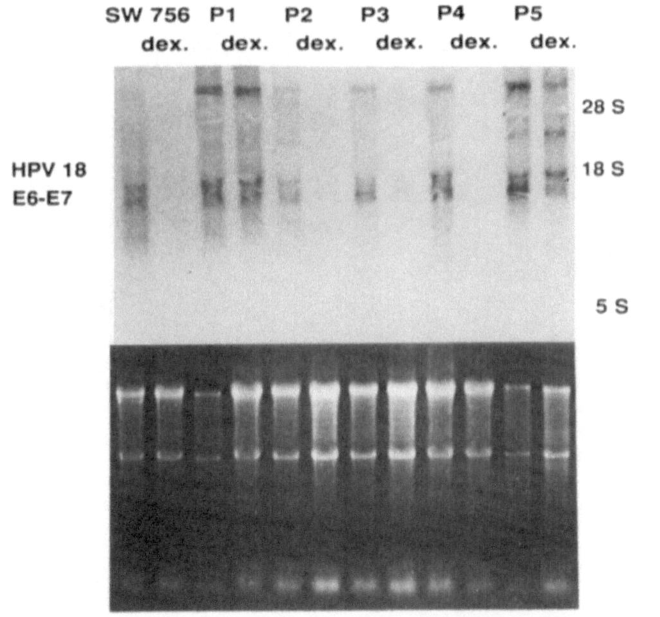

Figure 9. Northern blot analysis of SW 756 cells and subclones (P1-P5) partially reconstituted for HPV 18 E6-E7 expression in the presence of dexamethasone. Total cellular RNA was extracted (Chomczynski and Sacchi, 1987) and upon electrophoresis in 1% agarose gel transferred onto nylon membranes. The filters were hybridized with a ^{32}P-labeled HPV 18 E6-E7 probe. Clones P1 and P5 showed complete and clones P2 and P3 partial reconstitution of viral gene expression upon hormone treatment. The parental cells SW 756 and the clone P4 had no detectable levels of viral transcripts upon hormone treatment.

The growth properties of the cells exactly reflected the rate of viral gene expression as indicated by their ^{3}H-thymidine incorporation rate, growth in semi solid medium and their plating efficiency (Figure 10). The morphological phenotype also correlated well with the level of viral gene expression, since cells with reconstituted E6-E7 expression upon hormone treatment resembled untreated SW 756 cells (Figure 11).

These experiments strongly suggest that the continuous expression of the viral E6-E7 genes is an essential prerequisite for the proliferative phenotype and, therefore, for the malignant growth potential of cervical cancer cells.

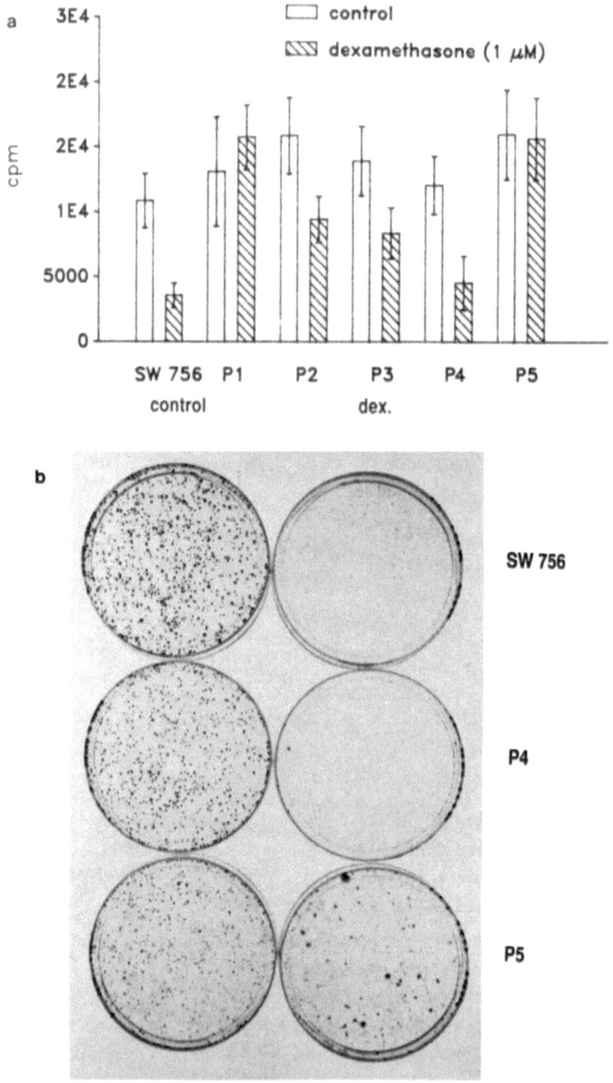

Figure 10. Growth characteristics of SW 756 cells with different levels of viral gene expression. a) ^{3}H-thymidine incorporation rates in SW 756 clones with complete (P1 and P5), partial (P2 and P3) and without (SW 756 and P4) reconstituted E6-E7 expression. b) plating efficiency of the parental SW 756 cells, the P4 clone and the P5 clone either with or without dexamethasone treatment.

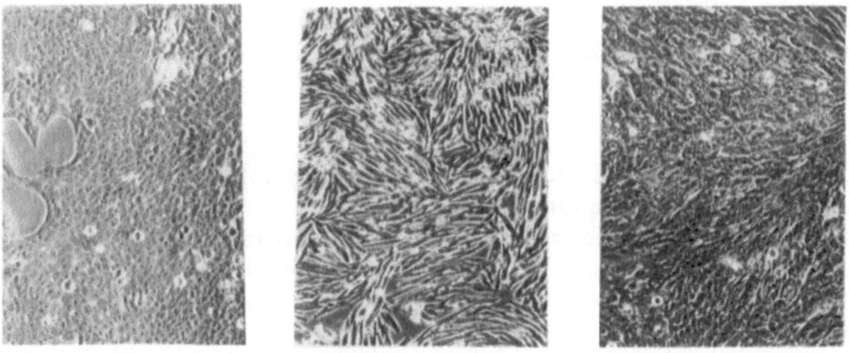

Figure 11. Morphology of SW 756 cells without (P4) and with (P5) reconstituted viral gene expression upon dexamethasone treatment. Untreated SW 756 cells are shown as the control.

How might the growth regulating properties of the viral E6-E7 genes work? How might these genes influence the physiological growth regulation? The groups of Peter Howley and Ed Harlow could demonstrate that the papillomavirus proteins can bind like the transforming proteins of other DNA tumor viruses to cellular proteins as, for example, the retinoblastoma gene product and the p53 protein, which are both regarded as cancer suppressor factors (Dyson et al. 1989; Werness et al. 1990). These results suggest that papillomavirus proteins might bind to specific cellular proteins involved in the regulation of the cell cycle. By binding to these proteins, they might interfere with the intracellular concentration of a functional active protein and thus, disturb the homeostasis of controlled cell growth. Whether this model holds true has to be demonstrated in the future.

INTEGRATION OF HUMAN PAPILLOMAVIRUS DNA MODIFIES THE REGULATION OF VIRAL GENE EXPRESSION

The heterogenous regulation of the integrated viral genes in different cervical carcinoma cell lines was an unexpected observation. Depending on the clonal origin of the cell, dexamethasone had a different influence on the expression level of these genes. Nuclear run on experiments demonstrated that the dexamethasone induced modification of the expression level are accompanied by significantly modified transcription rates of the integrated viral genes (von Knebel Doeberitz et al., in press).

These findings might be explained by the presence of a hormone induced transacting factor that interferes with the action of the activated glucocorticoid receptor on the integrated viral promoter in dexamethasone treated HeLa and SW 756 cells. Alternatively, the heterogenous regulation of the viral genes in some of the cell lines could be due to different cis acting response elements at the site of chromosomal integration, interfering with the glucocorticoid response of the viral promoter. These response elements should differ among various cell lines, since the viral sequences are integrated at different chromosomal loci in the different cell lines.

To test which of both possibilities is correct, we isolated the viral promoter sequences from genomic clones of HeLa and SW 756 cells and tested their activity in a non-integrated form in a transient chloramphenicol-acetyltransferase assay (Gorman et al. 1982). Both the HeLa and the SW 756 URR were significantly activated by dexamethasone in these cell lines (Figure 12) (von Knebel Doeberitz et al., in press).

This finding indicates that cis acting response elements probably of cellular origin, might significantly interfere with the physiological regulation of the viral promoter in the integrated state.

This might not be restricted to the glucocorticoid regulation, but might also be true for other regulatory features that modulate the activity of the viral promoter. Indirect evidence for this is given by the following experiment. The E2 orf inhibits the activity of the viral promoter when tested in transient assays. If this gene is transfected into a variety of different cervical carcinoma cell lines, some are inhibited in their growth properties, as for example, HeLa or SW 756 cells. Others, however, like C4-1 cells do not show a marked difference in cell growth if they are transfected with an intact E2 gene or an inactive mutant. Growth inhibition here is most likely due to down-regulation of the viral genes. Since it is not observed in all cell lines, the integrated viral promoter might be unresponsive to regulatory features of the E2 gene product in the integrated state in these particular lines.

It thus appears that integration of the viral genes renders the activity of the E6-E7 genes independent from regulatory features acting on the viral promoter in the transient state. This might represent one factor in the course of tumor progression, since uncontrolled E6-E7 expression emerges to be an essential prerequisite for the growth potential of malignant cervical cancer cells (von Knebel Doeberitz et al. 1988, in press).

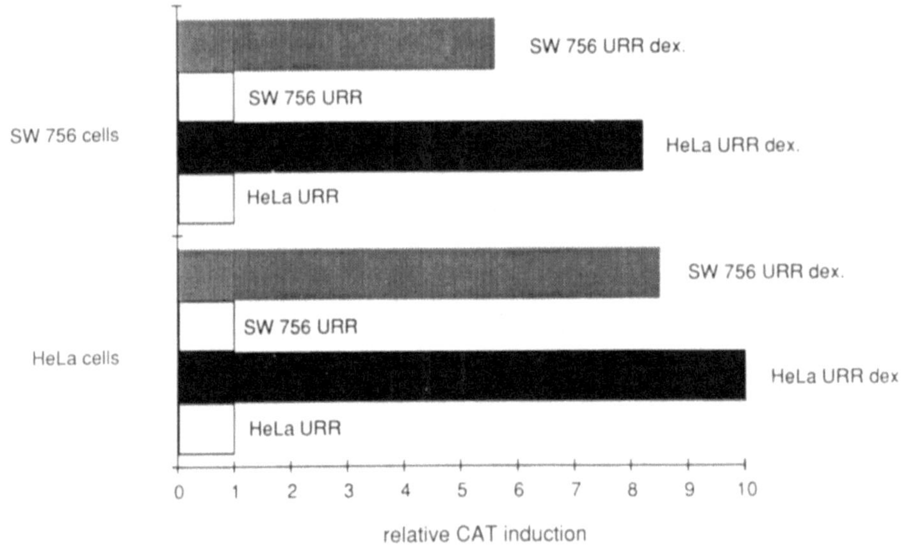

Figure 12. Chloramphenicol-acetyl transferase activity mediated by the HeLa and SW 756- HPV 18 URR controlled plasmids in HeLa and SW 756 cells with and without dexamethasone treatment. Dexamethasone significantly activated both isolated viral promoter regions in both cell lines when tested in the non-integrated form in transient expression assays.

EVIDENCE FOR CELLULAR FACTORS INTERACTING WITH PAPILLOMAVIRUS GENE EXPRESSION

If viral gene expression in a normal undifferentiated keratinocyte is strictly controlled, cellular factors are likely to contribute to this inhibition of viral gene expression. The functional loss of such factors could provide an alternative pathway leading to uncontrolled viral early gene expression (zur Hausen, 1986, 1989). This mechanism is suggested by experiments using hybrid cells between a cervical cancer cell line (HeLa) and a non-malignant fusion partner (either primary human fibroblasts or keratinocytes) established by Stanbridge et al. (1976, 1982). These hybrids have a non-malignant phenotype upon inoculation into athymic nude mice, but show all characteristics of transformed cells in tissue culture. Segregates of these hybrids converted to the malignant phenotype in nude mice. In tissue culture, both hybrid types express the HPV E6-E7 genes. However, upon implantation into nude mice, the non-malignant hybrids show selective down-regulation of the papillomavirus genes, whereas *in vivo*, the malignant cells also actively express viral genes (Bosch et al. 1990).

These findings are compatible with the concept of selective control system for viral genes in the non-malignant hybrids, probably provided by the non-malignant fusion partner (zur Hausen, 1986, 1989). In the malignant segregates, this system is apparently out of function, leading then to uncontrolled viral gene expression. It could be activated either by direct deficiency of the regulatory factor in the cell or alternatively, by functional loss of the receptor that mediates that activity of the system under *in vivo* conditions, since it appears from the experimental data that this system needs activation by presumably soluble factors provided by the living organism, but not by cell culture conditions (Bosch et al. 1990; zur Hausen, 1989).

All experimental data available so far support the concept that deregulated expression of the viral E6-E7 genes is one very important factor in the progress of carcinogenesis. When does deregulation occur in this multistep mechanism? Where in the progression sequence from a latently infected keratinocyte to the fully malignant cell occurs integration of the viral genes and/or inactivation of the regulatory features that control the papillomavirus gene expression under physiological conditions?

The expression, although in the deregulated form of the E6-E7 genes, is obviously not sufficient for the malignant conversion of human keratinocytes. Other uncharacterized factors contribute to carcinogenesis. Still, preliminary experiments based on *in situ* hybridization studies indicate that cervical lesions with a low grade of dysplasia still have regulated viral gene expression in the basal cell layer. However, lesions with more severe signs of dysplasia show a deregulated pattern of viral gene expression. E6-E7 expression can be demonstrated in these lesions over the whole area of the dysplastic epithelium and is not restricted to the differentiated cell layers as in the lesions with a low grade of dysplasia (Dürst et al., in preparation).

Therefore, integration of the viral genes or inactivation of the putative viral control system might occur during the course of progression of the premalignant lesion. Since the cells have a significant growth advantage by the uncontrolled expression of the viral genes, this probably marks a point from which the regression of the respective lesion is less likely.

CONCLUSIONS

Taken together, the following model for HPV associated cervical carcinogenesis may be proposed: In a latently infected keratinocyte, papillomavirus replication can only take place if the keratinocytes have reached a specific differentiation status. In the basal cell, the virus genome persists obviously without significant genetic activity. However, if by a molecular accident, the structure of the viral genes is modified in a sense that permits active expression of the E6-E7 genes, or if control of them by putative cellular factors is lost, one out of several essential prerequisites which lead the cell into the malignant phenotype is fulfilled. Expression of the E6-E7 genes per se, is not sufficient, however, it provides a selective advantage of the respective cells so that their survival is favored and further molecular events may then occur, pushing these cells forward into the more dysplastic state. Once the cells have reached the malignant phenotype and grow as invasive carcinoma, they still depend on the deregulated expression of the viral genes, since these obviously provide essential factors that interfere significantly with the regulation of cell growth.

REFERENCES

Almeida JD, Howatson AF, Williams MG (1962) Electron microscopic studies of human warts: sites of virus production and nature of inclusion bodies. J Invest Dermatol 38:337-345.

Androphy EJ, Hubbert NL, Schiller JT, Lowy DR (1987) Identification of the HPV 16 E6 protein from transformed mouse cells and human cervical carcinoma cell lines. EMBO J 6:989-992.

Auersperg N, Hawryluk AP (1962) Chromosome observations on three epithelial cell structures derived from carcinomas of the human cervix. J Natl Cancer Inst 28:605-627.

Baker CC, Phelps WC, Lindgren V, Braun MJ, Gonda MA, Howley PM (1987) Structural and transcriptional analysis of human papillomavirus type 16 sequences in cervical carcinoma cell lines. J Virol 61:962-971.

Bernard B, Baily C, Lenoir MC, Darmon M, Thierry F, Yaniv M (1989) The human papillomavirus type 18 (HPV18) E2 gene product is a repressor of the HPV18 regulatory region in human keratinocytes. J Virol 63:4317-4324.

Boshart M, Gissmann L, Ikenberg H, Kleinheinz A, Scheurlen W, zur Hausen H (1984) A new type of papillomavirus DNA and its presence in genital cancer and in cell lines derived from cervical cancer. EMBO J 3:1151-1157.

Chan WK, Klock G, Bernard HU (1989) Progesterone and glucocorticoid response elements occur in the long control region of several papillomaviruses involved in anogenital neoplasia. J Virol 63:3261-3269.

Chomczynski P, Sacchi N (1987) Single step method of RNA isolation by acid guanidinium thiocyanate-phenolchloroform extraction. Anal Biochem 162:156-159.

Ciuffo G (1907) Innesto positivo con filtrato di verruca volgare. G Ital Mal Venerol 48:12-17.

Coggin JR, zur Hausen H (1979) Workshop on papillomaviruses and cancer. Cancer Res 39:545-546.

Cripe TC, Haugen TH, Turk JP, Tabatabai F, Schmid PG, Dürst M, Gissmann L, Roman A, Turek L (1987) Transcriptional regulation of the human papillomavirus-16 E6-E7 promoter by a keratinocyte dependent enhancer and by viral E2 transactivator and repressor gene products: implications for cervical carcinogenesis. EMBO J 6:3745-3753.

de Villiers E-M, Gissmann L, zur Hausen H (1981) Molecular cloning of viral DNA from human genital warts. J Virol 40:932-935.

de Villiers E-M (1989) Heterogeneity of the human papillomavirus group. J Virol 63:4898-4903.

diPaolo JA, Woodwarth CD, Popescu NC, Notario V, Doninger J (1989) Induction of human cervical squamous cell carcinoma by sequential transfection with human papillomavirus 16 DNA and viral Harvey ras. Oncogene 4: 395-399.

Dürst M, Dzarlieva-Petrusevska RT, Boukamp P, Fusenig N, Gissmann L (1988) Molecular and cytogenic analysis of immortalized human primary keratinocytes obtained after transfection with human papillomavirus type 16 DNA. Oncogene 1:251-256.

Dürst M, Gallahan D, Jay G, Rhim JS (1989) Glucocorticoid enhanced neoplastic transformation of human keratinocytes by human papillomavirus type 16 and inactivated ras oncogene. Virology 173:767-771.

Dürst M, Croce CM, Gissmann L, Schwarz E, Hübner K (1987) Papillomavirus sequences integrate near cellular oncogenes in some cervical carcinomas. Proc Natl Acad Sci USA 84:1070-1074.

Dyson N, Howley PM, Munger KM, Harlow E (1989) The human papillomavirus-16 E7 protein is able to bind to the retinoblastoma gene product. Science 243:934-937.

Fasel N, Pearson K, Buetti E, Digglemann H (1984) The region of mouse mammary tumor virus DNA containing the long terminal repeat includes a long coding sequence and signals for hormonally regulated transcription. EMBO J 1:3-7.

Freedman RS, Bowen JM, Leibovitz A, Pathak S, Siciliano MJ, Gallager HS, Giovanella BC (1982) Characterization of a cell line (SW 756) derived from a human squamous cell carcinoma of the uterine cervix. In Vitro 18:719-726.

Gasser CS, Simonsen CC, Schilling JW, Schimke RT (1982) Expression of abbreviated mouse dihydrofolate reductase genes in cultured hamster cells. Proc Natl Acad Sci USA 79:6522-6526.

Gey GO, Coffman, WD, Kubiciek MT (1952) Tissue culture studies of the proliferative capacity of cervical carcinoma and normal epithelium. Cancer Res. 12:264-265.

Gissmann L, Schwarz E (1985) Cloning of papillomavirus DNA. In Becker Y (ed) Recombinant DNA Research and Virus. pp. 173-197. Martinus Nijhoff Publishing, Boston, USA.

Gissmann L, zur Hausen H (1980) Partial characterization of viral DNA from genital warts (condylomata acuminata) Int J Cancer 25:605-609.

Gissmann L, Diehl V, Schultz-Coulon HJ, zur Hausen H (1982) Molecular cloning and characterization of human papillomavirus DNA derived from a laryngeal papilloma. J Virol 44: 393-400.

Gissmann L, Pfister H, zur Hausen H (1977) Human papillomavirus (HPV): characterization of four different isolates. Virology 76:569-580.

Gloss B, Bernard HU, Seedorf K, Klock G (1987) The upstream regulatory region of the human papillomavirus-16 contains an E2 protein-independent enhancer which is specific for cervical carcinoma cells and regulated by glucocorticoid hormones. EMBO J 6:3735-3743.

Gorman C, Moffat LF, Howard B (1982) Recombinant genomes which express chloramphenicol acetyltransferase in mammalian cells. Mol Cell Biol 2:1044-1051.

Hakama M, Chamberlain J, Day NE, Miller AB, Prorok PC (1985) Evaluation of screening programmes for gynaecological cancer. Brit J Cancer 52:669-673.

Hawley-Nelson P, Voudsen KH, Hubbert NL, Lowy D, Schiller JT (1989) HPV 16 E6 and E7 proteins cooperate to immortalize human foreskin keratinocytes. EMBO J 8:3905-3910.

Hudson JB, Bedell MA, McCance D, Laiminas LA (1990) Immortalization and altered differentiation of human keratinocytes *in vitro* by the E6 and E7 open reading frames of human papillomavirus type 18. J Virol 64:519-526.

Inagaki Y, Tsunokawa Y, Takebe N, Nawa H, Nakanishi S, Terada M, Sugimura T (1988) Nucleotide sequences of cDNAs for human papillomavirus type 18 transcripts in HeLa cells. J Virol 62:1640-1646.

Inouye M (1988) Antisense RNA: its functions and applications in gene regulation. A review. Gene 72:25-34.

Izant IG, Weintraub H (1985) Constitutive and conditional suppression of exogenous and endogenous genes by antisense RNA. Science 229:345-352.

James GK, Kalousek DK, Auersperg N (1989) Karyotypic analysis of two related cervical carcinoma cell lines that contain human papillomavirus type 18 DNA and express divergent differentiation. Cancer Genet Cytogenet 38:53-60.

Jones HW, McKusick VA, Harper PS, Wuu KD (1971) The HeLa cell and a reappraisal of its origin. Obstet Gynecol 38:945-949.

Le JY, Defendi V (1988) A viral-cellular junction fragment from a human papillomavirus type 16-positive tumor is competent in transformation of NIH 3T3 cells. J Virol 62:4420-4426.

Lutzner MA (1978) Epidemodysplasia verruciformis: An autosomal recessive disease characterized by viral warts and skin cancer. Bul Cancer 65:169-182.

Münger K, Phelps WC, Bubb V, Howley PM, Schlegel R (1989) The E6 and E7 genes of the human papillomavirus type 16 together are necessary and sufficient for transformation of primary human keratinocytes. J Virol 63:4417-4421.

Munoz N, Bosch X, Kaldor JN (1988) Does human papillomavirus cause cervical cancer? The state of the epidemiological evidence. Br J Cancer 57:1-5.

Oltersdorf T, Seedorf K, Röwekamp W, Gissmann L (1987) Identification of human papillomavirus type 16 E7 protein by monoclonal antibodies. J Gen Virol 68:2933-2938.

Orth G, Jablonska S, Jarzabek-Chorzelska M, Obalek S, Rzesa G, Favre M, Croissant O (1979) Characteristics of the lesions and risk of malignant conversion associated with the type of papillomavirus involved in epidermodysplasia verruciformis. Cancer Res 39:1074-1082.

Papanicolaou, G, Traut HF (1943) Diagnosis of the uterine cancer by the vaginal smear. In: The Commonwealth Fund, New York.

Pater MM, Pater A (1985) HPV 16 and 18 sequences in carcinoma cell lines of the cervix. Virol 145:313-322.

Pecoraro G, Morgan D, Defendi V (1989) Differential effects of human papillomavirus type 6, 16 and 18 DNAs on immortalization and transformation of human cervical epithelial cells. Proc Natl Acad Sci USA 86:563-567.

Pirisi L, Yasumoto S, Feller M, Doninger JK, DiPaolo JA (1987) Transformation of human fibroblasts and keratinocytes with human papillomavirus type 16 DNA. J Virol 61:1061-1066.

Popescu NC, Amsbaugh SC, DiPaolo JA (1987) Human papillomavirus type 18 DNA is integreated at a single chromosome site in cervical carcinoma cell line SW 756. Virol 51:1682-1685.
Rous P, Beard JW (1935) The progression to carcinoma of virus-induced rabbit papillomas (Shope) J Exp Med 62:523.
Schneider A, Hotz M, Gissmann L (1987) Increased prevalence of human papillomaviruses in the lower genital tract of pregnant women. Int J Cancer 40:198-201.
Schneider-Gädick A, Schwarz E (1986) Different human cervical carcinoma cell lines show similar transcription patterns of human papillomavirus type 18 early genes. EMBO J 5:2285-2292.
Schwarz E, Freese UK, Gissmann L, Mayer W, Roggenbuck B, zur Hausen H (1985) Structure and transcription of human papillomavirus sequences in cervical carcinoma cells. Nature 314:111-114.
Seedorf K, Oltersdorf T, Krämmer G, Röwekamp W (1987) Identification of early proteins of the human papillomaviruses type 16 and 18 in cervical carcinoma cells. EMBO J 6:139-144.
Shirasava H, Tomita Y, Sekiya S, Takamizawa H, Simizu B (1987) Integration and transcription of human papillomavirus type 16 and 18 sequences in cell lines derived from cervical carcinomas. J Gen Virol 68:583-591.
Smotkin D, Wettstein FO (1987) The major human papillomavirus protein in cervical cancer is a cytoplasmic phosphoprotein. J Virol 61:1686-1689.
Smotkin D, Prokoph H, Wettstein FO (1989) Oncogenic and nononcogenic human genital papillomaviruses generate the E7 mRNA by different mechanisms. J Virol 63:1441-1447.
Southern PJ, Berg P (1982) Transformation of mammalian cells to antibiotic resistance with a bacterial gene under control of the SV40 early region promotor. J Mol Appl Genet 1:327-341.
Stanbridge EJ (1976) Suppression of malignancy in human cells. Nature 260:17-20.
Stanbridge EJ, Der C-J, Doersen CJ, Nishimi RY, Peehl DM, Weissmann BE, Wilkinson JE (1982) Human cell hybrids: Analysis of transformation and tumorgenicity. Science 215:252-259.
Stoler MH, Wolinsky SM, Whitebeck A, Broker TR, Chow LT (1989) Differentiation-linked human papillomavirus type 6 and 11 transcription in genital condylomata revealed by *in situ* hybridization with message-specific RNA probes. Virol 172:331-340.
Syverton JT, Berry GP (1935) Carcinoma in the Cottentail rabbit following spontaneous virus papilloma (Shope) Proc Soc Exp Biol Med 33:399-400.
Thierry F, Yaniv M (1987) The PBV 1-E2 transacting protein can be either an activator or a repressor of the HPV 18 regulatory region. EMBO J 6:3391-3397.
von Knebel Doebertiz M, Oltersdorf T, Schwarz E, Gissmann L (1988) Correlation of modified human papillomavirus early gene expression with altered growth properties in C4-I cervical carcinoma cells. Cancer Res. 48:3780-3786.
von Knebel Doeberitz M, Drzonek H, Koch S, Becker CM (1989) A simplified solid phase assay for quantitation of native membrane proteins: Application to the measurement of EGF receptor induction by dexamethasone. J Immunol Meth 122:259-264.
von Knebel Doeberitz M, Gissmann L, zur Hausen H (1990) Growth regulating functions of human papilomavirus early proteins in cervical cancer cells

acting dominant over enhanced epidermal growth factor receptor expression. Cancer Res. 50:3730-3736.
von Knebel Doeberitz M, Bauknecht T, Bartsch D, zur Hausen H (1990) Influence of chromosomal integration on glucocorticoid regulated transcription of growth-stimulating papillomavirus E6-E7 genes in cervical carcinoma cells. Proc Natl Acad Sci USA (in press).
Werness BA, Levine A, Howley PM (1990) Assocation of human papillomavirus type 16 and 18 E6 proteins with p53. Science 248:76-79.
Woodwarth CD, Doniger J, DiPaolo JA (1989) Immortalization of human foreskin keratinocytes by various human papillomarvirus DNAs corresponds to their association with cervical carcinoma. J Virol 63:159-164.
Yee C, Krishnan-Hewlett I, Becker CC, Schlegel R, Howley PM (1985) Presence and expression of human papillomavirus sequences in human cervical carcinoma cell lines. Am J Pathol 119:361-366.
zur Hausen H (1976) Condylomata acuminata and human genital cancer. Cancer Res. 36:530.
zur Hausen H (1986) Intracellular surveillance of persisting viral infections: Human genital cancer results from deficient cellular control of human papillomavirus gene expression. Lancet 2:489-491.
zur Hausen H (1989) Papillomaviruses as carcinomaviruses. Adv Virol Oncol 8:1-26.

DISCUSSION

Kilbourne E (Mount Sinai School of Medicine, New York, NY):

I just wondered whether you have revisited the problem of cultivation of the virus using corticosteroid hormones, in view of the findings you have? Also, could you relate the effects you've seen with dexamethasone to those of Harold Varmus with the mouse mammary tumor virus?

von Knebel-Doeberitz M:

To answer the second question first, there has been a glucocorticoid responsive element characterized within the papilloma virus promoter only in those papilloma viruses that are associated with mucosal infections. This kind of glucocorticoid response element is present in HPV-6, 11, 16 and 18, but it is not present in the cutaneous types. No, we haven't attempted to establish a replicative system. I doubt it is possible since dexamethasone per se doesn't lead cells into a differentiated status that is necessary for the viral replication.

Minocha H (Kansas State University, Manhattan, KS):

We have done some studies using glucocorticoids in herpes viruses, and with β–estradiol, using normal concentrations present in the *in vivo* system, that's 10^{-7}, 10^{-8} molar concentration, we have seen that there is just one or two times more production of virus. We have done some other studies where we see the morphogenesis is much more prevalent than in β–estradiol treated cells. This is in a tissue culture system. However, when we test dexamethasone concentrations the same, or even larger concentration like 10^{-4} molar, I think the production of virus is about ten times more than the non-treated cells. My question is what concentrations of dexamethasone did you use?

von Knebel-Doeberitz:

We have tried to do this with estradiol, however, the cells turned out not to have an active receptor for estrogens. The same was true for progesterones. The concentrations we used for these experiments ranged between 10^{-8} and 10^{-6} for dexamethasone.

Levy J (University of California Irvine, Irvine, CA):

You touched on my question already, but I was going to ask about the structural features that might correlate with the functional similarities of the different HPV strains. Can you comment any further on that?

von Knebel-Doeberitz:

It has been shown that the binding activity of the retinoblastoma and E7 protein is obviously restricted to those types that are associated with the malignant conversion. The E6 and E7 gene products of the 6 and 11, obviously are not capable of binding to the retinoblastoma or the p53 protein. Maybe that is one of the major differences between these proteins. However, this needs to be clarified. In addition, it's not yet known what the binding of these two viral proteins to the cellular proteins mean. Maybe there are other cellular proteins that are much more important in this type of cellular regulation.

DETECTION AND ELIMINATION OF BLOOD-BORNE VIRUSES TRANSMITTED BY TRANSFUSION

Girish N. Vyas

Department of Laboratory Medicine
University of California, San Francisco
San Francisco, California, USA

INTRODUCTION

A decade-long epidemic of AIDS has continued to raise an unprecedented concern about the safety of our blood supply. Considering that nearly four million American patients receive the life-saving benefits of blood transfusion every year, we need to recognize several of the inherent biological complications incidental to receiving transfusions (Vyas, 1988). The underlying phenomena of allogenic stimulation and suppression of the transfusion recipient's hemopoetic and immune systems facilitate graft-vs-host disease (Thaler et al. 1989; Vogelsang, 1990), as well as transmission of a variety of blood-borne viruses (BBVs) harbored by apparently healthy blood donors (Rawal et al. 1990a, b). The BBVs include hepatitis B viruses (HBV), hepatitis C viruses (HCV), hepatitis delta virus (HDV), human immunodeficiency virus (HIV-1/2), human T-cell leukemia virus (HTLV-I/II) and cytomegalovirus (CMV). Transfusion-transmitted BBVs cause a spectrum of multisystem diseases, including chronic hepatitis, liver cancer, AIDS, lymphoma-leukemia and myelopathy.

TRANSFUSION-TRANSMITTED BBVs

The current estimates of transfusion-transmitted viral infections and consequent disease in the United States shown in Table 1 provide a perspective of infectious risks of transfusion. Viral hepatitis and HIV-1 infections have been recognized as the most serious complications in transfusion medicine (Bove, 1987; Lenfant and McCurdy, 1990; Moore et al. 1987; Vyas, 1988). For transfusion recipients, the threat of widespread AIDS infection has raised the ominous specter of virus-induced suppression of the immune system itself. Two notable viral pathogens (CMV and HIV-1) are capable of bringing about such immunosuppression (Southern and Oldstone, 1986).

Medical Virology 10, Edited by L.M. de la Maza and
E.M. Peterson, Plenum Press, New York, 1991

TABLE 1. Estimated Risk of Transfusion-Transmitted Viral Infection and Associated Disease in the United States in 1990[a]

	Asymptomatic Infection	Clinical Disease
NANB-PTH[b] (Pre-HCV test)	5%	2.5%
HBV-PTH	0.5%	0.125%
HCV-PTH (Post-HCV test)	0.5%	0.25%
CMV	9%	<1%
HTLV-I	0.3%	Rare Myelopathy
HIV	0.002%	0.002%

a These figures represent estimates of a transfusion recipient's risk of infection, based on nationally reported data (Moore et al. 1987). They are in part based on known or estimated virus incidence rates among donors and on the assumption that a typical transfusion recipient receives five units of blood from different donors.
b NANB-PTH: Non A, non B post transfusion hepatitis.

Profound changes in blood banking services in the United States followed the discovery of Australia antigen or HBsAg. The introduction of HBsAg screening, combined with the changeover from paid to voluntary blood donors, significantly reduced post-transfusion hepatitis (PTH). The remainder of PTH, characterized as a disease cluster termed non-A, non-B hepatitis (NANB), is predominantly caused by HCV, and is expected to decline with the recently introduced anti-HCV screening of donor blood. Infectious agents, such as HTLV-I/II and CMV, which are exclusively associated with leukocytes are avoidable risks of transfusion.

The impact of legal liabilities due to transmission of HIV-1 (causing AIDS in patients not in the known high-risk groups), has been the most dramatic driving force for the improved safety of blood transfusion. An unprecedented set of serologic assays is currently employed to screen every one of the millions of units of blood with tests for syphilis, HBsAg, alanine aminotransferase (ALT), anti-HBc, anti-HIV-1/2, anti-HTLV-I/II, and anti-HCV. This effort is 60-90% effective in preventing transmission of BBVs. For example, HBV transmission has been reduced by 70% as a result of HBsAg screening (i.e. PTH due to HBV dropped from 30% to 10% of all PTH; Vyas and Blum, 1984). Similarly, anti-HCV screening is only 60% effective in preventing PTH (Esteban et al. 1990). Although specific effectiveness of anti-HIV-1 screening in preventing posttransfusion AIDS is not clearly established, the current incidence of anti-HIV-1 positive blood donors ranges from 1:10,000-20,000. In contrast, the estimates of HIV-1 infection in seronegative blood donors range between 1:50,000-150,000 (Busch et al. 1990). Thus, we can estimate that the anti-HIV-1 screening of donor blood is currently 80-90% effective. Similarly, HTLV screening is 65% effective (Kamahira et al. 1987). To further improve the safety of our blood supply, two distinct but complimentary approaches to research in screening and leukocyte filtration are necessary. Serologic screening of each unit of blood, coupled with a routine removal of unwanted leukocytes from each unit of blood soon after it is collected by blood centers, could significantly reduce the bioburden of transfusion-transmitted BBVs. The list of pretransfusion screening tests of donor blood shown in Table 2 is steadily growing, and consequent to the performance of these multiple assays at blood collection centers, there is an increase in the number of human errors. Such errors have become a source of serious liability issues increasingly encountered by the blood centers.

The current technology of enzyme-linked immunoassays (EIA) has gradually superseded the radioimmuno assays (RIA) and has proven to be very sensitive, rapid, simple and economical for large-scale screening of donated blood. However, false-positive and equivocal EIA observations are not uncommon (but at an acceptable level), requiring specificity testing using confirmatory assays. The technology of polymerase chain reaction (PCR) for direct detection of infectious agents is rapidly becoming a useful confirmatory test. Serological screening and confirmation of seropositive reactions has been time-consuming and costly. Therefore, research and development of alternative technologies, including PCR, which can be fully automated has been a subject of our major research effort. Before describing this effort, it is essential to describe each test as it is individually performed at the present time in blood collection facilities.

SEROLOGICAL SCREENING OF BLOOD FOR BBVs

Syphilis Serology

The serologic test for syphilis (STS), introduced more than four decades ago, has been the first test employed for preventing transfusion-transmitted infection (Felman, 1989). Paradoxically, the STS test routinely performed on donor blood does not prevent transmission of syphilis because the test does not become positive until long after the brief period of spirochaetemia and the potential infectivity of *Treponema pallidum*. The test has a high incidence of biological false positive reactions. However, the test is not removed from the list of screening assays because it has a possible clinical benefit of identifying donors with high risk of sexually transmitted diseases, including HBV, that may occasionally not be detectable by the HBsAg screening assays.

TABLE 2. Screening tests for blood donations mandated by the U.S. Food and Drug Administration

Antibodies to *Treponema pallidum* (Syphilis)
Hepatitis B surface antigen (HBsAg)
Hepatitis B core antibodies (Anti-HBc)
Hepatitis C virus antibodies (Anti-HCV)
Alanine aminotransferase (ALT)
Antibodies to human immunodeficiency virus (Anti-HIV-1)
Antibodies to human T-cell lymphotropic virus (Anti-HTLV)
Antibodies to cytomegalovirus (Anti-CMV, optional for selected recipients)

HBsAg

The envelope protein of HBV originates from the S gene encoding 226 amino acids which in turn determine the HBsAg epitopes. The antigenic epitopes of HBsAg have a common 'a' determinant and mutually exclusive d/y and w/r determinants (Tiollais et al. 1981). Following HBV infection most humans produce anti-HBs directed against the 'a' epitope determined by at least 9 amino acid residues (#139-147) of the sequence of 226 amino acids (Tiollais et al. 1981; Vyas and Blum, 1984). By a sandwich type of immunoassay HBsAg/a is determined by human or animal antibodies produced against HBsAg. In fact, HBsAg in serum is the primary means of diagnosing acute or chronic HBV infection in man. The absence of immune response or immunologic tolerance specific for the HBsAg/a determinant leads to persistence of HBsAg in chronic HBV infection. There are over 200 million chronic carriers of HBsAg in the world, and at least a million Americans are chronic carriers of HBV. Thus, HBsAg was the first and the most important test, introduced in 1971 for routine screening of donated blood. The HBsAg test is used worldwide for the detection of carriers of HBV infection and for the prevention of its transmission from apparently healthy blood donors to transfusion recipients. An array of immunoassays, beginning with simple Ouchterlony's immunodiffusion, counterelectropheresis, complement fixation, hemagglutination, RIA and finally EIA have been employed for the detection of HBsAg in human serum. Currently, EIA is most widely used for the detection of HBsAg by blood centers across the U.S.A.

Anti-HBc

Anti-HBc appear in acute phase of HBV infection and persist long after recovery. It is a secondary serological marker of acute or chronic HBV infection. Because of its significant association with NANBH and HIV-1 infection in homosexual men, it was introduced as a surrogate test for the prevention of transmitting HIV-1 and HCV infection in the U.S.A. In view of the fact that specific tests are now available for anti-HIV-1 and anti-HCV, it would be propitious to consider retraction of the routine testing for anti-HBc performed on every unit of blood collected in the U.S.A.

Anti-HCV

A principal etiologic agent of NANBH has been the newly discovered HCV (Kuo et al. 1989). A serological assay for anti-HCV is expected to reduce non-A, non-B PTH by 60-80%. The etiologic agent is a flavi-like RNA virus that encodes the structural genes for HCV core and envelope proteins (Okamoto et al. 1990), and nonstructural proteins analogous to NS2, NS3, NS4 and NS5 of flaviviruses (Choo et al. 1989). An anti-HCV test for screening blood has been introduced recently (Kuo et al. 1989), and appears to reduce PTH by 60% (Esteban et al. 1990). Using the 5'-noncoding sequence preceding the structural part of HCV genome, we have designed a PCR assay that may very well serve as a direct means for confirming the presence of HCV in seropositive and seronegative blood donors transmitting HCV infection through transfusion (Romeo et al. submitted; Ulrich et al. 1990). Current EIA tests require recombinant immunoblot (RIBA) assay as a confirmatory test. Anti-HBc, anti-HIV-1 and AIDS-related measures for exclusion of high-risk blood donors has coincidentally reduced NANB PTH by at least 50% (from 10% to 5% of transfused patients). After anti-HCV testing, at least 0.5% of residual HCV is expected to be a major problem in transfusion medicine.

ALT

ALT elevation is a biochemical indicator or hepatocellular necrosis. In the absence of a specific test for agent(s) of NANB PTH, ALT elevation 2.0 X above the upper limit of the normal range has been used as a surrogate test. Elevated ALT has been used as a means to exclude units of blood that can potentially transmit NANB PTH. The assay employs a kinetic enzymatic conversion of NADH to NAD. The determination of ALT level suffers from methodologic variations. In the absence of definite and universal standards, elevation above 70 I.U. (the normal range is 5-35 I.U.) is considered a basis for exclusion of the unit from transfusion. Because it is a simple liver function test the ALT abnormality varies, yet is likely to remain as a marker of blood unacceptable for transfusion. A new predonation spot-test for ALT is under clinical evaluation and may prove to be useful at blood centers.

Anti-HIV-1

Antibodies to HIV-1 envelope glycoproteins gp 41, gp120 and gp 160 are most commonly found in the serum of asymptomatic, apparently healthy individuals infected with HIV-1. The antibodies to the envelope protein reach high levels in early HIV-1 infection and remain persistently elevated throughout the spectrum of HIV-1 disease. In addition, HIV-1-p24 antigen and antibodies are variably found during the natural history of HIV-1 infection in man. In fact, prior to the appearance of any anti-HIV-1, the HIV-1-p24 antigen appears transiently in blood and disappears with the appearance of anti-HIV-1-p24. Later during HIV-1 infection when overt disease is manifested, the anti-HIV-1-p24 declines with concomitant reappearance of the p24 antigen. At this stage an HIV-1-infected person may be sufficiently ill to be unable to donate blood. While HIV-1-p24 antigen has proven to be of little or no value in screening donated blood, anti-HIV-1-gp41 is most useful in identifying infected units of blood (Scillian et al. 1989). Routine EIA reactive specimens are required to be confirmed by immunoblot or Western blot (WB) assays. Both HIV-1 and HIV-2 (uncommon in the U.S.) share their core antigens and are known to cause HIV-1 disease (AIDS and ARC). A PCR-based confirmatory assay may be useful in the future (Kwok et al. 1987; Ou et al. 1988; Simmonds et al. 1990).

Anti-HTLV

Antibodies to HTLV I/II are more commonly encountered among blood donors than anti-HIV-1 (Menitove, 1989; Okochi and Sato, 1986). Because HTLV-I causes adult T-cell leukemia and is associated with tropical spastic paraperesis or myelopathy, the tests for anti-HTLV-I/II were introduced in 1988 for routine screening of donated blood. The routine EIA results must be confirmed by immunoblot or WB. A PCR-based confirmatory assay for HTLV may also be useful in the future.

Anti-CMV

Anti-CMV test is positive in 30-50% of the blood donors. A positive test is considered a marker of potential infectivity. Thus, blood for transfusion to seronegative immunocompromised patients and premature neonates must be selectively screened and only seronegative units of blood transfused in order to avoid the common CMV infection that produces severe disease in immunocompromised patients (Adler, 1983). An IgM anti-CMV test is diagnostic of acute infection with CMV.

IMMUNOREACTIVE BEAD ASSAY FOR BBVs

The problems of multiple EIA screening assays introduced in blood banks led us to innovate the next generation of screening assays using anti-HIV-1 detection by fluorescent antiglobulin test as a prototype (Scillian et al. 1989). A flow cytometric immunoreactive bead (IRB) assay was developed for quantitative, simultaneous, and early detection of antibodies to HIV-1. Polystyrene beads of four diameters, each size coated with a different HIV-1 recombinant DNA-produced protein (p24, p31, gp41, or gp120), bound anti-HIV-1 antibodies that were detected with fluorescent antiglobulin (Figure 1). The IRB assay was performed on a panel of blood donor samples, many giving consistently false positive EIA and indeterminate WB results. The IRB assay proved as sensitive and more specific than currently licensed EIA and WB tests. Results on serial samples from 8 HIV-1-infected individuals indicated that quantitation of anti-p24 by IRB assay may be useful in monitoring disease progression. Sequential pre- and post-EIA seroconversion sera from 35 HIV-1-infected homosexual men were tested by the IRB assay using IgM- and IgG-specific fluorescent antibody probes. All 35 cases were positive by IRB assay for at least one rDNA-p either prior to (17/35, 49%) or at the time of EIA positivity. Eleven cases (31%) initially had only IgM anti-HIV-1, primarily to gp41 (17%). In two individuals

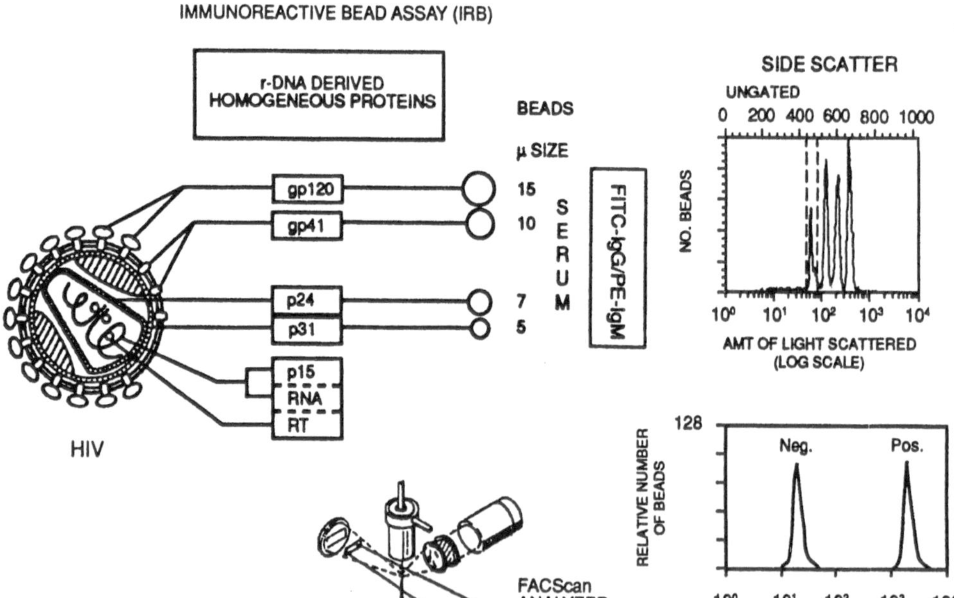

Figure 1. Flow cytometric immunoreactive bead assay for automated blood screening.

the IgM response was detected at least 18 months prior to EIA seroconversion. The results of IgM anti-HIV-1 are illustrated in Figure 2. The IRB assay is a widely applicable analytical procedure, not only useful in pretransfusion anti-HIV-1 screening of blood, but also useful for simultaneous detection of the multiple analytes listed in Table 2.

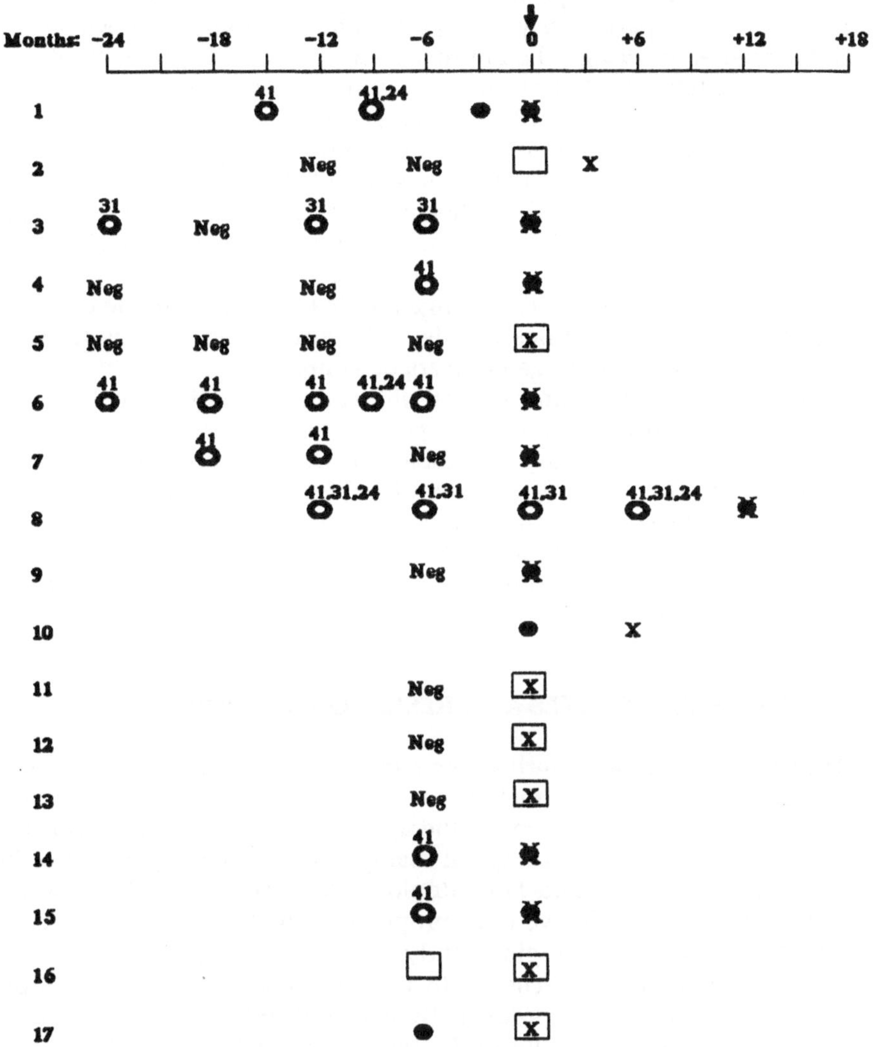

Figure 2. IgM and IgG anti-HIV-1 rDNA-p response on sequential sera from individuals, pre- and post-EIA seroconversion. Sera were assayed for IgM and IgG anti-HIV-1 rDNA-p reactivity by the IRB assay using FITC-anti-IgM or -IgG, Fc specific reagents. The origin (▼) marks the relative points in time at which sera were positive for HIV-1 antibodies by EIA. The months before (-) and after (+) EIA seroconversion are indicated on the scale (top). Relative times at which sera were confirmed positive by WB or immunofluorescence assay (X), when positive by IRB for one or more antigens [IgM only (O), IgM and IgG (●), IgG only (□)], or when negative for all study assays (Neg) are also indicated. Above the symbols for samples positive for IgM only (O) are shown the specific anti-HIV-1 rDNA-p reactivities found (p31 = 31, gp41 = 41, p24 = 24).

The IRB study by Scillian et al. (1989) demonstrates the importance of detecting gp41 specific antibodies in HIV-1 serodiagnosis. All HIV-1 positive individuals studied were IRB reactive for anti-gp41, generally at high levels. In 10/35 seroconvertors, IgM anti-gp41 was either the earliest antibody detected (6 cases), or was found with other IgM anti-HIV-1 specificities. Gaines et al. (1987) showed that the order of appearance of anti-HIV-1 antibody specificities was assay dependent with the earliest reactivity detected by radioimmunoprecipitation to gp160 and by WB to p24. Allain et al. (1986) showed that anti-gp41 appeared before anti-p24 in an rDNAp EIA assay, as was seen in this study. Others have shown that false positive reactivity to gp41 is unusual compared to other HIV-1 proteins, particularly p24 (Jackson and Balfour, 1988; Kleinman et al. 1988; Mitchell et al. 1988). The improved sensitivity and specificity of newer assays, which utilize selected rDNA-p or synthesized gp41 determinants is becoming increasingly apparent (Burke et al. 1987; Gnann et al. 1987).

The use of microspheres for flow cytometric based immunoassays has been previously described (Horan and Wheeless, 1977; McHugh et al. 1986, 1988; Saunders et al. 1985; Wilson and Wotherspoon, 1988). This is the first demonstration of the utility of the IRB assay for the simultaneous analysis of more than two specific antibody responses to an infectious agent (HIV-1). The procedure can be adapted to detect HIV-1 antigen, HIV-1-specific immune complexes, and other blood-borne infectious agents. The automation of this procedure will permit its use in testing large numbers of samples.

The principles of simultaneous determination of multiple analytes using the IRB procedure and homogenous antigens or antibodies, and the data from the foregoing studies, are broadly applicable to each of the tests listed in Table 2. A collaborative effort between University of California, Trans-Med Biotech Inc., and the National Heart, Lung and Blood Institute is envisaged to validate the practicality of an automated system of instrumentation and IRB reagents designed for simultaneous performance of the multiple assays required for screening our blood supply.

FILTRATION OF LEUKOCYTES AS A RESERVOIR OF BBVs

Blood components and derivatives such as packed erythrocytes, platelet concentrates, plasma and its fractions, e.g. the antihemophilic factor (Factor VIII) among others, are an enormous therapeutic resource and their functional integrity must be preserved through all manipulations for the removal or inactivation of BBVs (Prodouz and Fratantonio, 1988). To render these products safer than currently possible, we have proposed a combinatorial strategy to remove/inactivate BBVs. The original study design and progress made are illustrative of our research approaches with potential for practical application in providing safer cellular blood products (Rawal et al. 1989). Because leukocytes are the principal reservoir of the BBVs, their removal from blood is a logical approach to reducing the infectious bioburden in transfusion. Further, the uncommon but fatal transfusion-associated graft-versus-host (GvH) disease, due to HLA haplotype sharing (Atkinson et al. 1990; Thaler et al. 1989), has accentuated the controversial need for routine removal of leukocytes from blood donated at blood banks (Atkinson et al. 1990; McCarthy and Baldwin, 1989; Meryman, 1989; Rawal et al. 1990a, b; Sirchia et al. 1987; Vogelsang, 1990; Vyas, 1988). Because leukocytes in transfused blood have no therapeutic value, their removal alleviates the immunological risks of alloimmunization and consequent febrile reactions (Vogelsang, 1990). Incentive for removing leukocytes stems from the possibility that we could accrue a dual benefit of reducing both immunological and virologic risks of transfusion (Rawal et al. 1990a). Any process aimed at reducing transfusion-transmitted virus infection must include elimination of vector leukocytes to minimize the contaminating virus load in blood for transfusion. Using a new generation of filters, we have shown that leukocytes can be removed below the level of < 4 cells/ml blood using adsorptive filtration (Rawal et al. 1990b). We are also pleased to learn that our work has led two major companies (Pall Corporation and Cutter Laboratories) to jointly produce a new blood collection and component preparation set with built in filters, enabling blood banks to provide leukocyte-free blood products as a clinical choice. Leukocyte depletion also enhances erythrocyte survival during storage (Greenwalt, personal communication, Davey et al. 1989). Therefore, leukocyte removal not only reduces the immunological/virological risks of transfusion, but improves red cell preservation.

RECEPTOR-MEDIATED REMOVAL AND INACTIVATION OF BBVs

For absolute removal of any residual leukocytes, the leukocyte-depleted blood may be treated with specific MoAbs against human leukocyte CD45 (Becton Dickinson Immunocytometry System: Antileukocyte CD45 monoclonal antibodies Cat # 7465) using the principle of affinity chromatography (Lupien, 1976). Cell-free HIV-1 and circulating immune complexes (CIC) of HIV-1 could be similarly complexed with homologous MoAbs, including anti-HIV-1 (gp120), and/or anti-C1q. Alternatively, purified C1q itself may be used as a receptor for CIC. Since HBV selectively binds to polymerized human serum albumin (pHSA), it may be a candidate for receptor-mediated removal of HBV (Milich et al. 1981; Pontisso et al. 1983). We intend to determine if HCV can be removed by envelope-specific antibodies, and monitor its removal by PCR (Ulrich et al. 1990). We anticipate that molecularly cloned and expressed receptor molecules may become available for receptor-mediated removal of several BBVs by immunoaffinity. If needed, a final step for neutralizing the infectivity of BBVs could be provided by a biocompatible chemical of proven antimicrobial activity (Bissell et al. 1980, Rawal, 1978).

SUMMARY AND PROJECTIONS

Detection of infection with BBVs transmitted by the blood donated by apparently healthy volunteers is facilitated by several laboratory tests employed in screening out infected blood. Because several of the screening assays are immunological markers of host response to the infectious agents, the detection of viral nucleic acids remains a major goal that may possibly be fulfilled by the advent of PCR.

A conceptual framework of the complex host-virus interaction is finally depicted in Figure 3. Leukocytes in homologous blood transfusions are known to transiently suppress the hemopoetic and immune systems for a period of 2-3 weeks (Perkins, 1988; Vyas, 1988). Further, leukocytes are the primary vectors of human retroviruses and herpes viruses (Rawal et al. 1990a, b). Virus-induced suppression of the immune system may have an additive influence in establishing persistent infection (Southern and Oldstone, 1986). Latent viruses

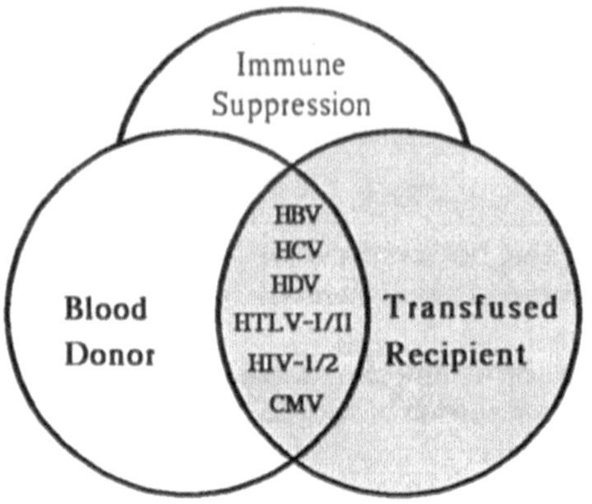

Figure 3. Conceptual framework of the host-virus interaction of transfusion-transmitted viral infections.

in leukocytes of blood donors may become actively expressed as a consequence of allogenic stimulation. The transiently induced immunosuppression in recipients of blood facilitates infection of the host by viruses activated from donor's leukocytes. Thus, it stands to reason to completely remove leukocytes that are therapeutically not needed and potentially harmful. Because of their roles both as vectors of cell-associated viruses and as mediators of immune suppression, their removal can only improve the safety of blood transfusion. This approach to the removal of vector leukocytes can be augmented with receptor-mediated removal and chemical inactivation of BBVs to make blood transfusions safer than ever before.

REFERENCES

Adler S (1983) Transfusion-associated cytomegalovirus infections. Rev Infect Dis 5:977-993.

Allain J-P, Laurian Y, Paul DA, Senn D and Members of the AIDS-haemophilia French study group (1986) Serological markers in early stages of human immunodeficiency virus infection in haemophiliacs. Lancet 2:1233-1236.

Atkinson K, Horowitz MM, Gale RP, van Bekkum DW, Gluckman E, Good RA, Jacobsen J, Kolb HJ, Rimm AA, Ringden O (1990) Risk factors from chronic graft-versus-host disease after HLA-identical sibling bone marrow transplantation. Blood 75:2459-2464.

Bissell MJ, Hatie C, Farson DA, Schwarz RI, Soo W (1980) Ascorbic acid inhibits replication and infectivity of avian RNA tumor virus. Proc Natl Acad Sci USA 77:2711-2715.

Bove JR (1987) Transfusion-associated hepatitis and AIDS. What is the risk? N Engl J Med 317:242-245.

Burke DS, Brandt BL, Redfield RR, Lee T, Thorn RM, Beltz GA, Hung C (1987) Diagnosis of human immunodeficiency virus infection by immunoassay using a molecularly cloned and expressed virus envelope polypeptide. Ann Intern Med 106:671-676.

Busch MP, Eble B. Heilbron D, Vyas G (1990) Risk associated with transfusion of HIV-antibody-negative blood (Letter). N Eng J Med 322:322-323.

Choo Q-L, Kuo G, Weiner AJ, Overby LR, Bradley DW, Houghton M (1989) Isolation of a cDNA clone derived from a blood-borne non-A, non-B viral hepatitis genome. Science 244:359-362.

Davey RJ, Carmen RA, Simon TL, Nelson EJ (1989) Preparation of white cell-depleted red cells for 42-day storage using an integral in-line filter. Transfusion 129:496-499.

Esteban JI, Gonzalez A, Hernandez JM, Viladomiu L, Sanchez C, Lopez-Talavera JC, Lucea D, Martin-Vega C, Vidal X, Esteban R, Guardia J (1990) Evaluation of antibodies to hepatitis C virus in a study of transfusion-associated hepatitis. N Engl J Med 323:1107-1112.

Felman YM (1989) Syphilis: from 1495 Naples to 1989 AIDS. Arch Derm 125:1698-1699.

Gaines H., Von Sydow M, Sonnerborg A, Albert J, Czajkowski J, Pehrson PO, Chiodi F, Moberg L, Fenyo EM, Asjo B, Forsgren M (1987) Antibody response in primary human immunodeficiency virus infection. Lancet 1:1249-1253.

Gnann JW, Schwimmbeck PL, Nelson JA, Truax AB, Oldstone MBA (1987) Diagnosis of AIDS by using a 12-amino acid peptide representing an immunodominant epitope of the human immunodeficiency virus. J Infect Dis 156:261-267.

Horan PK, Wheeless LL Jr (1977) Quantitative single cell analysis and sorting. Science 198:149-157.

Jackson JB, Balfour HH, Jr. (1988) Practical diagnostic testing for human immunodeficiency virus. Clin Microbiol Rev 1:124-138.

Kamihira S, Nakasima S, Oyakawa (1987) Transmission of HTLV-I by blood transfusion before and after mass screening of sera. Vox Sang 53:43-44.

Kleinman S, Fitzpatrick L, Secord K, Wilke D (1988) Follow-up testing and notification of anti-HIV Western blot atypical (indeterminant) donors. Transfusion 28:280-282.

Kuo G, Choo Q-L, Alter HJ, Gitnick GL, Redeker AG, Purcell RH, Miyamura T, Dienstag JL, Alter MJ, Stevens CE, Tegtmeier GE, Bonino F, Colombo M, Lee W-S, Kuo C, Berger K, Shuster JR, Overby LR, Bradley DW, Houghton M (1989) An assay for circulating antibodies to a major etiologic virus of human non-A, non-B hepatitis. Science 244:362-364.

Kwok S, Mack DH, Mullis K, Poiesz B, Ehrlich G, Blair D, Friedman-Kien A, Sninsky JJ (1987) Identification of human immunodeficiency virus sequences by using *in vitro* enzymatic amplification and oligomer cleavage detection. J Virol 61:1690-1694.

Lenfant C, McCurdy PR (1990) The National Heart, Lung and Blood Institute and Transfusion Medicine (Editorial). Transfusion 30:97-100.

Lupien P-J, Moorjani S, Awad J (1976) A new approach to the management of familial hypercholesterolaemia: Removal of plasma - cholesterol based on the principle of affinity chromatography. Lancet 1:1261-1264.

McCarthy LJ, Baldwin ML (eds.) (1989) Controversies of leukocyte-poor blood and components. Arlington, VA: American Association of Blood Banks.

McHugh TM, Stites DP, Casavant CH, Fulwyler MJ (1986) Flow cytometric detection and quantitation of immune complexes using human C1q-coated microspheres. J Immunol Meth 95:57-61.

McHugh TM, Miner RC, Logan LH, Stites DP (1988) Simultaneous detection of antibodies to cytomegalovirus and herpes simplex virus using flow cytometry and a microsphere based fluorescent immunoassay. J Clin Micro 26:1957-1961.

Menitove JE (1989) The decreasing risk of transfusion-associated AIDS. N Eng J Med 321:966-968.

Meryman HT (1989) Transfusion-induced alloimmunization and immunosuppression and the effects of leukocyte depletion. Transfusion Med Rev 3:180-193.

Milich DR, Gottfried TD, Vyas GN (1981) Characterization of the interaction between polymerized human albumin and hepatitis B surface antigen. Gastroenterology 81:218-225.

Mitchell R, Dow BC, Barr A, Follett EAC (1988) False positive anti-HIV tests on blood donations. Lancet 1:297-298.

Moore DJ, Bucens MR, Holman CD, Ott AK, Wells JI (1987) Prenatal screening for markers of hepatitis B in aboriginal mothers. Med J Australia 147:557-558.

Okamoto H, Okada S, Sugiyama Y, Yotsumoto S, Tanaka T, Yoshizawa H, Tsuda F, Miyakawa Y, Mayumi M (1990) The 5'-terminal sequence of the hepatitis C virus genome. Japan J Exp Med 60:167-177.

Okochi K, Sato H (1986) Transmission of adult T-cell leukemia virus (HTLV-I) through blood transfusion and its prevention. AIDS Res 3:3157-3161.

Ou C-Y, Kwok S, Mitchell SW, Mack DH, Sninsky JJ, Krebs JW, Feorino P, Warfield D, Schochtetman G (1988) DNA amplification for direct detection of HIV-1 in DNA of peripheral mononuclear cells. Science 239:295-297.

Perkins HA (1988) Transfusion-induced immunologic unresponsiveness. Transfusion Med Rev 2:196-203.

Pontisso P, Alberti A, Bortolotti F, Realdi G (1983) Virus-associated receptors for polymerized human serum albumin in acute and in chronic hepatitis B virus infection. Gastroenterol 84:220-226.

Prodouz KN, Fratantonio JC (1988) Inactivation of virus in blood products: Editorial. Transfusion 28:2-3.

Rawal BD (1978) Bactericidal action of ascorbic acid on *Pseudomonas aeruginosa* - alteration of cell surface as possible mechanism. Chemotherapy 24:166-172

Rawal BD, Busch MP, Endow R, Garcia-de-Lomas J, Perkins HA, Schwadron R, and Vyas GN (1989) Reduction of human immunodeficiency virus-infected cells from donor blood by leukocyte filtration. Transfusion 29:460-462.

Rawal BD, Davis RE, Busch MP, Vyas GN (1990a) Dual reduction in the immunologic and infectious complications of transfusion by filtration/removal of leukocytes from donor blood soon after collection. Transfusion Med Rev 4:36-41.

Rawal BD, Schwadron R, Busch MP, Endow R, Vyas GN (1990b) Evaluation of leukocyte removal filters modelled by use of HIV-infected cells and DNA amplification. Blood 76:2159-2161.

Saunders GC, Jett JH, Martin JC (1985) Amplified flow-cytometric separation-free fluorescence immunoassays. Clin Chem 31:2020-2023.

Scillian JJ, McHugh TM, Busch MP, Tam M, Fulwyler MJ, Chien DY, Vyas GN (1989) Early detection of antibodies against rDNA-produced HIV proteins with a flow cytometric assay. Blood 73:2041-2048.

Simmonds P, Belafe S, Peutherer J, Ludlam CA, Bishop JO, Brown AJL (1990) Human immunodeficiency virus-infected individuals contain provirus in small numbers of PBMCs and at low copy number. J Virol 64:864-872.

Sirchia G, Rebulla P, Paravicini A, Carnelli V, Gianotti GA, Bertolini F (1987) Leukocyte depletion of red cell units at the bedside by transfusion through a new filter. Transfusion 27:402-405.
Southern P, Oldstone MBA (1986) Medical consequences of persistent viral infections. N Engl J Med 314:359-366.
Thaler M, Shamiss A, Orgad S, Huszar M, Nussinovitch N, Meisel S, Gazit E, Lavee J, Smolinsky A (1989) The role of blood from HLA-homozygous donors in fatal transfusion-associated graft-versus-host disease after open heart surgery. N Engl J Med 32:25-28.
Tiollais P, Charnay P, Vyas GN (1981) Biology of hepatitis B virus. Science 213:406-411
Ulrich PP, Romeo JM, Lane PK, Kelly I, Daniel LJ, Vyas GN (1990) Detection, semiquantitation, and genetic variation in hepatitis C virus sequences amplified from the plasma of blood donors with elevated alanine aminotransferase. J Clin Invest 86:1609-1614.
Vogelsang GB (1990) Transfusion-associated graft-versus-host disease in non-immunocompromised hosts. Transfusion 30:101-103.
Vyas GN, Blum HE (1984) Hepatitis B virus infection: Current concepts of chronicity and immunity. West J Med 140: 754-762.
Vyas GN (1988) Transfusion-associated infections and immune response: Guest editorial. Transfusion Med Rev 4:193-195.
Wilson MR, Wotherspoon JS (1988) A new microsphere based immunofluorescence assay using flow cytometry. J Immunol Meth 107:225-230.

DISCUSSION

Mendelson E (The Chaim Sheba Medical Center, Tel-Hashomer, Israel):

The fact that you find a little piece of DNA or RNA in a blood product by PCR, does it mean that there is an infectious agent there?

Vyas G:

That is a very good question, I appreciate it. PCR can amplify dead DNA which is incapable of replicating. However, the way we are performing, it's actually the immunoabsorbtion of a virion. For example, anti-HBs will only adsorb intact virion, that has an HBsAg on the envelope. When you get the virus, the nucleic acid inside is protected and stable to storage conditions, etc. The ultimate test is a biological test for HBV infectivity in a chimpanzee model. We have shown in my lab (Journal of Infectious Disease) that PCR and chimpanzee infectious dose are exceptionally correlated. Actually, it's slightly more sensitive than biological infectivity dose. We know that in biology, if you introduce one virus, the chance of that causing infection is not 100%. We know that we can detect 10-20 virions by these methods, and those virions have been shown to be infectious. The way we detect it suggests that these are infectious virions. Also, in plasma, if there was naked DNA or RNA floating around, there are endonucleases that are naturally present in plasma that would chew it up. It's the envelope and the core of the virus that protects it from being chewed up. What we are doing is establishing correlation between biologic infectivity and PCR. When we employ PCR as a substitute for chimpanzee infectious dose, can we do it as a substitute for human research or deliberate infection? Balayan is a very famous Russian virologist who decided to establish the hepatitis E virus infectivity by preparing a soup of the fecal matter. He drank it himself and produced infection. Now these kind of experiments, I think, in 1990 we don't need to do. Your point is well taken that ultimately, biologic infection within some animal models has to be established.

Ehrlich G (University of Pittsburgh, Pittsburgh, PA):

I just wanted to make a couple of comments. We've recently shown that HTLV-II and HIV-1 can be found, at least PCRable, in cell-free plasma and serum from infected individuals. Also, in regards to the problem that you had with protein inhibitors when you're doing lysis of the samples, we found that if you boil these samples for an extended period of time which, admittedly, is a little work but less work than doing DNA extractions, that you can get just as good, if not better, sensitivity by doing a 15 minute boiling after you do the lysis rather than have to go through the organic extractions.

Vyas G:

I totally agree with you that boiling is a simple method. For extraction usually of purified DNA, boiling is a very simple method of denaturing the proteins. The problem you run into is when you take 2 or 3 ml serum and boil it, and actually if you quantitate then compare the virion titers by DNA extraction versus boiling, there is a 60-70% loss of HBV. It doesn't matter, 60-70% of three million particles doesn't matter. This boiling method we have been using routinely for three or four years. I learned about it in Sardinia. But, your point is well taken. The second point I want to make in response to Dr. Mendelson, if you found HTLV-I or II in plasma, I can guarantee that it is not infectious. The reason for that is there is an enormous amount of data in Japan that factor VIII and fresh frozen plasma are not infectious for HTLV I and II. So, when you collect blood, you lyse some cells and the DNA that comes out will still be floating around. So a piece of naked DNA that is there for HTLV-I in plasma, you could amplify and get a true PCR positive, but that is biologically not significant. The two questions are complimentary.

Ehrlich G

What we were finding was viral associated RNA. Now, whether it was infectious or not, I don't know. But what we did was to DNAse the samples and do a reverse transcriptase and a retroviral type of pelleting.

Vyas G:

If you had shown that it was endogenously reverse transcriptase positive, I would be quite convinced.

Ehrlich G:

I'm not saying that it was reverse transcriptase positive. What we did is we DNAsed samples to make sure that there was no cellular DNA there. Then we used reverse transcriptase to make a DNA copy of, I presume, the RNA which was there and then PCR'ed that.

Vyas G:

I think that the viruses such as HTLV-I and II are cultureable so that we have a biological system to evaluate infectivity, which is not possible with hepatitis B virus.

Kilbourne E (Mount Sinai School of Medicine, New York, NY):

Going back to the matter of biological sensitivity in the assay. In the first place, it may be arguable whether that is a definitive objective. To use an unelegant story, if there is horse feces around, there must be a pony around here somewhere. In the sense that if one does get an authentic fragment of viral RNA or DNA identified, then one would have to assume that somewhere in the body, replication is going on. The other point is simply to ask directly, how good the evidence is that, say one human ID-50 equals one chimpanzee ID-50. Isn't it possible that there is an insensitivity even at that biological system, that it may take 100 human infectious doses to infect a chimpanzee.

Vyas G:

There are no absolute data available, but the data are available that what is infectious in humans is infectious in chimpanzee in a qualitative way. That is, 100% correlation between chimps and human infectivity has been demonstrated for hepatitis B virus. What Dr. Krugman did in the children at the Willowbrook hospital was also infectious in chimpanzee model. Quantitatively, you may be right. The infectious dose could be 100 times more in humans than in chimpanzee, and I don't think we have any real data to make any comment on that. Now we have very powerful molecular biology tools with which you can almost quantitate the number of virion particles per microliter. The correlation between chimp infectious dose and the number of virus particles is fairly good. That is 10 virus particles produce an infection in chimp.

Minocha H (Kansas State University, Manhattan, KS):

Thank you for your very inspiring talk. In the past I have had two friends, one who is working with me at Kansas State University, who had a bypass about six months ago. Before he had a bypass, he asked the doctors if he needed blood transfusions, and he was told that there was no need of blood transfusions. It happened that he had to have blood transfusions during his bypass surgery. They gave him whatever blood they had, and he was in the hospital for about two months because he ran temperature and fever. It was presumed that it was due to the viruses in the blood. What are those particular viruses in the blood that physicians think of these days? This is not the only incidence, probably. There must be many. I know of two of those.

Vyas G:

Fortunately, I am not a physician, so I don't have an answer to your question. I'll pass it on to Dr. Lennette.

Lennette E (California Public Health Foundation, Berkeley, CA):

I'll pass it on to Kilbourne.

Vyas G:

I can only quote from my humble background. The fear of unknown is so enormous, that it is best not to be afraid of it. In other words, there is so much that we don't know, we don't even have the foggiest notion of, what we don't know. I frankly think that the number of viruses, and that is where I left off, by saying that non-specific methods of inactivating viruses, if it is compatible with blood cells and so forth, would be the ultimate rescue. I'm not sure a) that it is realistic, b) that it is needed. As Dr. Kilbourne said very correctly, one has to put these things into perspective as to how far we can afford to go as a society. Today we may not have the budget and tomorrow half of our government employees may be jobless. What I'm saying is, that one has to put into perspective right now, that the country's worry is driven by HIV phenomenon. One in a million risk is not acceptable to this society. That to me is crazy. I am free as an American citizen to think that is crazy, and I say so. But the truth is, that the rest of society does want and expect that if it can be prevented, at any cost, do so. The next thing we worry about is can we reduce the cost so that it is practical? That is the mood of the country.

Minocha H:

I was thinking that since this thing must have happened a number of times, it's just a matter of checking the blood before the transfusion is given, and after the transfusion, what kind of antibodies appear in that blood so that they could be related to certain viruses if they were there. That's what I was thinking about.

Vyas G:

I think that known agents are there. The list that I gave you, that is the known transfusion-transmitted infections. But unknowns, there must be a lot of endogenous retroviruses, some of them may be pathogenic, others may not be pathogenic.

Lennette E:

Wearing my clinical hat, I have a very simple answer to the question that you posed. If the patient was fearful of reaction to the transfusion just like he anticipated, his fears could have been allayed by autologous transfusion. It's very simple.

Vyas G:

That is correct. But the clinical syndrome of FUO, fear of unknown origin, still remains important in clinical medicine. I have been Chief of blood bank for 20 years at UCSF, and when the AIDS litigation era started, and after two law suites, I decided that I didn't need this job. I resigned as Chief of Blood Bank.

Berlin B (Michigan Department of Public Health, Lansing, MI):

I would like to go back to your particle sorter machine and its concept. The use of latex particles and other particles for detection of antigens or antibodies is an old one. One of the things we have noticed in doing these, is that agglutination of these particles occur. How do you avoid agglutination? If it does occur, do you do it by means of a particular window in your cell sorting system? If you do it by window, that would imply that there are clumps of particles off there somewhere. What could be in those?

Vyas G:

Agglutination phenomenon is a function of the antigen density per particle. For example, the reason anti-Rh does not agglutinate cells is simply because there are 50-100 thousand sites per erythrocyte. Whereas anti-A and anti-B agglutinates red cells, it's because there are 5-6 million sites per erythrocyte membrane. By engineering the particle protein at a level, what we call in a functional sub-agglutinating dose, you can avoid completely, agglutination reaction. Then, by anti-globulin reaction, come back and detect 200-300 molecules per particle. We know enough about quantitating immunochemistry in this regard. You are absolutely right that these concepts have existed. But polystyrene uniform size beads, these are the things that were not available. I was in Cleveland at Western Reserve where I trained, and I remember that the same company, which is in Norway, offered me those particles for any experimental immunochemical work that I could think of. I said, it's a waste of time. How naive I was. After 25 years, wisdom dawned on me.

Lennette E:

I would like to make a comment a propos to the statement you made. Dr. Vyas, in passing remarked that Blood Bankers are not Microbiologists, they're Immunologists. I mention this only because that has been the basis of my running feud with the FDA. At a consensus meeting in Kansas City two years ago, maybe three years ago, we had a group sitting at a table like this, and Jay Epstein, who is the Chief of the Blood Products Group in the Division of Biologics was on the Panel. Every participant had his chance to say things and I started out by saying "I am not here to praise FDA, I'm here to attack it". Which I did. My attack was on the concept that if you are dealing with a diagnostic test in a clinical laboratory, why in God's name do you have to follow on a full furrah that goes to show that a blood product is acceptable. You have to have 10,000 normal controls to show that your test is workable. You have to go through a lot of other antics. I say, instead of sending everybody in this direction, why don't we split off the microbiologist in the diagnostic lab and go in this direction. You know, I'm still fighting that battle. I think I am winning it though. I don't see why diagnostic labs should have to make this a poor diagnostic test. You should have to meet the same criteria that you have for a blood product.

Vyas G:

I totally agree with you. I can only say from personal experience that I was the first example, I think in the country, of a Ph.D. Microbiologist, being Chief of the Blood Bank at University of California Medical Center.

Lennette E:

Since you've got all this money, you're deluged with money left and right, why not spend a few bucks on artificial blood?

Vyas G:

Actually, there is a very important program at National Heart, Lung and Blood Institute, as well as what the Army has in place. The FDA committee on Blood and Blood Products had a sort of, I wouldn't call it a consensus conference, but a review of the state of the art, seven or eight countries had products and all the data were reviewed. I think that there is enough effort in that area. Fortunately, the all mighty dollar draws this. There is a biotechnology company in Denver where an M.D., Ph.D. wanted to come and be a resident at our place, and decided to become the vice president of this company. When recombinant DNA derived hemoglobin is made in *E. coli,* they turn pink. The colony turns pink because of oxygen in the air. The technology like that where recombinant hemoglobin is now produced *in vitro,* without stem cells and so forth, is very reassuring to me that it is a matter of time and effort. I think by putting in a billion dollars, we did not cure cancer. I don't think that by putting in another billion dollars, we'll have a blood substitute, but there is a reasonable intense program, both in the private sector and in the public sector, that is currently going on.

Lennette E:

Long before most of you were born, we used to have in this country what was called the Chitauqua Circuit where speakers would go from city to city delivering the talks on various subjects. We now have a Chitauqua Circuit in science, because Dr. Vyas is due to leave here shortly and speak this afternoon to the Clinical Chemists up in San Francisco. So, Godspeed. I hope you get there on time.

Vyas G:

I will.

PROGRAMMED ANTIGENIC STIMULATION - A NEW APPROACH TO INFLUENZA VACCINATION

Edwin D. Kilbourne and Bert E. Johansson

Mount Sinai School of Medicine
New York, New York, USA

INTRODUCTION: CATEGORICAL AND GENERIC OBJECTIVES OF ARTIFICIAL IMMUNIZATION

Although much is currently written about "new vaccines" and "new approaches to immunization" most contemporary approaches represent technologic developments in the synthesis or presentation of antigens rather than departure from the strategies of the past. The old categories of live and inactivated vaccines can readily accommodate present day recombinant (replicating) viruses (live) or "subunit" or oligopeptide vaccines, perhaps better characterized as non-replicating antigens. The primary and hallowed objective of most vaccines has been the prevention of disease by the prevention of infection. Exceptions are such vaccines as rabies or hepatitis B vaccine which may be given post-exposure and with which suppression of infection already acquired is the only possible objective.

Of course, in practice the goal of complete inhibition of infection is not always achieved and in fact disease may occur in some vaccinated subjects and asymptomatic infection in others.

Underlying the two generically different approaches to immunization are two challengeable assumptions: 1) that optimal immunity will be engendered by simulation of natural infection (i.e., with living agent vaccines) or 2) that viral attenuation is difficult and unreliable so that immunity must be stimulated with antigens incapable of replication and infection (inactivated or subunit vaccines). A combination of the two approaches has seldom been used and then usually without deliberate intent. A single example that comes to mind is the sequential immunization of adults with Salk (inactivated) and Sabin (live virus) vaccines to reduce the risk of paralytic poliomyelitis from the live virus vaccine.

In the present paper, we shall present evidence to suggest that 1) the prevention of infection need not be the primary objective of immunization and 2) antigen presented out of the context of association with other virion antigens may be more effective than when presented in the intact virion and 3) the sequential presentation of a single antigen followed by infection may eventually prove superior to either infection alone or whole inactivated virus vaccine followed by infection in the achievement of definitive immunity. Our studies have been specifically on the use of influenza virus neuraminidase (NA) in the prevention of influenza in man and experimental animals.

WHAT ARE THE PROBLEMS WITH CURRENT INFLUENZA VACCINES?

In the United States, formalin inactivated whole virus or subunit vaccines have been the bulwark of influenza prevention for 40 years. When matched with antigenically identical challenge virus, such vaccines are 75-90% effective (Couch et al. 1986; Hoskins et al. 1979; Kendal and Patriarca, 1986). Immunity induced with these vaccines, mediated principally by an IgG response to the hemagglutinin (HA) antigen, however, is only briefly sustained. Adding to the problem is the notorious antigenic variability of influenza A virus, the principal target of immunization (Kilbourne, 1987). It should be noted here that when contemporary influenza vaccines work, and they work very well, indeed, for nine to twelve months, they do so principally by the traditional strategy of prevention of infection. A further practical problem in the manufacture of contemporary inactivated virus vaccines is the need not only for the replacement of vaccine strains by new antigenic variants every one or two years, but the necessity for genetic reassortment of these strains with laboratory viruses for the production of higher yielding reassortant viruses containing the surface antigens of the new variants and internal proteins of the laboratory strain, which confer high yield properties.

Aggressive and imaginative research has led to the development of attenuated live virus vaccines for influenza, none of which is currently licensed in the United States. These experimental vaccines make use of temperature sensitive (ts) or cold-adapted (ca) mutants or reassortment with avian influenza A viruses to achieve attenuation (Kilbourne, 1987). It is fair to say that there is no present consensus concerning the superiority of live virus vaccines or non-replicating antigens either in short-term efficacy or duration of immunity. Nor do live virus vaccines obviate the problem of annual vaccine revision. Any assumption that the transfer of attenuating genes to new strains of virus will guarantee attenuation of new vaccine reassortants is untenable in view of the intrinsic mutability of the virus and evidence that the effects of new gene combinations are unpredictable, even leading to virulence greater than that of either parent (Scholtissek et al. 1979). Therefore, it seems likely that each new reassortant will require new trials of safety and efficacy before wide scale use.

As infective agents introduced into the respiratory tract, live virus vaccines have the potential for increasing susceptibility to secondary infections or of compromising pulmonary function in the chronically ill. The latter concern has been somewhat allayed by a recent study of cold-reassortant vaccines in asthmatics (Atmar et al. 1990). A further concern not adequately addressed by either of the present vaccine approaches is the lack of a non-reactogenic vaccine for infants and young children who are not only vulnerable to the effects of the virus but who serve as active transmitters of infection during epidemics.

Clearly a new approach to influenza immunization is needed that provides for the induction of more durable and broader gauged immunity by a safe procedure that is available to people of all ages.

IMMUNITY TO INFLUENZA

Immunity to influenza is principally the composite effect of exposure to the two external glycoprotein antigens of the influenza virus, the HA (the major antigen) and the NA (the minor antigen) (Figure 1). The probable and potential role of viral internal proteins in influenza immunization, especially as related to T cell participation, is discussed later in this chapter.

The separate and separable functions of HA and NA antigens have been clearly identified through the use of genetically engineered recombinant virus "antigenic hybrids" into which either antigen can be segregated (Figure 2).

Antibody to the HA neutralizes the virus and inhibits its entry into cells. When anti-NA antibody binds to a virus particle it does not neutralize the virus or prevent its attachment to the target host cells. Rather, anti-NA antibody acts in subsequent cycles of infection to aggregate virus particles, to bind them to cells or cell associated antigen, or to inhibit viral enzymatic activity directly (Figure 3). The result is a reduction in the total amount of virus produced. If sufficient antibody is present, then virus production may be reduced below the level required for causation of disease. The contrasting effects of anti-NA and anti-HA antibody are illustrated in a cell culture plaque reduction system. (Figure 4).

The infection-permissive but disease-limiting effect of anti-NA immunity has been demonstrated in mice (Schulman et al. 1968), chickens (Allen et al. 1971; Rott et al. 1974), ferrets (McLaren et al. 1974) and man (Beutner et al. 1979; Couch et al. 1974; Kilbourne, 1987; Murphy et al. 1972).

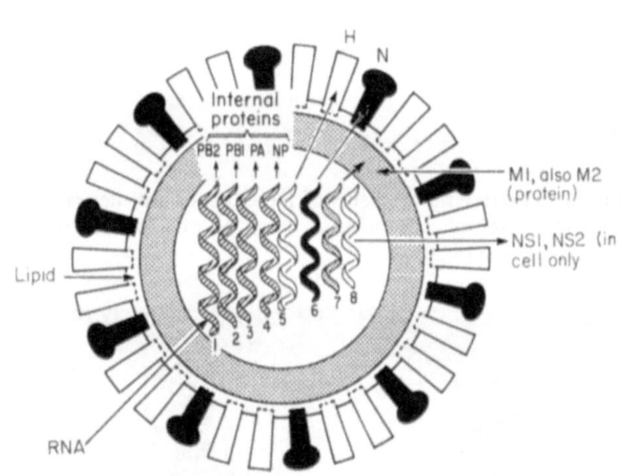

Figure 1. Schematic representation of the influenza virion, indicating the eight viral genes and their polypeptide products. The relative sizes of the RNA genes are only approximate, and the structural details of their relationship to one another and to the internal proteins P1-P3 and NP are not known. H, hemagglutinin glycoprotein; N, neuraminidase glycoprotein. Modified from Kilbourne (1979).

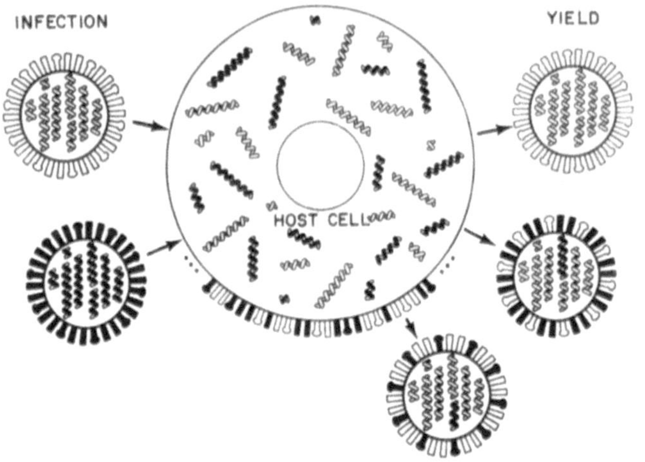

Figure 2. Intratypic genetic reassortment of influenza viruses, illustrating diagrammatically three of 256 possible reassortant progeny from infection with two parental viruses. Reprinted with permission from Kilbourne (1987).

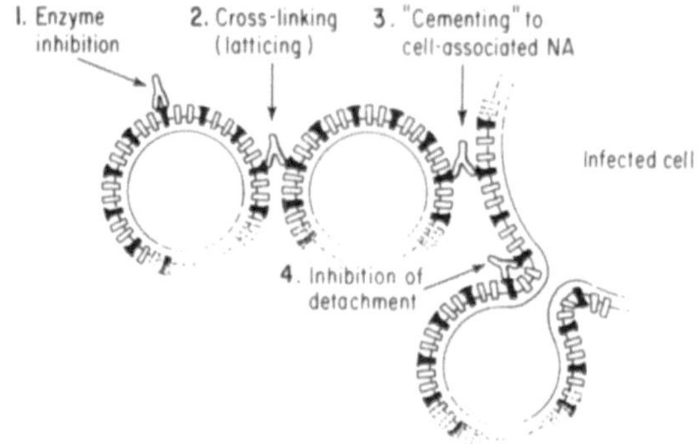

Figure 3. Putative mechanisms of the effect of anti-neuraminidase antibody. Reprinted with permission from Kilbourne et al. (1975).

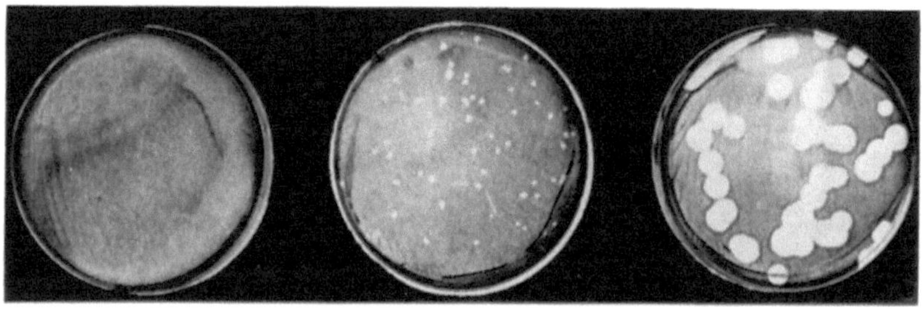

Figure 4. Differing effects of anti-HA and anti-NA antibodies on influenza virus plaque formation in clone 1-5C-4 monolayer cultures. Antiserum is incorporated into the agar overlay. Complete inhibition of plaque formation occurs with anti-HA (left); reduction in plaque size (PSR) with no reduction in plaque number occurs in the presence of anti-NA. Plaque development in the absence of specific antibody is shown at the right. Reprinted with permission from Kilbourne et al. (1975).

IF HA-INDUCED ANTIBODY PREVENTS INFECTION, WHY SETTLE FOR PARTIAL IMMUNIZATION WITH NA?

Conventional inactivated influenza virus vaccines contain both HA in and NA, and stimulate antibody production against both antigens. With conventional immunization the effect of anti-HA immunity is overriding and infection usually is prevented.

If infection is prevented, then disease is prevented, which is a good result in the "short-term" and represents the classical objective of most immunization procedures. But following conventional influenza vaccination, immunity is brief, ordinarily less than one year, and the virus, often in changed form, returns yearly as a recurrent threat. Hoskins et al. (1979) in studies in a boys' school, observed no difference in the cumulative attack rates of vaccinated and unvaccinated subjects over a period of years, and concluded that "annual vaccination with inactivated influenza A vaccines confers no long-term advantage."

It is the objective of NA-specific immunization to achieve both the short term advantage afforded by conventional HA-containing vaccine and the long-term gain which should result from infection-induced immunity. Preliminary evidence suggests that the proper dose of NA antigen can indeed prevent disease and yet allow immunizing infection to occur.

Naturally acquired anti-NA antibody has been correlated with resistance to challenge (Kilbourne, 1987; Kim et al. 1976) or natural (Kim et al. 1976) infection. The consensus that the Hong Kong (H3N2) pandemic of 1968 was less severe than the Asian (H2N2) pandemic of 1957 (Kilbourne, 1975) may well reflect the relative resistance of a population naturally immunized with the N2 NA antigen common to both causative viruses during the decade 1957-1968. (If this conclusion is correct, it also follows that natural immunization with NA was inadequate to forestall a major epidemic with a virus with a radically different HA antigen. However, infection with H2N2 virus in the year prior to exposure to H3N2 virus seemed to spare certain localities from severe epidemics (Gill et al. 1971; Rodin, 1969). Eickoff and Meiklejohn (1969) found that an adjuvant H2N2 vaccine gave 40% protection against the new H3N2 virus.

INFECTION-PERMISSIVE IMMUNIZATION

Our original concept of infection-permissive immunization was based on the premise that immunity conferred by natural infection is superior to that induced by either non-replicating antigens or genetically attenuated virus. Although our recent studies in mice suggest that we might have to modify this view, the assumption that infection is a definitive immunizing step remains central to our concept. This concept is outlined in Figure 5. As indicated, an NA-specific vaccine induces the formation of antibody which does not prevent infection following the first natural challenge but reduces viral replication below the pathogenic threshold; i.e., without disease. This first infection results in a secondary response to the NA antigen contained in the challenge virion and a primary response to the virion HA and other viral proteins. Subsequent challenge with the same virus will be met with solid immunity to infection as a result of NA priming and modulated prior infection with a wild-type virus.

VACCINES FOR INDUCTION OF NA-SPECIFIC IMMUNITY

Whole or Split Virus Vaccines

Whole or split virus vaccines in which the NA antigen is associated with an HA antigen which is "irrelevant" to human experience are readily produced as reassortant "antigenic hybrids" (e.g., H7N2, H7N1) (Figure 2) that incorporate HA antigens from animal strains with NA(s) of human viruses. Such vaccines already have been used in the prevention of experimental infection in extensive field trials (Beutner et al. 1979; Couch et al. 1974; Kilbourne, 1987) and require no "new drug" approval because they are produced exactly like conventional influenza vaccines.

Isolated Purified NA Antigen

NA is a large (56K) tetrameric glycoprotein that is separable from the influenza virion by detergent disruption of the particle and ion exchange chromatography (Gallagher et al. 1984; Johansson et al. 1989). In aggregated form (after removal by dialysis of virus splitting detergents) it is highly immunogenic in rabbits and mice. The gene for the NA has been cloned (Lai et al. 1980) and the protein in an enzymatically and antigenically active form can be expressed in large quantities by a baculovirus vector in insect larvae (*Trichoplusia ni*) or in insect cell culture (Price et al. 1989).

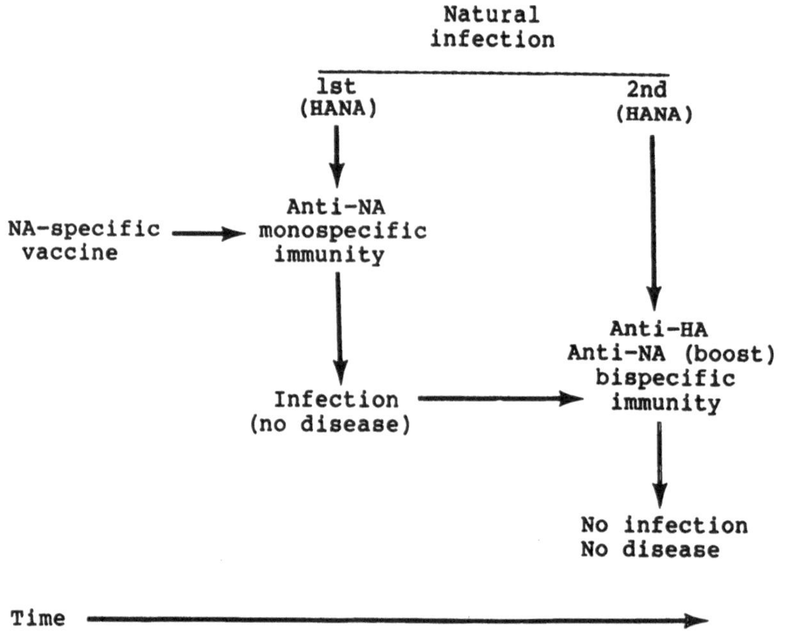

Figure 5. NA-specific vaccine allows initial infection to occur without disease. Boosting effect of infection then creates definitive immunity. Infection presents both HA and NA antigens (HANA). Modified from Couch et al. (1974).

NA Specific Immunization with Viral Antigenic Hybrids as Inactivated Whole Virus Vaccines

Couch et al. (1974) clearly demonstrated reduction of viral replication and protection by an H7N2 virus vaccine in volunteers challenged with H3N2 virus and also showed solid resistance to reinfection and illness in subjects challenged six months later with the original virus.

Ogra and his colleagues (1977) in a two year study of 875 Buffalo school children, compared the effects of two vaccines under conditions of natural exposure and successive challenge by the Port Chalmers and Victoria variants of H3N2 virus. The two vaccines, X-41 and X-42, were manufactured under identical conditions. X-41 contained both the HA and NA antigens of the Port Chalmers strain and hence was comparable to conventional inactivated vaccines. X-42 was an H7N2 antigenic hybrid possessing the irrelevant HA of an equine virus.

X-41 (H3N2) induced seroconversion for HA antibody in 90% of 270 seronegative children. X-42 (H7N2) converted 279 seronegative children with respect to anti-NA antibody. (Sixteen per cent of X-42 immunized subjects had some heterotypic anti-H3 response to the H7 HA.) In a prior study of immunogenic efficacy of a comparable vaccine this antibody had been demonstrated to be non-neutralizing, however. The outcome of natural infection, first with Port Chalmers virus then with Victoria/75 virus in two successive winters, is shown in Table 1.

The infection rate in the first winter was shown to be the same in control and NA-specific vaccine groups although the incidence of illness was reduced in the latter. As expected, X-41 (the conventional vaccine) caused a greater initial reduction of illness. In the second winter, the hypothesis was further substantiated in that the infection as well as the disease rate in the X-42 vaccinees was now reduced. The only anomaly was with respect to the conventional X-41 vaccine in that: 1) 23% were infected in the first year, a higher rate than expected, and 2) the effect in the second year approximated that following X-42 immunization, which should not have been the case if infection had been adequately prevented the first year by the conventional vaccine. Thus, the apparently equal effect of X-41 and X-42 vaccines might reflect a combination of unanticipated vaccine-induced residual heterotypic immunity from X-41, combined with immunity from the preceding Port Chalmers virus infection in subjects not adequately protected by vaccine the first year. Results of this large study are perplexing mainly with respect to the unexpected efficacy of conventional vaccine in the second year. However, in this study the severity of clinical illnesses was not ascertained or defined; neither was it possible to follow the fate of individual vaccinees through the two year period as was done with the Hoskins study (Hoskins et al. 1979) and, as acknowledged by the authors, their test for the determination of anti-Victoria antibody might not have employed a maximally sensitive reagent.

In studies in Czechoslovakia by Vonka et al. (1977), an inactivated whole-virus vaccine was prepared from the influenza virus reassortant, H1N2, and administered to two groups of subjects. A total of 1,200 subjects were vaccinated; comparable groups of subjects served as controls. From a portion of the vaccinees sera were obtained prior to vaccination and 3-4 weeks afterwards. Serological tests revealed development of or increase in antibody against H1 in a great majority of the vaccinated subjects and against N2 in slightly more than

TABLE 1. Comparison of NA-specific (X-42) and Conventional (X-41) Vaccines in School Children

Natural Infection I
(A/Port Chalmers H3N2 Virus in 1975)

Vaccine	Serologic Infection Rate	Ill	Protection Ratio[a]
X-41 (H3N2)	70/300 (23%)	28 (9%)	69
X-42 (H7N2)	119/300 (40%)	56 (18%)	37
Placebo	123/275 (45%)	82 (29%)	0

Natural Infection I
(A/Victoria/75 virus in 1976)

Vaccine	Serologic Infection Rate	Ill	Protection Ratio[a]
X-41 (H3N2)	35/220 (16%)	9 (4%)	80
X-42 (H7N2)	45/201 (22%)	12 (6%)	73
Placebo	73/185 (40%)	38 (20%)	0

[a] $\frac{(\%\text{ ill in placebo group} - \%\text{ ill in vaccine group}) \times 100}{\%\text{ ill placebo}}$

Modified from Beutner et al. (1979).

half of them. Antibody response to H3 antigen was only rarely encountered. Approximately three months after vaccination, an influenza epidemic caused by Victoria-like (H3N2) viruses broke out in Czechoslovakia. Numerous influenza cases occurred in the two populations followed. Morbidity was significantly lower among the vaccinated than among the control subjects, indicating a protective effect of the NA vaccine under field conditions.

In a two year study of NA-specific immunization of 2,000 college students, which involved a number of campuses on the Eastern seaboard, the vaccine used was an H7N1 vaccine. During the first year, problems were encountered with the stability of the NA in this vaccine, but Lederle Laboratories had success in stabilizing the antigen in later batches. Results of this study indicated that the vaccine was of borderline potency but was apparently protective in at least one college epidemic (Cerini et al. 1981; Kilbourne et al. 1987b).

Revelation of HA-NA Antigenic Competition: The Superiority of H7N2 Over H3N2 Vaccine in Stimulation of Anti-N2 Antibody Response in H3N2-Primed Subjects

Kilbourne (1976) made the surprising observation that whereas 25% of those given a conventional H3N2 vaccine had significant anti-NA antibody response, a 69% response occurred in those given H7N2 (NA-specific) vaccine containing an irrelevant HA antigen, although both vaccines had been standardized to contain equivalent amounts of NA antigen. A similar result was seen in subsequent studies of antigenically hybrid NA-specific vaccines in man (Beutner et al. 1979; Cerini et al. 1981; Ogra et al. 1977). Furthermore, anti-NA response to conventional vaccine was poor (18%) in a population naturally primed to H3N2 virus (Kendal et al. 1980). Therefore, it seemed important to explore the use of biochemically pure NA vaccine devoid of associated HA antigen. It also appeared likely that higher levels of anti-NA antibody might be attained by such artificial immunization than through natural infection.

Preliminary Study of NA Vaccines in Humans - a Summary

The studies summarized above in which the NA specific vaccines were essentially conventional inactivated whole virus vaccines containing a substituted "irrelevant" HA were encouragingly confirmatory of our expectations for infection permissive immunization. As predicted, these vaccines permitted but reduced viral replication and as a consequence prevented disease. Reduction of virus shedding appeared to be directly related to pre-infection titers of NA antibody (Couch et al. 1974) and infection-inhibiting immunity to reinfection was also demonstrated. However, important questions remained to be answered:

1) Is purified NA sufficiently antigenic to be useful for vaccination? If so, what is the best source of NA? It seems desirable to use a single polypeptide in order to reduce the reactogenicity intrinsic in whole virus vaccines and to avoid possible antigenic competition with HA antigen.
2) What is the mechanism of HA - NA antigenic competition?
3) If purified NA is immunogenic, how does it compare with purified HA and with whole virus vaccines?
4) Does NA vaccine address the problem of influenza virus antigenic variation more effectively than conventional vaccine?
5) What are the time/dose factors in the programming of NA vaccine followed by infection boosting?

We have now answered most of these questions by exploitation of a mouse model system in which our initial studies had been conducted.

Studies of NA-Specific Vaccination in BALB/c Mice - Sequential Infection of Mice Simulates the Human Experience

In order to simulate a background of immune experience characteristic of the human population, mice were infected sequentially with H3 variant viruses bearing N1 (irrelevant) or N2 NAs, then were vaccinated with hybrid vaccines of H3N2 or H7N2 phenotype. As had been the case in infection-primed humans, vaccine stimulated post-infection anti-N2 booster response was four-fold greater with H7N2 vaccine and was reciprocal to the magnitude of anti-H3 response stimulated by H3N2 vaccine. Thus HA-influenced suppression of immunologic response to NA was not unique to man and its mechanism could be investigated in the murine model (Johansson et al. 1987b).

IMMUNOLOGIC RESPONSE TO INFLUENZA VIRUS NA IS INFLUENCED BY PRIOR EXPERIENCE WITH THE ASSOCIATED VIRAL HA

Mechanism of HA-NA Antigenic Competition

In BALB/c mice primed to H3 HA and N2 NA by infection, the presentation of N2 in association with a heterotypic HA resulted, as expected, in production of a greater amount of N2 antibody than was found with homologous (H3N2) reimmunization. Titration of primed helper T (Th) cell activity by adoptive transfer of purified T cells to athymic mice given H6N2 vaccine demonstrated a lesser number of N2-specific Th cells in mice subjected to homologous (H3N2-H3N2) reimmunization. These observations demonstrated the participation of Th cells in the mediation of intermolecular intravirionic antigenic competition between influenza virus HA and NA (Johansson et al. 1987a).

In further exploration of the mechanism, the *in vitro* reaction of purified splenic B and T lymphocytes from mice immunized by various immunization schedules was investigated. Assay of the proliferation response of T cells in B/T cell mixtures stimulated by HA specific and NA specific reassortant viruses *in vitro* enabled differentiation of cellular responses to HA and NA antigens. It was shown that: 1) intravirionic HA is dominant over NA in both B and T cell priming; 2) memory B cells functioned as antigen presenting cells and interacted with memory helper T cells in the mediation of intravirionic HA/NA antigenic competition in favor of HA. Thus, the damping of response to the NA in favor of HA prohibited balanced response to the two virion antigens under ordinary conditions of immunization, whether by whole virus vaccine or by infection (Johansson et al. 1987c).

Although resolution of the question of HA-NA antigenic competition is a contribution to the understanding of antigenic competition in general, its relevance to the definitive goal of NA immunization was limited to reassortant whole virus vaccines. It also indicated, of course, that in natural infection immunologic response to the viral glycoproteins is perpetually skewed in favor of HA.

The Comparative Immunogenicity of HA and NA Glycoproteins in Mice

BALB/c mice immunized with graded doses of chromatographically purified HA and NA antigens derived from A/Hong Kong/1/68 (H3N2) influenza virus demonstrated equivalent responses when HA-specific and NA-specific serum antibodies were measured by enzyme-linked immunosorbent assays (ELISAs) (Figure 6). Antibody responses measured by HA inhibition or NA inhibition titrations showed similar kinetic patterns, except for more rapid decline in hemagglutination inhibiting antibody (Figure 7).

Injection of mice with either purified HA or NA resulted in immunity manifested by reduction in pulmonary virus following challenge with virus containing homologous antigens. However, the nature of the immunity induced by the two antigens differed markedly. While HA immunization with

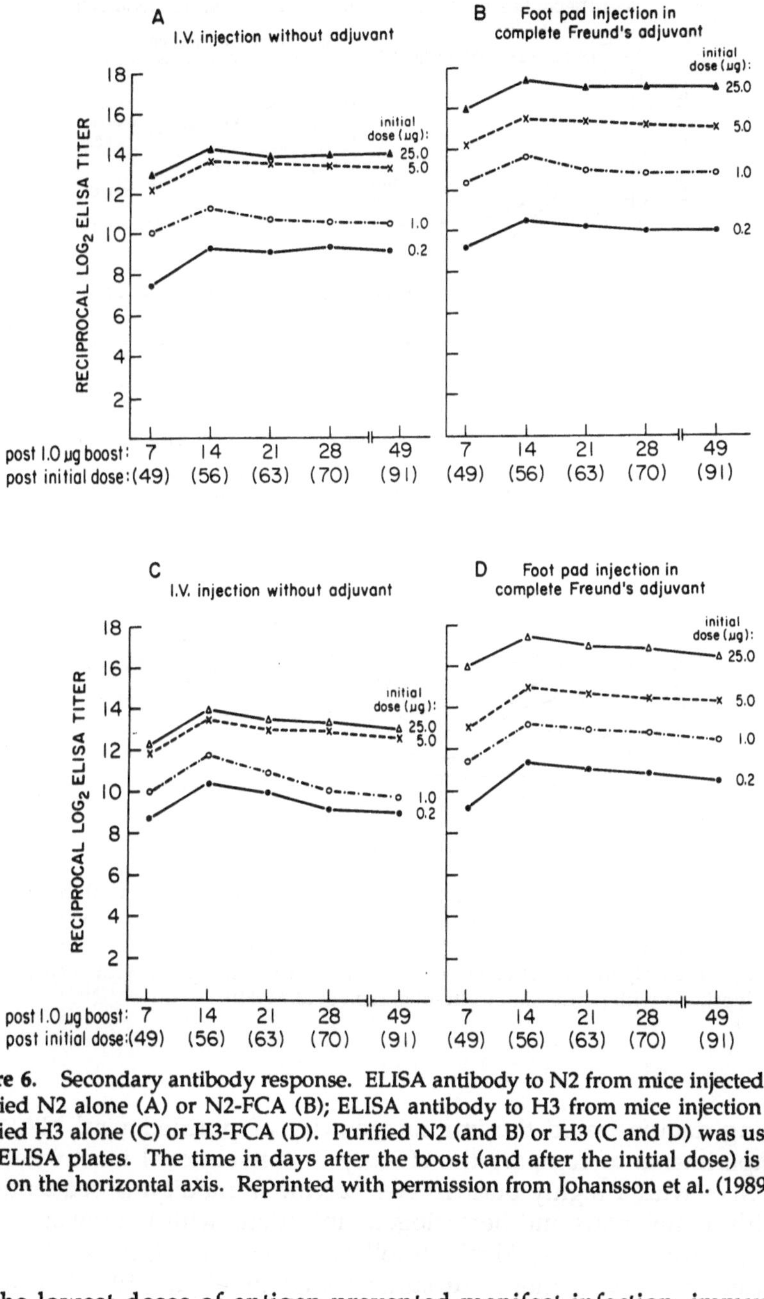

Figure 6. Secondary antibody response. ELISA antibody to N2 from mice injected with purified N2 alone (A) or N2-FCA (B); ELISA antibody to H3 from mice injection with purified H3 alone (C) or H3-FCA (D). Purified N2 (and B) or H3 (C and D) was used to coat ELISA plates. The time in days after the boost (and after the initial dose) is indicated on the horizontal axis. Reprinted with permission from Johansson et al. (1989).

all but the lowest doses of antigen prevented manifest infection, immunization with NA was infection-permissive at all antigen doses (Figure 8), although reduction in pulmonary virus was proportional to the amount of antibody induced (Figure 9).

The immunizing but infection-permissive effect of NA immunization over a wide range of doses is in accord with results of earlier studies with mice in which single doses of NA and antigenically hybrid viruses were used. The demonstrable immunogenicity of highly purified NA as a single glycoprotein without adjuvant appeared to offer a novel infection-permissive approach with potentially low toxicity for human immunization against influenza (Johansson et al. 1989).

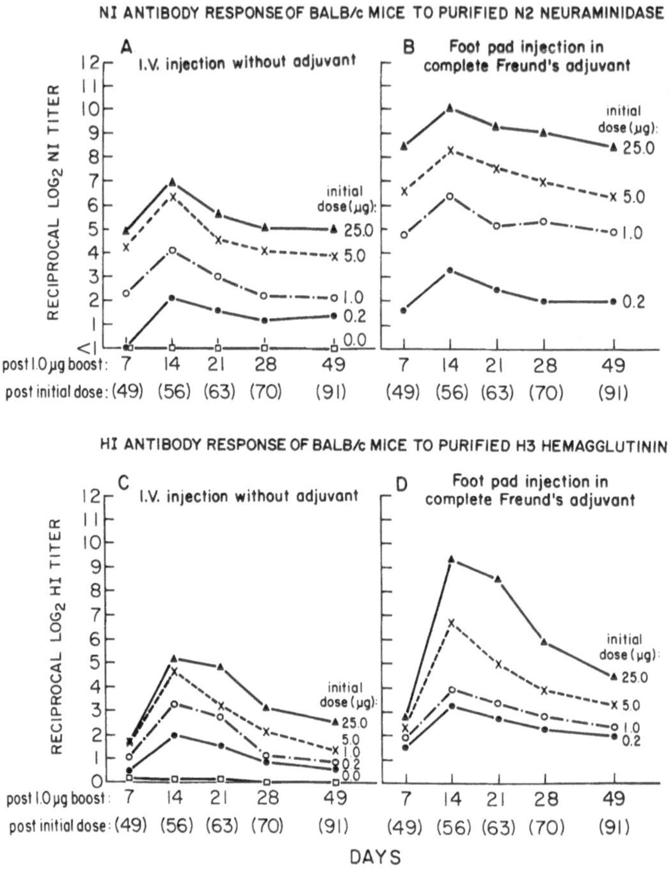

Figure 7. NI = neutralization inhibition and HI = hemagglutination inhibition antibody responses. Antibody to N2 from mice injected with purified N2 alone (A) or N2-FCA (B) was measured by NI with whole H1N2 virus. Antibody to H3 from mice injected with purified H3 alone (C) or H3-FCA (D was measured by HI with whole H3N2 virus. Reprinted with permission from Johansson et al. (1989).

Having demonstrated the immunogenicity of purified viral NA, we designed experiments to compare the efficacy of conventional inactivated whole virus vaccines with a highly purified NA vaccine in modifying sequential challenges with homologous and heterologous infections with heterovariant H3N2 influenza A viruses, i.e., A/Aichi/68 followed by A/Philippines/82. The results of one such experiment are summarized in Figure 10. The essential points to be made are 1) the initial superiority of whole virus vaccine during first infection (day 59) as manifested by reduction in pulmonary virus titer and conversely 2) the greater inhibition by NA vaccine of both homologous and heterologous challenges on second infection (day 93).

In the above experiment, the intervals between vaccination and infectious challenge had been somewhat arbitrarily chosen. With reference to modeling the human situation, it was important to define the optimal interval and program for antigenic stimulation by the NA vaccine and initial challenge (boosting) infection and to gain some insight into the duration of subsequent immunity.

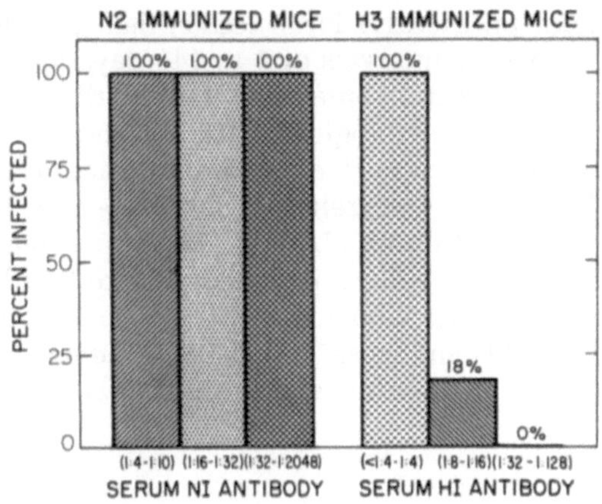

Figure 8. Incidence of infection in mice immunized with N2 (left) or H3 antigen (right) and infected with H3N2 virus containing homologous antigen. Infection was defined by the detection of egg-infective virus in normal lungs in a 10^{-2} dilution of lung suspension. (Even at highest antibody levels infection occurs in N2 vaccinated mice.) Reprinted with permission from Johansson et al. (1989).

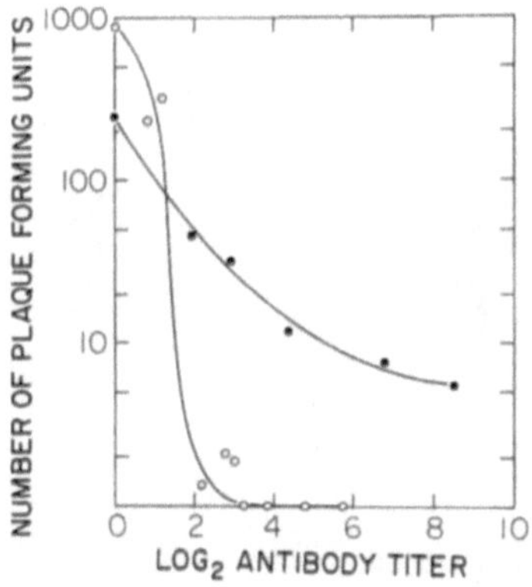

Figure 9. Reduction in pulmonary virus levels. Plot of PFU versus amount of NI antibody in N2-immunized mice (●) and of HI antibody in H3-immunized mice (○) challenged with 100, 50% mouse infective doses of H3N2 virus. Reprinted with permission from Johansson et al. (1989).

TIMING AND KINETICS OF THE IMMUNE RESPONSE TO CHALLENGE INFECTION OF MICE PRIMED WITH INACTIVATED WHOLE VIRUS NA VACCINE

Mice were immunized with either inactivated whole virus influenza A (H3N2) virus vaccine or purified N2 NA vaccine, then challenged with mouse-adapted homologous virus at intervals of 1 to 141 days later in order to ascertain the optimal vaccine infection interval for induction of resistance to still later infection. Measured by serological (data not shown) and infection-suppressing response, this interval was 15 days for both vaccines.

A summary of this study correlating vaccine - infection intervals and prechallenge antibody titers to both NA and HA is presented in Table 2. The results presented here represent the experience of mice that had first been vaccinated with two doses of either whole virus of NA vaccine, then boosted by intranasal infection. The table shows as "initial PVT" the log 10 titers of infective virus in mice in each group three days after the initial boosting infection. As

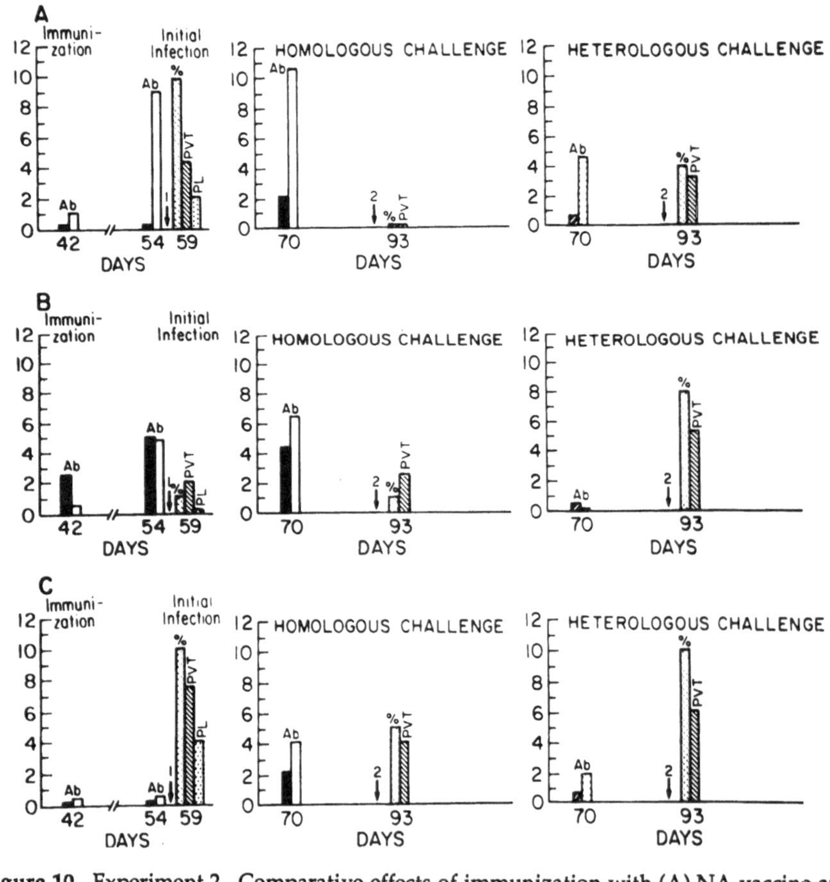

Figure 10. Experiment 2. Comparative effects of immunization with (A) NA vaccine and (B) whole virus vaccine on homologous and heterologous influenza virus challenge infections. C and D, results of infection of unvaccinated control animals. The score represents an estimate of the mean percentage of total lung involvement by grossly apparent pulmonary consolidation. Modified from Johansson and Kilbourne (1990).

TABLE 2. Correlation of Vaccine-Infection Intervals and Pre-Challenge Antibody Titers with Immunity to Final Infection

	Interval Days		Group A					Group B				
			NA Vaccine					WV Vaccine				
				Pre-Final Challenge					Pre-Final Challenge			
Sub-group	Vaccine to Inf. 1	Inf. 1 to Inf. 2	Init. PVT[a]	HI[b]	NI[b]	PVT[c]	Reduction in Infection (%)[d]	Init. PVT[a]	HI[b]	NI[b]	PVT[c]	Reduction in Infection (%)
1	1	156	6.2	<1	4.8	3.1	40	5.9	2.5	<2	2.6	20
2	15	141	4.4	3.3	7.6	4.1	60	<2	2.0	3.4	1.7	20
3	43	113	4.5	3.8	7.5	3.9	60	2.3	2.0	2.6	1.9	20
4	57	99	5.5	4.0	5.7	5.3	80	3.7	3.1	3.4	3.3	40
5	85	71	5.2	3.5	5.6	7.6	100	4.6	4.2	4.0	7.6	100
6	141	15	5.9	3.2	5.5	7.6	100	6.0	3.8	4.1	7.6	100
			Mean	3.0	6.0	5.3[d]	80		3.0	3.0	4.1	56
						(4.1)[e]					(2.4)[e]	

a $\log_{10}$ pulmonary virus titer
b $\log_2$ antibody titer following vaccination and first infection
c $\log_{10}$ reduction in PVT, Groups 1-6
d Reduction of number of mice infected
e $\log_{10}$ reduction in PVT, Groups 1-4

From Johansson and Kilbourne (submitted)

expected, these initial titers were higher in NA vaccinated groups. Not shown are titers of unimmunized control mice that exceeded the NA group titers by 1.3 to 4.0 logs. With both vaccines, PVTs in first and second infection were reciprocal reflecting respectively the greater antigenic stimulation provided by higher PVT in initial vaccination of NA vaccinated mice or the lower levels of replication of virus in first infection in mice primed with whole virus vaccine. This effect is graphically illustrated in Figure 11. This figure depicts only the results in mice in groups 1-4 of Table 2 because it was found that when the inter-infection interval was shortened to 71 or 15 days, reinfection did not occur (ie., 100% reduction of infection in subgroups 5 and 6 in Table 2).

ANTIGENIC DRIFT OF NA IS LESS RAPID THAN THAT OF VIRAL HA

The HA and NA external glycoprotein antigens of H1N1 and H3N2 subtypes of epidemiologically important influenza A viruses prevalent during recent decades were subjected to intensive antigenic analysis by four different methods. Prior to serological analysis with polyclonal rabbit antisera, HA and NA antigens of four viruses of each subtype were segregated by genetic reassortment to forestall nonspecific steric hindrance during antigen-antibody combination. This analysis demonstrated that with respect to antigenic phenotype, HA and NA proteins have evolved at different rates (Figure 12). With H1N1 viruses, an arrest of significant evolution of the NA discordant with the continuing antigenic drift of HA was found in the 1980-1983 period. It is probable that the different and independent rates of evolution of HA and NA reflect the greater selective pressure of HA antibodies, which forces the more rapid emergence of HA escape mutants. The slower antigenic change found for NA further supports the potential for NA-specific infection-permissive immunization as a useful stratagem against influenza (Kilbourne et al. 1990).

SUMMARY AND CONCLUSIONS

Under the carefully controlled conditions possible with experiments with large numbers of inbred mice and precisely defined viral challenges we have shown that:

1) Highly purified influenza virus NA is immunogenic in small quantities in mice in the absence of adjuvant as manifested by the stimulation of humoral antibody production and the priming of B and T lymphocytes.
2) Immunization with purified NA reduces virus replication in mice in proportion to NA dose over a wide range of doses. Consequently, pulmonary lesions and mortality are reduced.
3) Purified influenza virus NA and HA are equivalent in the stimulation of antibody response but antibodies to each have contrasting effects on the suppression of virus replication; virus replication is prevented in HA vaccinated mice but only partially suppressed in mice given NA vaccine.
4) Priming of mice given either whole virus or NA vaccine prior to infection is followed by immunity superior to that induced by infection(s) alone.
5) The initial superiority of whole virus vaccine with respect to the prevention of infection is lost at the time of the second infectious challenge. At that time mice initially primed with NA antigen are more resistant to homologous or heterologous infection.

6) NA antigen evolves more slowly than HA antigen in nature and consequently can offer more sustained protection.
7) Programmed antigenic stimulation with initial priming with the NA antigen has great potential value in the immunization of children before their initial encounters with influenza viruses.

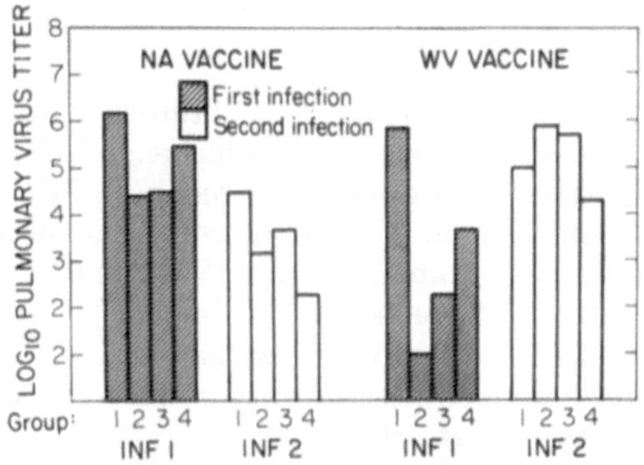

Figure 11. Inverse relationship of pulmonary virus titers (PVT) on first and second infection. Groups A 1-4, B 1-4 (Table 1). From Johansson and Kilbourne (Vaccine, in press).

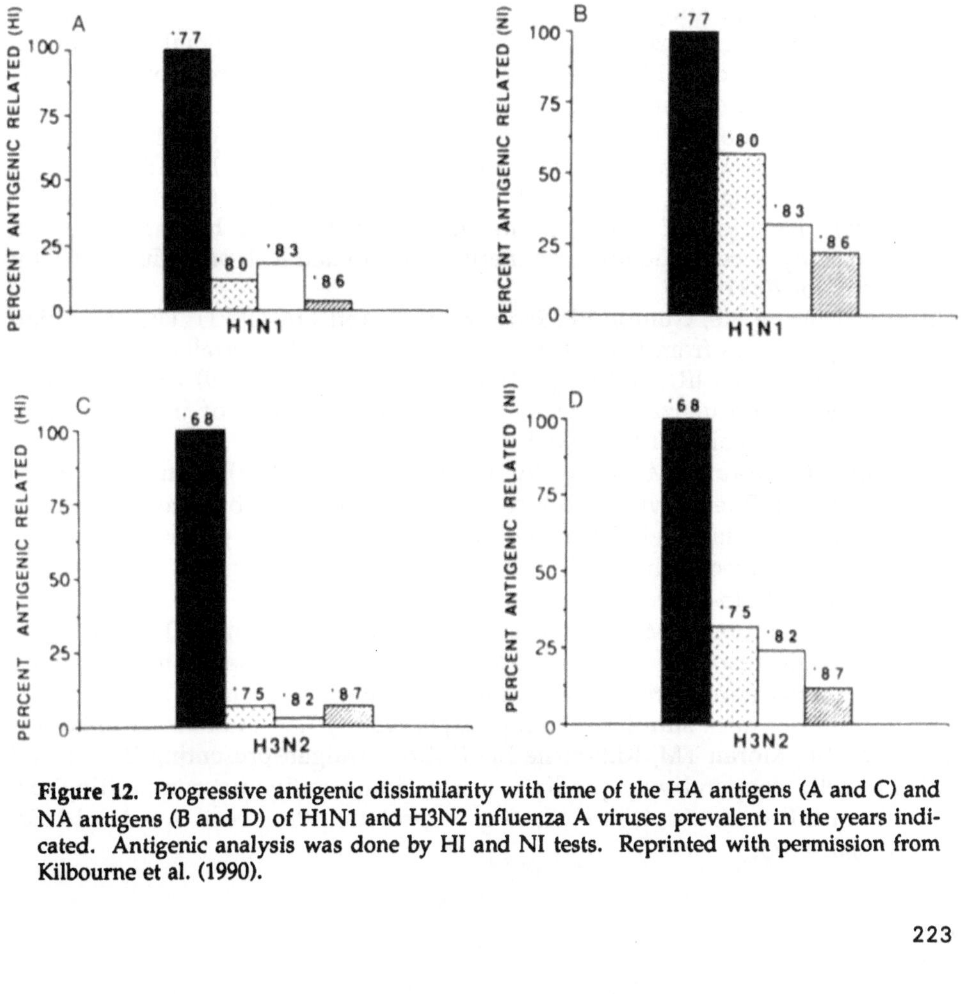

Figure 12. Progressive antigenic dissimilarity with time of the HA antigens (A and C) and NA antigens (B and D) of H1N1 and H3N2 influenza A viruses prevalent in the years indicated. Antigenic analysis was done by HI and NI tests. Reprinted with permission from Kilbourne et al. (1990).

ACKNOWLEDGEMENTS

We wish to thank Ms. Barbara A. Pokorny for excellent technical assistance. This work was supported by Public Health Service Grant AI 09304 and the Aaron Diamond Foundation.

REFERENCES

Allen WH, Madelye CR, Kendal AP (1971) Studies with avian influenza A viruses: cross protection experiments in chickens. J Gen Virol 12:79-84.

Atmar RL, Bloom K, Keitel W, Couch RB, Greenberg SB (1990) Effect of live attenuated, cold recombinant (CR) influenza virus vaccines on pulmonary function in healthy and asthmatic adults. Vaccine 8:217-224.

Beutner KR, Chow T, Rubi E, Strussenberg J, Clement J, Ogra PL (1979) Evaluation of a neuraminidase-specific influenza A virus vaccine in children: antibody responses and effects on two successive outbreaks of natural infection. J Infect Dis 140:844-850.

Cerini C, Joseph A, Kalish G, Kahn M, Kutzner W, Stebbins M, McCoy D (1981) Efficacy of neuraminidase specific H7N1 influenza vaccine in an influenza H1N1 outbreak. Curr Chemotherapy & Immunotherapy, Proc. 12th International Congress of Chemotherapy, Florence, Italy, July, 1981.

Couch RB, Kasel JA, Gerin JL, Schulman JL, Kilbourne ED (1974) Induction of partial immunity to influenza by a neuraminidase-specific influenza A virus vaccine in humans. J Infect Dis 129:411-420.

Couch RB, Kasel JA, Glezen WP, Cate TR, Six HR, Taber H, Frank AL, Greenberg SB, Zahradnik M, Keitel WA (1986) Influenza, its control in persons and populations. J Infect Dis 153:431-441.

Eickhoff TC, Meiklejohn G (1969) Protection against Hong Kong influenza by adjuvant vaccine containing A2/Ann Arbor/67. Bull WHO 41:562.

Gallagher M, Bucher DJ, Dourmashkin R, Davis JF, Rosenn G, Kilbourne ED (1984) Isolation of immunogenic neuraminidases of human influenza viruses by a combination of genetic and biochemical procedures. J Clin Microbiol 20:89-93.

Gill PW, Babbage NF, Gunton PE, Flower W, Garrett DA (1971) Did the Asian virus protect us from Hong Kong influenza? Med J Australia 2:53-54.

Hoskins TW, Davies JR, Smith AJ, Miller CL, Allchin A (1979) Assessment of inactivated influenza-A vaccine after three outbreaks of influenza-A at Christ's Hospital. Lancet 1:33-35.

Johansson BE, Moran TM, Bona CA, Kilbourne ED (1987a) Immunologic response to influenza virus neuraminidase is influenced by prior experience with the associated viral hemagglutinin. III. Reduced generation of neuraminidase-specific helper T cells in hemagglutinin-primed mice. J Immunol 139:2015-2019.

Johansson BE, Moran TM, Bona CA, Popple SW, Kilbourne ED (1987b) Immunologic response to influenza virus neuraminidase is influenced by prior experience with the associated viral hemagglutinin. II. Sequential infection of mice simulates human experience. J Immunol 139:2010-2014.

Johansson BE, Moran TM, Kilbourne ED (1987c) Antigen-presenting B cells and T_H cells cooperatively mediate intravirionic antigenic competition between influenza A virus surface glycoproteins. Proc Natl Acad Sci USA 84:6869-6873.

Johansson BE, Bucher DJ, Kilbourne ED (1989) Purified influenza virus hemagglutinin and neuraminidase are equivalent in stimulation of antibody response but induce contrasting types of immunity to infection. J Virol 63:1239-1246.

Johansson BE and Kilbourne ED (1990) Comparative long term effects in a mouse model system of influenza whole virus and purified neuraminidase vaccines followed by sequential infections. J Infect Dis 162:800-809.

Kendal AP, Patriarca PA (1986) Options for the control of influenza. UCLA Symp Mol Cell Biol 36:139-183.

Kendal AP, Bozeman FM, Ennis FA (1980) Further studies of the neuraminidase content of inactivated influenza vaccines and the neuraminidase antibody responses after vaccination of immunologically primed and unprimed populations. Infect Immun 29:966-971.

Kilbourne ED (1975a) Epidemiology of influenza. In Kilbourne ED (ed), The Influenza Viruses and Influenza. New York: Academic Press.

Kilbourne ED, Palese P, Schulman JL (1975b) Inhibition of viral neuraminidase as a new approach to the prevention of influenza. In Pollard M (ed) Perspectives in Virology IX. New York: Academic Press, pp. 99-113.

Kilbourne ED (1976) Comparative efficacy of neuraminidase-specific and conventional influenza virus vaccines in the induction of anti-neuraminidase antibody in man. J Infect Dis 134:384-394.

Kilbourne ED (1979), The Harvey Lectures, Series 73, pp. 225-258, Academic Press, New York.

Kilbourne ED (1987) Influenza. New York: Plenum Medical Book Co., pp. 119-122.

Kilbourne ED, Cerini CP, Khan MW, Mitchell JW, Jr., Ogra PL (1987) Immunologic response to the influenza virus neuraminidase is influenced by prior experience with the associated viral hemagglutinin. I. Studies in human vaccinees. J Immunol 138:3010-3013.

Kilbourne ED, Johansson BE, Grajower B (1990) Independent and disparate evolution in nature of influenza A virus hemagglutinin and neuraminidase. Proc Natl Acad Sci USA 87:786-790.

Kim W, Arrobio J, Brandt C, Parrott R, Murphy B, Richman D Chanock R (1976) Temperature-sensitive mutants of influenza A virus: response of children to the influenza A/Hong Kong/68-ts-1[E] (H3N2) candidate vaccine viruses and significance of immunity to neuraminidase antigen. Pediatr Res 10:238-242.

Lai C-J, Markoff LJ, Zummerman S, Cohen B, Berndt JA, Chanock RM (1980) Cloning DNA sequences from influenza viral RNA segments. Proc Natl Acad Sci USA 77:210-214.

McLaren C, Potter CW, Jennings R (1974) Immunity to influenza in ferrets. Med Microbiol Immunol 160:33-45.

Murphy BR, Kasel JA, Chanock RM (1972) Association of serum anti-neuraminidase antibody with resistance to influenza in man. New Engl J Med 25:1329-1332.

Ogra PL, Chow T, Beutner KR, Rubi E, Strussenberg J, DeMello S, Rizzone C (1977) Clinical and immunologic evaluation of neuraminidase-specific influenza A virus vaccine in humans. J Infect Dis 135:499-506.

Price PM, Reichelderfer CG, Johansson BE, Kilbourne ED, Acs G (1989) Complementation of recombinant baculoviruses by coinfection with wild type virus facilitates production in insect larvae of antigenic proteins of HBV and influenza virus. Proc Natl Acad Sci USA 86:1453-1456.

Rodin AT (1969) National experience with Hong Kong influenza in the United Kingdom, 1968-69. Bull WHO 41:375-380.
Rott R, Becht H, Orlich M (1974) The significance of influenza virus neuraminidase in immunity. Virol 22:35-41.
Scholtissek C, Vallbracht A, Flehmig B, Rott R (1979) Correlation of pathogenicity and gene constellation of influenza A viruses. II. Highly neurovirulent recombinants derived from non-neurovirulent or weakly neurovirulent parent virus strains. Virology 95:492-500.
Schulman JL, Khakpour M, Kilbourne ED (1968) Protective effects of specific immunity to viral neuraminidase on influenza virus infection of mice. J Virol 2:778-786.
Vonka V, Zavadova H, Bruj J, Skocil V, Janout V, Uvizl M, Kotikova J. (1977) Small-scale field trial with neuraminidase vaccine. Dev Biolog Stand 39:337-339.

DISCUSSION

Peterson E (University of California Irvine Medical Center, Orange, CA):

The protective antibodies you get when you inject your N2 purified antigen, are they more to a conformational epitope or a linear epitope?

Kilbourne E:

We have begun to do a little work with some linear oligopeptides, but I don't think we're prepared to answer that yet.

Minocha H (Kansas State University, Manhattan, KS):

Dr. Kilbourne, thank you for a very enlightening review of influenza viruses. I want to ask, where are we on the interferon situation? There is a model system for parainfluenza in veterinary medicine. When we inject calves and let them go in feed locks where there is a lot of influenza virus going on, in the first seven days, they don't develop any antibodies. There is humoral immune response. There is data available that due to interferon production, they'll be protected in the first six or seven days, before the antibodies come along. In humans, if there was an outbreak going on, I was thinking in the same terms, would there be a vaccine which would protect initially and still the antibodies are made?

Lennette E (California Public Health Foundation, Berkeley, CA):

Are you talking about shipping fever?

Minocha H:

Partly. Shipping fever is a complex of many viruses. No doubt about it. I'm talking about mostly, either IBR which is infectious bovine trachea or parainfluenza-3. In both cases, there is interferon production.

Kilbourne E:

I'm not exactly sure where you're going with the question. It would seem to me that if you are looking for interferon production as a control mechanism, you're probably better off with an infective virus vaccine. I think that there is some evidence now that Julie Younger has, that through defective interfering particle formation with certain cold adapted mutant influenza virus, it's not interferon, but defective interfering particles that are actually damping infection with wild type virus. If you want to go that particular route, I think that would be related more to the use of live virus vaccine.

Minocha H:

I was wondering if a purified neuraminidase or hemagglutinin, has the production of interferon been checked?

Kilbourne E:

No, I don't think you would expect this from what we know about interferon induction. These are purified proteins and contain no nucleic acid.

Lennette E:

You mentioned in one of your slides, over and under attenuation problems. I think you might take polio vaccine as a possible example of under attenuation, where you get cases of poliomyelitis, rare, but nevertheless, they occur. A better example, perhaps, is yellow fever vaccine. In the original work with the chick embryo passage of the virus, you could make so many passages and then the virus was attenuated, and actually could be used in the field. But, if you went beyond a certain number, the vaccine virus would still kill mice, but would produce no immunity in man. Everything is held at one passage level and then aliquots are taken from that, and the vaccines are made. That is one way to get around the problem. That is a difficult way to go, though. Expensive.

Kilbourne E:

The problem there, Ed, is you have only one yellow fever vaccine. You're not changing the antigen every two years and you don't have to go through it repetitively.

Lennette E:

I was emphasizing over-attenuation. If you go too far, then you can make things worse.

Kilbourne E:

Actually, the closest model that I can think of, to what we're trying to do, involves polio. That is, the use of Salk vaccine before Sabin. What you're really trying to do, is to do something like this. You're trying to create a floor of immunity so you don't get the viremia and the paralytic manifestations, but you do allow infection of the gut to occur as a definitive immunizing step. That's really one of the few models I can think of in contemporary vaccinology. The other model is, again, what was mentioned this morning. That is the modification of measles with specific immune globulin. Here again, one could modify the infection, get a few lesions, and yet have rather durable immunity thereafter.

Berlin B (Michigan Department of Public Health, Lansing, MI):

Just to start off, are you intending to give us a childhood vaccine with X to the Nth power number of new antigens? Is that what you propose for the future?

Kilbourne E:

I think it should be a trivalent vaccine with N2, N1 and influenza B neuraminidases.

Berlin B:

Some of the other people who believe in other antigens would be a little disappointed you didn't include those too, but let me ask you something else. It isn't entirely clear to me as to whether the response you see with N2 in the presence of the hemagglutinin is due to antigenic dampening because of the antigenic load, or whether there is some other mechanism. If you were to do this with say, and H0-N2 whole virus, would you see the same phenomenon as you do with N2 alone?

Kilbourne E:

That was exactly how we first observed this. We saw this when we were using an H6 or an H7-N2, and found the response was damped in that situation compared to the response in an H3 primed animal or man with an H3-N2. So, the original observations were made with antigenic hybrids. Later, we found out that we got even a greater release from the tyranny of the HA by using purified N2.

Lennette E:

I forgot why you used the animal hemagglutinins rather than the human, H7.

Kilbourne E:

We used it as sort of a carrier for the neuraminidase. We picked H7 which is an avian strain or an equine strain. That has very little crossreactivity with the contemporary human strains. We chose this deliberately as a carrier for neuraminidase. I would like to forget all that because we're now into using the purified antigen. I think it was a useful device in the past because we were able to get vaccine made easily, but now we're focusing on using the purified antigen.

Lennette E:

OK, totally forgotten.

LESSONS FROM VARICELLA VACCINE

Anne A. Gershon

Columbia University
College of Physicians & Surgeons
New York, New York, USA

INTRODUCTION

The Oka strain of live attenuated varicella vaccine was developed in the early 1970's by Takahashi and his colleagues in Japan (Takahashi et al. 1974). When this vaccine was first being tested, it was thought to have the greatest potential utility for use in immunocompromised children such as those with leukemia, in whom the mortality rate from varicella (chickenpox) had been shown to approach 10% (Feldman et al. 1975). While at that time it was considered contraindicated to administer a live virus vaccine to immunocompromised patients in the United States, it was found in Japan that the vaccine could be safely administered to high risk immunocompromised children (Ha et al. 1980; Izawa et al. 1977).

Studies with this vaccine in the United States in patients at high risk to develop severe varicella such as leukemic children and healthy adults began in the early 1980's. It has now become clear from studies conducted by the Collaborative Varicella Vaccine Study Group of the National Institute of Allergy and Infectious Disease and others that this vaccine is not only safe but also highly protective against varicella for immunocompromised varicella-susceptible children and adults (Arbeter et al. 1986a, 1986b, 1990; Brunell et al. 1982; Gershon et al. 1984, 1989, 1990). It has also been realized, however, that healthy children too could derive important health benefits from varicella vaccine (Arbeter et al. 1986a, 1986b; Brunell et al. 1988; Johnson et al. 1988, 1989). Moreover since it became apparent that administration of varicella vaccine to children with leukemia results in a vaccine-associated rash in about 50% and almost half of these need to be treated with acyclovir (ACV) to control the rash (Brunell et al. 1987; Gershon et al. 1984, 1989), there is now a much greater interest in vaccinating healthy varicella-susceptible individuals rather than leukemic children. It is now recognized that if immunization with varicella vaccine were recommended for all 15 month old infants, most children who become immunosuppressed due to development of a disease such as leukemia would already have had a primary immune response to varicella-zoster virus (VZV) by virtue of having already been immunized. Moreover, it is hoped that eventually varicella vaccine will be administered routinely to healthy

Medical Virology 10, Edited by L.M. de la Maza and
E.M. Peterson, Plenum Press, New York, 1991

toddlers along with measles-mumps-rubella vaccine as a 4-valent product (Arbeter et al. 1986b; Brunell et al. 1988). Finally it is planned that varicella vaccine be distributed to leukemic children in remission under a "compassionate use" protocol rather than licensing the vaccine for them.

EFFICACY AND SAFETY OF LIVE VARICELLA VACCINE

The efficacy and safety of varicella vaccine in various groups to whom it has been given in the United States are presented in Table 1. The data are a summary of the findings of a number of studies (Arbeter et al. 1984, 1986a, 1986b, 1990; Brunell et al. 1982, 1987, 1988; Chartrand et al. 1985; Gershon et al. 1984, 1986, 1988, 1989, 1990; Johnson et al. 1989; Jones and Grose, 1988; Weibel et al. 1984, 1985). As can be seen, 95% of healthy children seroconvert to VZV after one dose of vaccine, and therefore second doses are not given. In contrast, less than 90% of leukemic children and healthy adults seroconvert against VZV after 1 dose of vaccine and therefore 2 doses are routinely given to insure that a seroconversion has occurred in these high risk patients. After a second dose of vaccine there is an improvement in the seroconversion rate for both groups, approximating the rate for healthy children after 1 dose of vaccine. Leukemic children vaccinated in the Collaborative Study have been in remission for at least 1 year prior to immunization. Their maintenance chemotherapy has been stopped for 1 week before and 1 week after the first dose of vaccine, and steroids are avoided for an additional week after vaccination. Even if there has been no seroconversion after the first immunizing dose in a leukemic child, the chemotherapy is not withheld for the second dose of vaccine.

Most vaccinees experience few if any side effects after vaccination; those that do most frequently develop a rash about 1 month later. Rashes occur almost entirely after the first dose of vaccine. The incidence of rash in immunologically normal children and adults is very low, from 5 to 10%. Its character is also mild; most frequently it is described as a few bumps resembling mosquito bites. In contrast, the incidence of rash is significantly greater in leukemic children, about 50%, and its nature is more pronounced than that seen in healthy vaccinees. Roughly 40% of leukemic children who develop a rash will develop over 50 papulovesicular lesions that require treatment with the anti-

TABLE 1. Varicella Vaccine in Various Groups

Group	Seroconversion Rate 1 dose	Seroconversion Rate 2 doses	Rash	Protection
Healthy children	95%	not given	5%	90%
Adults	88%	94%	10%	70%
Leukemic children	82%	98%	50%	85%

viral drug ACV for 7 to 10 days. This is usually administered by mouth; in about 30% of leukemics with rash, a 5-7 day course of ACV has been administered. No leukemic vaccinee in the Collaborative Study (of more than 500) has developed a disseminated VZV infection from the vaccine (Gershon et al. 1984, 1989). Thus even at its worst, the vaccine is less pathogenic for leukemic children than varicella caused by the wild type virus.

Varicella vaccine has been highly protective against subsequent chickenpox in all the groups to whom it has been administered. As can also be seen from Table 1, the protective efficacy is greater in children (85-90%), than it is in healthy adults (70%).

ANALYSIS OF VACCINE EFFICACY

The efficacy of the vaccine has been judged in two ways. In healthy children it was possible to perform a double blind placebo controlled study (Weibel et al. 1984). In this study 368 children were immunized and 446 were given placebo. Over the period of 9 months, there were 39 cases of varicella; all cases of chickenpox occurred in placebo recipients. While there were no breakthrough cases of varicella in this small, brief study, a few cases of breakthrough cases of varicella have been reported in healthy vaccinated children (Arbeter et al. 1986a; Asano et al. 1983, 1985; Johnson et al. 1989). The protective efficacy of varicella vaccine in healthy children is therefore estimated to be about 90% (Table 1).

It was not ethically possible to perform a placebo controlled study in leukemic children and varicella-susceptible adults who were considered to be at high risk to develop severe varicella, vaccine efficacy was judged in them by determination of the attack rate of varicella following a household exposure to chickenpox. It is well known that the clinical attack rate of chickenpox is exceedingly high after exposure of susceptible children in a household setting, ranging from 80% to 90% (Ross et al. 1962). In vaccinated leukemic children the attack rate after such an exposure was only 15% (Gershon et al. 1989), and in vaccinated adults it was only 30% (Gershon et al. 1990), indicating that the vaccine is highly protective for both high risk groups. Interestingly, the vaccine was more effective in children than in adults although it is clinically useful for adults despite the 30% rate of breakthrough varicella. This is because breakthrough cases of varicella that have occurred were almost always modified with few skin lesions and no systemic toxicity or disease complications.

TOWARDS LICENSURE IN THE UNITED STATES

Major questions impeding the licensure of varicella vaccine have been the duration of immunity after immunization and the potential of vaccinees for development of zoster (shingles), which has been postulated to be due to reactivation of latent VZV. While it will be many years before these two questions can be fully answered, the data thus far indicate that immunity to varicella persists for years after immunization (Asano et al. 1983, 1985; Gershon et al. 1989). This has been determined by a stable attack rate of varicella with time occurring after vaccination as well as stable antibody titers after 1 year following vaccination (Asano et al. 1983; Gershon et al. 1988, 1989, 1990; Johnson et al. 1989). It is of interest, however, that during the first year after immunization about 25% of both adult and leukemic vaccinees lose detectable VZV antibody titers

(Gershon et al. 1988, 1989, 1990). Many of these individuals regain detectable VZV antibody in the absence of symptoms, however, indicating protection from clinical disease (Gershon et al. 1989, 1990). Importantly, the incidence of zoster is not increased after immunization when compared to the natural infection, and if anything, may actually be decreased (see below) (Brunell et al. 1986; Lawrence et al. 1988).

Since the answers to these questions have been answered in a limited although very positive way, it is hopeful that we are close to licensure of varicella vaccine for use in healthy children in the United States. The vaccine has already been licensed for use in Japan, Korea, and several European countries. Although the vaccine has been used successfully in leukemic vaccinees, inducing a high level of protection against varicella and causing only mild to moderate adverse effects and no increase in the rate of relapse of leukemia despite the withholding of chemotherapy for 2 weeks, it now seems unlikely that the vaccine will be licensed for use in leukemic children. It is planned that the vaccine will be distributed to them under compassionate use with distribution in tertiary medical centers by specialists in infectious diseases who are familiar with the problems encountered such as vaccine-associated rash, use of ACV, and management of breakthrough varicella.

LESSONS FROM VARICELLA VACCINE

In addition to providing a potential benefit to varicella-susceptible individuals, studies of the live varicella vaccine have provided a means for learning a great deal about the natural history of VZV infections. Infection with this virus has been difficult to study in the absence of a reproducible practical animal model. As a result of various vaccine studies, a tremendous amount of information has emerged concerning not only varicella but also zoster. The advances made in our knowledge of VZV using illustrations from vaccine trials will now be reviewed. They are listed in Table 2.

Immunity to VZV is Not an All-or-None Phenomenon

Prior to efficacy studies of varicella vaccine, it was predicted that protection of vaccinees known to have experienced a "take" who were subsequently exposed to the virus would approach 100%. In healthy children this has been

TABLE 2. Lessons from Varicella Vaccine

- Immunity is not an all-or-none phenomenon
- The major source of transmissible VZV is the skin
- Specific immunosuppression begins early in adulthood
- Zoster is due to reactivation of latent virus

the case although it is clear that some children who are immunized against chickenpox may develop a mild attack of the natural infection some months to years later (Arbeter et al. 1986a; Asano et al. 1983, 1985; Johnson et al. 1988, 1989). The exact incidence of this occurrence in healthy immunized children is not yet known but is probably about 10 percent. Such breakthrough cases are almost always exceedingly mild with only a few lesions, suggesting that these children have partial immunity to varicella. As has been mentioned, the incidence of clinical reinfection in leukemic children who are immunized is somewhat greater, about 15% (Gershon et al. 1989). Clearly, although leukemic children who are reinfected with the wild type virus have a more pronounced breakthrough illness than healthy children, these immunized leukemics too seem to have partial immunity due to vaccination, with an average of 100 vesicles (Gershon et al. 1989). Interestingly, healthy adults immunized at an average age of 27 years have a breakthrough rate of clinical varicella after exposure of about 30%, but their illness too is usually very mild with only about 25 skin vesicles on the average (Gershon et al. 1988, 1990). To put these observations in perspective, it had been found in a natural history study of varicella that otherwise healthy unimmunized children develop 250-500 skin vesicles on the average (Ross et al. 1962). Another indication of partial immunity in recipients of varicella vaccine is that a boost of humoral or cellular immunity frequently follows exposure to the wild type virus (Asano et al. 1983, 1985; Gershon 1987, 1989, 1990). Interestingly, this type of boosting has also been observed following the natural infection (Arvin et al. 1983; Gershon et al. 1990).

While it was initially thought that the presence of detectable antibodies in the blood would correlate with protection against varicella in vaccinees, this has not necessarily been the case. The attack rate of varicella-susceptibles exposed to chickenpox in a household setting is 85-90% (Ross et al. 1962). It was observed in the Collaborative Study that the attack rate for varicella following a household exposure to chickenpox among seropositive leukemic vaccinees was 8%, while among those who had originally seroconverted but had become seronegative when exposed, the attack rate was 29% (Gershon et al. 1989). Thus antibodies are neither totally protective in themselves nor have those children who have lost detectable antibodies (after an initial seroconversion) reverted to susceptibility to varicella. Presumably cell-mediated immunity (CMI) was also acting in host defense against VZV in these vaccinees, again illustrating that partial immunity to VZV is possible and that immunity is not an all-or-none phenomenon but exists in various shades of grey.

The Major Source of Transmissible VZV is the Skin

From hospital airflow studies using tracer gases and/or involving varicella and zoster patients who were isolated from others, it has been shown that VZV spreads by the airborne route (Gustafson et al. 1984; Hyams et al. 1984; Josephson and Gombert 1988; Leclair et al. 1980). Possibly since the virus spreads in the air and infects a new host by invading the respiratory tract, it has been assumed that the source of infectious VZV must also be the respiratory tract (Nelson and St. Geme, 1966). There are reports in the literature of possible spread of varicella ostensibly during the pre-eruptive stages of varicella, but such reports are very few, involving at best less than 10 children (Brunell, 1989; Evans, 1940; Gordon and Meader, 1929; Nelson and St. Geme, 1966). Moreover, one cannot rule out the possibility that a few skin lesions were overlooked in these reported instances. Furthermore, it is difficult to isolate VZV from the pharynx, even in the early stages of varicella. In a recent report, VZV was isolated during the early stages of varicella from 5/117 (4.3%) of children tested (Ozaki et al. 1989). Others have reported an inability to isolate VZV from the pharynx (Nelson and St. Geme, 1966). In contrast, it is not difficult to isolate VZV from skin vesicles, particularly during the first few days after onset of varicella or zoster. It seems illogical to assume that the source of infectious VZV is the respiratory tract and not the skin when is it difficult or impossible to isolate virus from the throat but easy to isolate from the skin!

In our studies of vaccinated leukemic children, we were unable to isolate VZV from the throat of 30 immunized children in the Collaborative Study. In contrast, VZV was often isolated from the skin lesions of leukemic children with a vaccine-associated rash (Gershon et al. 1984, 1989). Recently in the Collaborative Study, for example, of 31 leukemics with a vaccine-associated rash that had a vesicular component, vaccine-type VZV was isolated from 10 (31%) children (Gershon et al. unpublished).

We also examined the sera of varicella-susceptible siblings of leukemic children to determine whether there was any spread of VZV to them. We did not find any evidence of spread of VZV to siblings if the leukemic vaccinee had no skin rash (Gershon et al. 1984). In contrast, we observed that there was serologic evidence of spread of vaccine virus to 15/88 (17%) of siblings when the leukemic vaccinee had a rash after immunization (Tsolia et al. 1990). There was also a direct relationship between the number of skin lesions in the leukemic child with a vaccine-associated rash and the chance of spread; the more skin lesions the more likely it was that a sibling would be infected. These observations indicate that the vaccine virus spreads to others from the skin, and it seems unlikely that the wild type virus would spread in a different manner since the two viruses are very similar. Presumably infectious VZV reaches the air from skin vesicles through skin trauma such as scratching. Varicella is well known for its pruritic nature; most children complain of terrible itching during the disease, and symptomatic therapy is often aimed at relief from the itching. Therefore in this author's opinion, the idea that virus reaches the air from the skin does not seem far-fetched.

The studies on spread of VZV also afforded another means of proof that the vaccine virus is attenuated. One indication that vaccine VZV is attenuated is the observation that when the vaccine was administered to a group of healthy children by the natural route, inhalation, no cases of varicella developed (Bogger-Goren et al. 1982). Another is that with an increasing number of tissue culture passages of a non-Oka VZV vaccine strain, there was a decreasing incidence and number of skin rashes after injection of the virus; however, this phenomenon was observed with an early strain of varicella vaccine that is no longer in use (Neff et al. 1981).

The observation that the contact cases of varicella caused by the vaccine virus were mild or subclinical is important evidence that the vaccine virus is indeed attenuated. That these infections were actually caused by the vaccine virus was verified by analysis of the VZV isolates by restriction endonuclease analysis of the VZV DNA (Gelb et al. 1987; Martin et al. 1982; Tsolia et al. 1990). This test has been used very successfully to differentiate between vaccine- and wild-type isolates of VZV. The average number of skin lesions in siblings exposed to a leukemic vaccinee with a rash was 38, in contrast to an average of 250-500 in children with natural varicella (Ross et al. 1962; Tsolia et al. 1990). The incidence of subclinical seroconversion to VZV in these siblings was 27%; in contrast subclinical varicella is estimated to occur in only 5% of instances (Ross et al. 1962; Tsolia et al. 1990). It is also of interest that in the one instance of tertiary spread of vaccine type VZV the infection was also very mild in the second sibling, with no significant change in the viral DNA seen on passage through 3 children in the family (Tsolia et al. 1990).

Specific Impairment of the Immune Response to VZV Begins in Early Adulthood

It has long been recognized that zoster is a disease associated with aging. Zoster is quite unusual in healthy children and young adults, but the incidence of zoster begins to increase at about 50 years of age and climbs steadily afterward (Hope-Simpson, 1965). A major factor in whether zoster develops is a decline in specific CMI to VZV (Arvin et al. 1978); this has been found to decrease significantly with advancing age (Berger et al. 1981; Burke et al. 1982). Thus it has come to be generally accepted that the increasing incidence of zoster with advancing years is due to a decrease in CMI to VZV that occurs as individuals age.

Studies on varicella vaccine have expanded the observation of impairment of immunity to VZV with increasing age. It has been recognized that varicella is 25 times more severe when it occurs in adults than when it occurs in children (Preblud, 1986). For this very reason we began, in the Collaborative Study, to immunize adults who had escaped varicella as children. We discovered that although adults tolerated varicella vaccine very well, with an incidence of mild vaccine-associated rash of only about 10%, their seroconversion rate was lower than that observed in children (Table 1) and therefore a two dose schedule was instituted for them. Even more surprising was the observation that following a household exposure to varicella, only 70% of immunized adults were completely protected from developing varicella. Moreover, immunized children, including the leukemic vaccinees, fared better with regard to vaccine efficacy, with 85-90% protection. In trying to explain the somewhat disappointing responses of adults to varicella vaccine, we found that the CMI response was significantly lower after vaccination of adults than it was in children, including those with leukemia (Gershon et al. 1990). The adults ranged in age from 17 to 57 years with a mean of 27 years at immunization. It appears that the primary immune response to VZV is impaired in young adults in comparison to the primary immune response to VZV in children, even including immunosuppressed children. Thus not only is there waning of immunity to VZV with advancing age as indicated by an increased incidence of zoster and low VZV CMI, but also a there is a poor ability of even young adults to respond to VZV when they encounter the virus for the first time as an adult rather than as a child.

Zoster is Caused by Reactivation of Latent VZV

Although viruses had not yet been identified, in 1909, von Bokay appreciated that the same agent that caused zoster was responsible for varicella; he observed that varicella often developed in children who were exposed to a patient with zoster (von Bokay, 1909). This was soon proven; a number of children were inoculated with vesicular fluid from zoster patients and they developed chickenpox about 2 weeks later (Bruusgaard, 1932; Kundratitz, 1925). Garland was the first to postulate that zoster developed as a result of reactivation of latent VZV in a manner analogous to that of herpes simplex virus (Garland, 1943). Hope-Simpson (1965) in a classic review of zoster, postulated that VZV reactivated from latency when there was a loss of immunity to the virus with time. That depression of VZV CMI rather than humoral immunity was a factor in development of zoster was determined by several investigators about 15 years ago (Arvin et al. 1978; Rand et al. 1977; Ruckdeschel et al. 1977). Epidemiologic studies showing that there was a seasonal increase of varicella but not of zoster in the winter and spring seemed to confirm the hypothesis that zoster was not due to reinfection with VZV. Nevertheless, publications continued to appear suggesting that zoster could be contracted following exposure to someone with a VZV infection (Morens et al. 1980; Palmer et al. 1985; Schimpff et al. 1972).

Studies on leukemic children immunized with varicella vaccine have yielded positive proof that zoster is caused by reactivation of latent virus. One child developed zoster in his left arm, at the site where a dose of varicella vaccine had been administered 21 months previously. Another developed grouped vesicular lesions on his right arm 16 months after he had been immunized in the left arm. In both of these children a vaccine-type VZV was isolated from the zoster rash, determined by analysis of the viral DNA with restriction enzymes (Gelb et al. 1987; Williams et al. 1985). Since the vaccine type virus is not circulating in the United States where these studies have been carried out, the only way in which these children could have been infected with this particular VZV was by their immunization. Therefore their zoster must have been due to reactivation of latent VZV.

In studies of others, two VZV viruses isolated from a child with Wiscott-Aldrich syndrome who had experienced natural varicella subsequently followed quickly by zoster, were similarly studied and the two viruses were shown to be identical (Straus et al. 1984). However, this finding is somewhat less conclusive than that described concerning zoster caused by the vaccine type virus due to the unique character of the vaccine virus. That VZV is latent in dorsal root ganglia has, also recently been proven by demonstration of the presence of viral RNA and DNA by hybridization methodology (Croen et al. 1988; Gilden et al. 1983, 1987; Hyman et al. 1983).

Studies of vaccinated leukemic children children also afforded an opportunity to determine the incidence of zoster after vaccination. One possible impediment to widespread use of varicella vaccine would be an observed increase in either the incidence or severity of zoster in vaccinees. In the Collaborative Study, therefore, the rate of zoster was compared in leukemic vaccinees and a control group of leukemic children who had had natural varicella either before or after developing leukemia. Leukemic children are a sentinel population for studying the incidence of zoster since those with past varicella have a higher rate of zoster than is normal after natural VZV infection; this occurs within a short period of time, within a few years of the diagnosis of leukemia. A group of 84 children in each category was therefore studied, each for a period of about 5 years. The rate of zoster was 2% in the vaccinees, and it was 14% in the controls (Lawrence et al. 1988). The zoster in vaccinees was identified as being caused by vaccine type virus in 1 case and wild type virus in another; in the other 3 cases the type of virus was not known since no virus had been isolated from the rash. A life table analysis of the two groups revealed no significant difference in the incidence of zoster, although follow-up over a longer period of time may reveal that the incidence of zoster after immunization is actually lower than after the natural disease (Lawrence et al. 1988). In any case, however, this study showed that the incidence of zoster is clearly no higher after vaccination than after the natural infection. Interestingly, other studies have also indicated that there is not an increase in the incidence of zoster after immunization (Brunell et al. 1986; Plotkin et al. 1989).

REFERENCE

Arbeter AM, Starr SE, Preblud S, Ihara T, Paciorek TT, Miller, D S, Zelson C M, Proctor EA and Plotkin S. (1984) Varicella vaccine trials in healthy children: a summary of comparative follow-up studies. Am J Dis Child 138:434-438.

Arbeter AM Baker L, Starr SE, Levine B, Books E and Plotkin S (1986a) Combination measles, mumps, rubella, and varicella vaccine. Pediatrics 78:742-742.

Arbeter A, Starr SE and Plotkin SA (1986b) Varicella vaccine studies in healthy children and adults. Pediatrics. 78 (suppl): 748-756.

Arbeter A, Granowetter L, Starr S, Lange B, Wimmer R and Plotkin S (1990) Immunization of children with acute lymphoblastic leukemia with live attenuated varicella vaccine without complete suspension of chemotherapy. Pediatrics 85:338-344.

Arvin AM, Pollard RB, Rasmussen L and Merigan T (1978) Selective impairment in lymphocyte reactivity to varicella-zoster antigen among untreated lymphoma patients. J Infect Dis 137:531-540.

Arvin A, Koropchak CM and Wittek AC (1983) Immunologic evidence of reinfection with varicella-zoster virus. J Infect Dis 148:200-205.

Asano Y, Albrecht P, Vujcic LK, Quinnan GV, Jr., Kawakami K and M Takahashi (1983) Five-year follow-up study of recipients of live varicella vaccine using enhanced neutralization and fluorescent antibody membrane antigen assays. Pediatrics 72:291-294.

Asano Y, Nagai T, Miyata T, Yazaki T, Ito S, Yamanishi K and Takahashi M (1985) Long-term protective immunity of recipients of the Oka strain of live varicella vaccine. Pediatrics 75:667-671.

Berger R, Florent G and Just M (1981) Decrease of the lympho-proliferative response to varicella-zoster virus antigen in the aged. Infect Immun 32:24-27.

Bogger-Goren S, Baba K, Hurley P, Yabuuchi H, Takahashi M and Ogra P (1982) Antibody response to varicella-zoster virus after natural or vaccine-induced infection. J Infect Dis 146:260-265.

Brunell PA (1989) Transmission of chickenpox in a school setting prior to the observed exanthem. Amer J Dis Child 143:1451-1452.

Brunell PA, Shehab Z, Geiser C and Waugh JE (1982) Administration of live varicella vaccine to children with leukemia. Lancet 2:1069-1072.

Brunell PA, Taylor-Wiedeman J and Geiser CF (1986) Risk of herpes zoster in children with leukemia: varicella vaccine compared with history of chickenpox. Pediatrics 77:53-56.

Brunell P, Geiser C, Novelli V, Lipton S and Narkewicz S (1987) Varicella-like illness caused by live varicella vaccine in children with acute lymphocytic leukemia. Pediatrics 79:922-927.

Brunell PA, Novelli VM, Lipton SV and Pollock B (1988) Combined vaccine against measles, mumps, rubella, and varicella. Pediatrics 81:779-784.

Bruusgaard, E (1932). The mutual relation between zoster and varicella. Brit J Derm Syph 44:1-24.

Burke BL, Steele RW, Beard OW, Wood JS, Cain TD and Marmer DJ (1982) Immune responses to varicella-zoster in the aged. Arch Intern Med 142:291-293.

Chartrand S, MBG, Steinberg S and Gershon A. Varicella vaccine in day care centers. Interscience Conference on Antimicrobial Agents and Chemotherapy. 1985.

Croen K, Ostrove J, Dragovic L and Straus S (1988) Patterns of gene expression and sites of latency in human ganglia are different for varicella-zoster and herpes simplex viruses. Proc Soc Nat Acad Sci USA 85:9773-9777.

Evans P (1940) An epidemic of chickenpox. Lancet 2:339-340.

Feldman S, Hughes W and Daniel C (1975) Varicella in children with cancer: 77 cases. Pediatrics 80:388-397.

Garland J (1943). Varicella following exposure to herpes zoster. N Engl J Med 228:336-337.

Gelb LD, Dohner DE, Gershon AA, Steinberg S, Waner JL, Takahashi M, Dennehy P and Brown AE (1987) Molecular epidemiology of live attenuated varicella virus vaccine in children and in normal adults. J Infect Dis 155:633-640.

Gershon A (1987) Live attenuated varicella vaccine. Ann Rev Med 38: 41-50.

Gershon AA, Steinberg S, Gelb L and NIAID-Collaborative-Varicella-Vaccine-Study-Group (1984) Live attenuated varicella vaccine: efficacy for children with leukemia in remission. JAMA 252:355-362.

Gershon AA, Steinberg S, Gelb L and NIAID-Collaborative-Varicella-Vaccine-Study-Group (1986) Live attenuated varicella vaccine: efficacy in immunocompromised children and adults. Pediatrics 78 (S):757-762.

Gershon AA, Steinberg S, LaRussa P, Hammerschlag M, Ferrara A and NIAID-Collaborative-Varicella-Vaccine-Study-Group (1988) Immunization of healthy adults with live attenuated varicella vaccine. J Infect Dis 158:132-137.

Gershon AA, Steinberg S and NIAID-Collaborative-Varicella-Vaccine-Study-Group (1989) Persistence of immunity to varicella in children with leukemia immunized with live attenuated varicella vaccine. N Engl J Med 320:892-897.

Gershon AA, Steinberg S and NIAID-Collaborative-Varicella-Vaccine-Study-Group (1990) Live attenuated varicella vaccine: protection in healthy adults in comparison to leukemic children. J Infect Dis 161:661-666.

Gilden D, Vafai A, Shtram Y, Bercker T, Devin M and Wellish M(1983). Varicella-zoster virus DNA in human sensory ganglia. Nature. 306:478-480.

Gilden D, Rozenman Y, Murray R, Devlin M and Vafai A (1987) Detection of varicella-zoster virus nucleic acid in neurons of normal human thoracic ganglia. Ann Neurol 22:337-380.

Gordon JE and Meader FM (1929) The period of infectivity and serum prevention of chickenpox. JAMA 93:2013-2015.

Gustafson TL, Shehab Zand Brunell P (1984) Outbreak of varicella in a newborn intensive care nursery. Am J Dis Child 138:548-550.

Ha K, Baba K, Ikeda T, Nishida M, Yabuuchi H and Takahashi M (1980) Application of a live varicella vaccine in children with acute leukemia or other malignancies without suspension of anticancer therapy. Pediatrics 65:346-350.

Hope-Simpson RE (1965) The nature of herpes zoster: a long term study and a new hypothesis. Proc Roy Soc Med 58:9-20.

Hyams P, Stuewe M and Heitzer V (1984) Herpes zoster causing varicella (chickenpox) in hospital employees: cost of a casual attitude. Am J Infect Contr 12:2-5.

Hyman RW, Ecker JR and Tenser RB (1983) Varicella-zoster virus RNA in human trigeminal ganglia. Lancet 2:814-816.

Izawa T, Ihara T, Hattori A, Iwasa T, Kamiya H, Sakurai M and Takahashi M (1977) Application of a live varicella vaccine in children with acute leukemia or other malignant diseases. Pediatrics 60:805-809.

Johnson CE, Shurin PA, Fattlar D, Rome LP and Kumar ML (1988) Live attenuated varicella vaccine in healthy 12- to 24- month old children. Pediatrics 81:512-518.

Johnson C, Rome L, Stancin T and Kumar M (1989). Humoral immunity and clinical reinfections following varicella vaccine in healthy children. Pediatrics 84:418-421.

Jones F and Grose C (1988) Role of cytoplasmic vacuoles in varicella-zoster virus glycoprotein trafficking and virion envelopment. J Virol 62:2701-2711.

Josephson A and Gombert ME (1988) Airborne transmission of nosocomial varicella from localized zoster. J Infect Dis 158:238-241.

Kundratitz K (1925) Experimentelle Ubertragung von Herpes Zoster auf den Mensschen und die Beziehungen von Herpes Zoster zu Varicellen. Monatss fuer Kinder 29:516-523.

Lawrence R, Gershon A, Holzman R, Steinberg S and NIAID Varicella Vaccine Collaborative Study Group (1988) The risk of zoster in leukemic children who received live attenuated varicella vaccine. N Engl J Med 318:543-548.

Leclair JM, Zaia J, Levin MJ, Congdon RG and Goldmann D (1980) Airborne transmission of chickenpox in a hospital. N Engl J Med 302:450-453.

Martin JH, Dohner D, Wellinghoff WJ and Gelb LD (1982) Restriction endonuclease analysis of varicella-zoster vaccine virus and wild type DNAs. J Med Virol. 9: 69-76.

Morens DM, Bregman DJ, West M, Green M, Masur M, Dolin R and Fisher R (1980) An outbreak of varicella-zoster virus infection among cancer patients. Ann Intern Med 93:414-419.

Neff BJ, Weibel RE, Villerajos VM, Buynak E, McLean A, Morton D, Wolanski B and Hilleman M (1981) Clinical and laboratory studies of KMcC strain of live attenuated varicella virus. Proc Soc Exp Biol Med 166:339-347.

Nelson A and St Geme J (1966) On the respiratory spread of varicella-zoster virus. Pediatrics 37:1007-1009.

Ozaki T, Matsui Y, Asano Y, Okuno T, Yamanishi K and Takahashi M (1989) Study of virus isolation from pharyngeal swabs in children with varicella. Am J Dis Child 143:1448-1450.

Palmer SR, Caul EO, Donald DE, Kwantes W and Tillett H (1985) An outbreak of shingles? Lancet 2:1108-1111.

Plotkin SA, Starr S, Connor K and Morton D (1989) Zoster in normal children after varicella vaccine. J Infect Dis 159:1000-1001.

Preblud, S. R. (1986). Varicella: complications and costs. Pediatrics 76 (suppl.): 728-735.

Rand KH, Rasmussen LE, Pollard RB, Arvin A and Merigan T (1977) Cellular immunity and herpesvirus infections in cardiac transplant patients. N Engl J Med 296:1372-1377.

Ross AH, Lencher E and Reitman G (1962) Modification of chickenpox in family contacts by administration of gamma globulin. N Engl J Med 267:369-376.

Ruckdeschel JC, Schimpff SC, Smyth AC and Mardiney MR (1977) Herpes zoster and impaired cell-associated immunity to the varicella-zoster virus in patients with Hodgkin's disease. Am J Med 62:77-85.

Schimpff S, Serpick A, Stoler B, Rumack B, Mellin H, Joseph JM and Block J (1972) Varicella-zoster infection in patients with cancer. Ann Intern Med 76:241-254.

Straus SE, Reinhold W, Smith HA, Ruyechan W, Henderson D, Blaese RM and Hay J (1984) Endonuclease analysis of viral DNA from varicella and subsequent zoster infections in the same patient. N Engl J Med 311:1362-1364.
Takahashi M, Otsuka T, Okuno Y, Asano Y, Yazaki T and Isomura S (1974) Live vaccine used to prevent the spread of varicella in children in hospital. Lancet 2:1288-1290.
Tsolia M, Gershon A, Steinberg S and Gelb L (1990) Live attenuated varicella vaccine: evidence that the virus is attenuated and the importance of skin lesions in transmission. J Pediatr 116:184-189.
von Bokay J (1909) Uber den aetiologischen zusammenhang der varizellen mit gewissen fallen von herpes zoster. Wein Klin Wochenschr 22:1323-1326.
Weibel R, Neff BJ, Kuter BJ, Guess HA, Rothenberger CA, Fitzgerald AJ, Connor KA, McLean AA, Hilleman MR, Buynak EB and Scolnick EM (1984) Live attenuated varicella virus vaccine: efficacy trial in healthy children. N Engl J Med 310:1409-1415.
Weibel R, Kuter B, Neff B, Rothenberger C, Fitzgerald A, Connor K, Morton D, McLean A and Scolnick E (1985) Live Oka/Merck varicella vaccine in healthy children: further clinical and laboratory assessment. JAMA 245:2435-2439.
Williams DL, Gershon A, Gelb LD, Spraker MK, Steinberg S and Ragab A (1985) Herpes zoster following varicella vaccine in a child with acute lymphocytic leukemia. J Pediatr 106:259-261.

DISCUSSION

Kilbourne E (Mount Sinai School of Medicine, New York, NY):

I think there is always a danger in extrapolating from the site where the virus is replicating in highest titer, so that the assumption is that is the site where the virus is always transmitted. I think this has been shown in influenza, for example, where the evidence on the basis that a small drop of nuclei are the primary infectious modality would indicate that primary alveolar deposition, where you don't really get striking lesions, is probably the most effective way of transmission. I wonder if you couldn't turn your data around and simply say that if you have many lesions, with much virus in the lesions in the skin, that at some unknown respiratory site you might also have more virus that would then be transmitted. I am not sure that I would follow or agree with you conclusion about the method of transmission. I wonder, in that connection, do you have any of the evidence that's been introduced for the common cold, for example, that close physical proximity is more important than, perhaps, aerosol transmission?

Gershon A:

Your points are well taken. I think it is good to have this discussion about it. In terms of the leukemic vaccinees, I think that if I had shown you the number of lesions that those children had, I might have convinced you a little bit more. Because, at least three of them had well over 200 skin lesions, at least three in which there was transmission. I think that because we are immunizing immunocompromised children, we wound up with producing a rash like that in which we saw transmission. So, clearly, it's not just that it is a child that has five or six skin lesions that goes on to transmit, but the more skin lesions that the children have, the higher the likelihood that there will be transmission. A number of them, because they were leukemic vaccinees, had a high number of skin lesions to begin with.

Kilbourne E:

I accept your data that with increasing numbers of lesions of severity, similarly, you should get higher transmissibility. But, I don't except your conclusion that it comes from the skin. It may be from respiratory sites.

Gershon A:

One answer to that would be to culture somebody's pharynx, nose, nasopharynx, or throat when they have chickenpox, and culture their skin, you can isolate the virus very easily from the skin. You can almost never isolate it from the respiratory tract. It's hard to walk away from that. We'll never resolve it. But, just for those people in the audience who are involved in infection control, I certainly don't want you to go back to your hospitals and say we don't have to isolate somebody if they were exposed to a child with early chickenpox that didn't have the rash yet. Because, in hospitals, you have to be conservative. I can't answer your question any further. I accept your doubts, but I still think it comes from the skin.

Lennette E (California Public Health Foundation, Berkeley, CA):

I can think of one possible exception. That is a paper, some 15 or so years ago, in which varicella lesions were described in the larynx. Are you familiar with that?

Gershon A:

The interesting thing is that even when there are lesions in the throat, you still can't grow the virus out of the throat. People have said that maybe it's the enzymes in saliva that are inactivating the virus, but it's just not really known. It's a very hard virus to get out of the respiratory tract. In contrast, you can get it out of the skin very easily. The way I think that it gets into the air from the skin, I'm postulating, as you know, chickenpox is a very itchy condition, and so children have the rash and they scratch it. That's how I think it gets out.

Kilbourne E:

I don't want to run this into the ground, but have you systematically cultured the throats of children with very severe and frequent skin lesions.

Gershon A:

People have done that for years, and you just can't get it out of the throat. Even if they have lots of skin lesions, and even it they have throat lesions present at the time you do your culture.

Lennette E:

A multifaceted question. Your last slide, you talked about zoster, and very few cases occurred in healthy children. With the passage of time, I would expect more cases to occur. According to all the available textbooks, currently post herpetic neuralgia appears in about 1-2% of the people that had zoster over the age of 60, let's say.

Gershon A:

I think it's higher than that. I would say more like 50% of elderly adults. Very high. The older you are when you get zoster, they more likely you are to get post herpetic neuralgia.

Lennette E:

Well, I've been one of these, and I can vouch for the problems it poses. When I was on the infectious disease service, I used to tell these people to go home and take some aspirin, till I got it. It took Demerol to ease the pain. What I'm driving at, you indicate zoster may appear in these younger age groups, 27 - 30?

Gershon A:

No. It's just that they don't respond. I think it's an interesting paradox that we have found that these adults who were immunized don't respond as well to the virus as we would like them to. When I first realized that, I began to think, oh, what if they start getting zoster. But, so far, we've only had one adult out of the 250 that we've immunized that has developed zoster. The interesting phenomenon there is that this is a physician who was one of those people who seroconverted, lost her antibody, was taking care of somebody with chickenpox in the hospital, re-seroconverted without any evidence of disease, and a year later, she got her zoster. Fortunately, we had a virus isolate from her, and it was wild type virus. So, she got zoster, but it wasn't from the vaccine. That's the only case that we've seen. I think that is what you were going to come around to asking. That was a worry once. We didn't know we were going to see adults not respond to the vaccine as well as we had hoped. But, we have only seen that one case of zoster that wasn't from the vaccine.

Ellis M (Medical Group Pathology Lab, Santa Barbara, CA):

Could you give an elderly adult the vaccine to try to boost their immunity? I watched my mother and father-in-law suffer terribly with it.

Gershon A:

People are beginning to do that now. There are studies that are organized in which elderly people are either being given vaccine or placebo to try to boost their cellular immunity.

Ellis M:

I think that is one of the worst things that can happen to an elderly adult.

Gershon A:

The incidence of it is low enough that you have to have a placebo controlled group, and you have to have a huge study to be able to demonstrate any efficacy of the vaccine. But, there are studies underway. I suppose that a subunit vaccine might be useful in that regard as well, but they're using a live vaccine.

Lennette E:

The other aspect would be cost/benefit ratio. With only 1 or 2% of these people acquiring zoster, is it worth all the money that's involved?

Ellis M:

I bet you would have said so when you had your zoster.

Gershon A:

I think it's going to take a long time before we ever get the data, but it's interesting to just speculate that because the vaccine virus is an attenuated agent, it might have a lower ability to reactivate, and it might also cause a less severe zoster, because it is attenuated. We don't know the answer to that. There haven't been, even in the leukemic vaccinees, enough cases to be able to analyze.

Berlin B (Michigan Department of Health, Lansing, MI):

Dr. Gershon, this is an interesting discussion. A very interesting phenomenon is the observation that you vaccinate somebody, he develops immune response, antibodies fall a little bit, and, presumably on reexposure to somebody, his antibodies increase. Is that increase due to direct stimulating effect of the dose of virus he got from that person, or is there replication in him at some particular spot that leads to enough virus to give rise to a secondary response?

Gershon A:

We don't know for sure, but certainly with that physician that I mentioned who developed zoster, she must have had some viral replication because she was latently infected with the wild type virus. By the way, she was definitely susceptible before she was immunized. There is no question about that. I couldn't tell you what percentage, but clearly some people have viral multiplication certainly, of those who develop a break-through disease, they have viral multiplication. We just don't know in what proportion the virus, say, hits the respiratory tract, stimulates it, and it never gets any further. It's very hard to demonstrate secretory immunity to VZ, by the way. You can't study CMI, because there is no way to study it. Antibodies have been demonstrated, IgA has been demonstrated after the natural disease as a very short lived response. So, it's hard to demonstrate.

Berlin B:

Given break-through that you observed, and given the possibility that virus can grow in a previously immunized person, would you rule out the possibility of a break-through zoster?

Gershon A:

I don't really understand your question. Are you saying a subclinical case of zoster.

Berlin B:

No, I'm saying that if one sees zoster in somebody who previously had zoster, a second attack, which presumably can occur in about 1%, is it possible that there could be break-through zoster acquired as a second infection? If one had chickenpox in childhood, is it possible that in some stage of immunity they could be re-exposed and therefore, develop a break-through zoster?

Gershon A:

I have trouble thinking about how that could happen. Because zoster is a localized infection, at least it begins that way. It's unilateral, it's dermatomal, it's not a systemic kind of a thing. So, you think about somebody who has had chickenpox or has had the vaccine, they come in contact with the virus, subsequently, they inhale it, how is it going to cause a dermatomal rash? What really makes sense is the reactivation hypothesis. I really don't think you get zoster from being re-exposed to the virus.

Minocha H (Kansas State University, Manhattan, KS):

Dr. Gershon, you mentioned about the attenuated virus or the vaccine virus develops rash one month later. Viremia with the virulent viral strain occurs, probably, earlier. I'm thinking that is a much more severe virus kind of attack. Then, the vaccine virus might be replicating at a much slower pace. That could be one of the reasons that you have the skin lesions later. During the course of that, maybe some antibodies are developing. So, while the virus is replicating, some of it might be inactivated, compounding the situation so that you have skin lesions much later, in the case of this particular virus.

Gershon A:

That is certainly possible. We do see in about 50% of the children who get a vaccine associated rash, that they've already made antibodies. So, it is certainly possible.

Lennette E:

In response to the first question that Dr. Berlin raised, Dorothy Horseman provided some of that information with rubella virus some years back. Up the pike here a few hundred miles, we have a place called Fort Ord, California training base. Rubella was not unusual in the recruits. People in Hawaii, despite the fact it's a crossroad of the world, probably, to some extent live in caves. Because there has been very little immunity in these males and recruits, to rubella. So, we decided to inoculate them with the rubella vaccine, and then let well enough alone. But, when they came to Fort Ord, some of them developed rubella and were found to be viremic. I'll leave it at that.

Galbraith J (University of Alberta Hospital, Edmonton, Alberta, Canada):

I was just wondering, based on the fact that you're looking at an 85% protection rate for leukemic children, what the role of VZIg will be. Of course with your study, you are following immunity, but would you advocate routine auditing of their immunity and then, if they were exposed, would they or would they not get VZIg. What would your protocol be?

Gershon A:

I think we are always going to need VZIg for those high risk individuals who haven't been vaccinated. I'm not convinced that if you know that the child seroconverted, that you need to give VZIg, even if they've lost their antibodies. In fact, we have not given VZIg in our studies. I should have mentioned that because obviously, that would have askewed the data very much. We did not give VZIg when we had household exposures, and as you saw, the children did just fine. However, there were a couple of situations in which we did give VZIg. One is if we knew that somebody hadn't seroconverted and we wanted to make sure they were protected. If we knew they had a household exposure, we gave them VZIg. Then, there were some people, since this was a collaborative study, who simply didn't go along with the protocol and when their kids who were vaccinated, even if they knew they had seroconverted, they gave them VZIg anyway. What is interesting, is that if you add VZIg on top of vaccine, it doesn't seem to decrease the break-through rate, very much anyway. There is certainly break-through. The infections are mild, but I am hard pressed to say that they are any milder. So, I'm not convinced that giving VZIg to children who have already been immunized successfully, has any advantage. In the real world, if this vaccine is out there for use and you have a leukemic child that has been vaccinated, and a sibling comes down with chickenpox, if you want to give VZIg on top of that, that is fine. You protect the kid, that is the major thing.

Lennette E:

We have some interesting things coming up. Those of you who watch TV this week will recognize this is the United Nations Child Week to protect children against infectious diseases, abuse or whatever, slave labor. And somebody came on the tube, representing WHO out of Washington, unheard of, I don't know what his name is or who he was, but the gist of his remark was, that we have to immunize these children against all of these infectious diseases which are numerous. His answer was that in a few years, we will have a vaccine, which is a cocktail, composed of all the important sequences out of different viruses, bacteria and parasites. All of this in one half of ml, I guess.

Gershon A:

I believe it.

Lennette E:

Do you believe it? There are other considerations. I would like to say something about antigenic mass, for one thing. Maybe Kilbourne doesn't agree with that. Those are important considerations too. Also, whether some of these oligopeptides will work. This was started years ago by Michael Sealess in Israel, when he talked about influenza and getting a vaccine which was composed only of the important immunizing portions. The rest of the ductresses you threw away. It comes back to the same thing.

Gershon A:

I think you can be lead down the garden path, though, when thinking about inactivated measles vaccine, and the trouble we got into with that because of the lack of certain antigens. Because it lacked certain antigens that were important for protection.

Kasal J (Baylor College of Medicine, Houston, TX):

Was the vaccine dose given to the adults and the children the same?

Gershon A:

Yes.

ANTISENSE OLIGONUCLEOTIDES AS ANTI-VIRAL AGENTS: NEW DRUGS FOR A NEW AGE

Jack S. Cohen

Pharmacology Department
Georgetown University Medical School
Washington District of Columbia, USA

INTRODUCTION

We need a new paradigm for the development of effective anti-viral agents. Knowledge of molecular biology, of how the cell and virus function, is now extensive. However, most drug development programs are still based on the old concepts: find an active natural product by screening, study its mode of action, and attempt to improve it by making analogs. In the more sophisticated cases the use of molecular modeling of small drug molecules to protein active sites is utilized. But, little progress has been made in developing effective anti-viral agents by these strategies.

One new strategy is that in which the drug molecules themselves are informational. The paradigm is that if you wish to disrupt a virus, use part of its genetic sequence, in other words use a gene-mimetic substance. In fact antisense mRNA's exist naturally (Inouye, 1988), and it is a logical step to introduce antisense messengers into the cell via plasmids or other vectors in order to counteract the expression of a particular gene (Izant and Weintraub, 1985; Melton, 1985). However, the control of synchronization and amount of endogenous anti-messenger expression in cell systems relative to those of the natural sense message is still very difficult (Kerr et al. 1988). In order to overcome this problem it is possible to synthesize small fragments of DNA, that contain sufficient number of bases to hybridize effectively with a single unique target site in the mRNA of the gene, yet are protected from the nucleases that are present in cells. Such a molecule is an oligodeoxynucleotide analog (Cohen, 1989a). This review will focus on the use of such substances as putative antiviral agents.

THE ANTISENSE APPROACH WITH OLIGONUCLEOTIDES

There are three basic approaches using oligonucleotides that have been developed in recent years. First, there is the true antisense effect in which a

Medical Virology 10, Edited by L.M. de la Maza and
E.M. Peterson, Plenum Press, New York, 1991

complementary oligodeoxynucleotide is used to target a specific (sense) sequence of a mRNA molecule to bring about *translation arrest* (Figure 1). Second is the use of an oligodeoxynucleotide to bind to duplex DNA to form a triple helix, or triplex structure, to cause *transcription arrest* (Figure 2) (Cooney et al. 1988). Third is the use of an oligoribonucleotide with catalytic properties, known as a ribozyme, to cleave specific target sequences in a mRNA (Haseloff and Gerlach, 1988; Zaug and Cech, 1986). In this review the focus will be on translation arrest, mainly because that is the approach that has so far yielded the most positive results. The other approaches will be considered in more detail in later sections.

The first antiviral application of the antisense method using an oligodeoxynucleotide was the suppression of Rous sarcoma virus by a synthetic 13-mer oligodeoxynucleotide targeted on the repeated terminal sequences of the 35S mRNA in chick embryo fibroblast cells (Zamecnik and Stephenson, 1978). The delay in exploiting this observation must be partly attributed to the difficulty at that time in synthesizing such compounds by chemical means. Subsequent development of the facile automated synthesis method (Caruthers, 1985) soon overcame that barrier.

A serious impediment to the realization of the goal of using oligodeoxynucleotides as inhibitors of gene expression is the susceptibility *in vivo*, of the natural phosphodiester linkage to nucleases. Clearly it would be a devastating limitation of this approach if the base sequence encoding the complement to the target could not reach that site intact. To overcome this problem chemically modified oligodeoxynucleotides have been developed that are nuclease resistant. Those used extensively to date include the phospho-triesters (Buck et al. 1990, these results have been withdrawn, see Moody et al. 1990; Miller et al. 1971), methylphosphonates (Agris et al. 1986), and thio-substituted analogs, or phosphorothioates (Eckstein and Gish, 1989; Stein et al. 1988) (Figure 3). The phospho-triesters are subject to de-esterification *in vivo* (Miller et al. 1977), but Buck et al. (1990) indicate that the methyl triesters are quite stable *in vivo*. It should be noted that the triesters and methylphosphonates are neutral molecules, and as such quite hydrophobic, while the phosphorothioate analog is charged like the natural phosphodiester. This makes a significant difference in the properties of these analogs, both in terms of solubility of long chain oligos, hybridization to the target sequence (i.e. stability of the duplex), and cellular uptake.

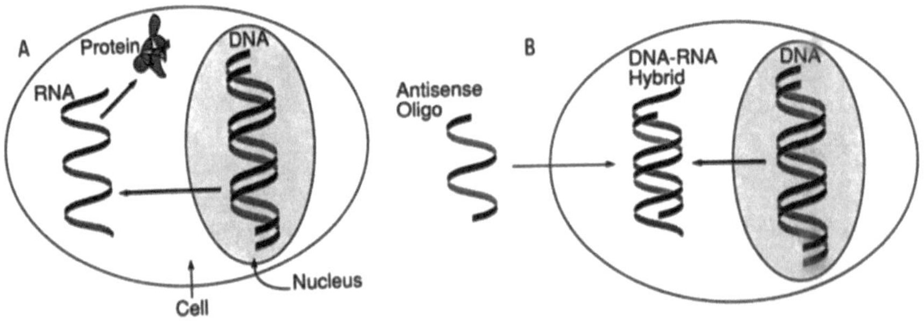

Figure 1. Diagrammatic representation of the steps involved in protein synthesis, i.e transcription from DNA to RNA, and translation from RNA to protein, and the interruption by an antisense oligo causing translation arrest.

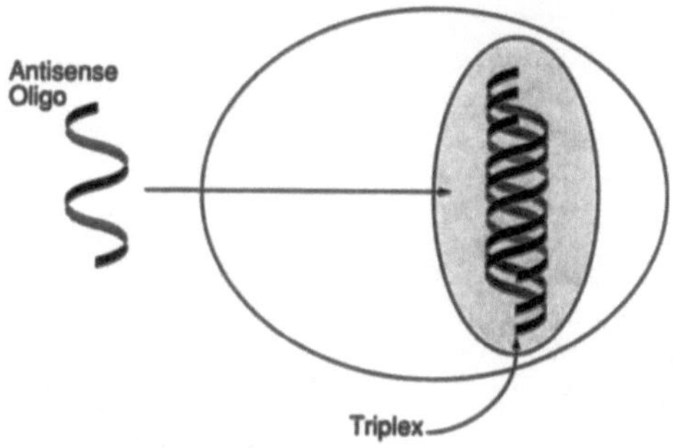

Figure 2. Diagramatic representation of the process of transcription from DNA to mRNA and its interruption by an antisense oligo causing transcription arrest.

INHIBITION IN CELL-FREE SYSTEMS

In order to determine the efficacy of an oligonucleotide or analog in regulating the expression of a particular gene, it is necessary to carry out studies in cell free systems, such as the rabbit reticulocyte and wheat germ. Usually rabbit β-globin mRNA is used, and the extent of inhibition of the incorporation of ^{35}S-methionine into globin is determined. Another approach is to microinject oligos and non-native mRNA into *Xenopus* oocytes (Kawasaki, 1985; Melton, 1985). This has the advantage that differential uptake into cells by the different oligo analogs is avoided. In a comparative study of a 17-mer sequence antisense to a segment of the β-globin mRNA of a normal phosphodiester, a phosphorothioate, and an α-oligo analog in which the glycosidic bond has the non-natural configuration (Figure 4), the differences that were observed could be rationalized in terms of the ability of the oligo to activate ribonuclease-H (Cazenave et al. 1989; Walder and Walder, 1988) (Figure 5). In this respect, the natural and phosphorothioate analogs were active, while the α-oligo (Cazenave et al. 1989; Gagnor et al. 1987) and methylphosphonate analog (Walder, 1988) were inactive. However, oocytes contain high levels of RNase-H, while this is not true for most mammalian cells. Consequently this convenient endogenous increase in effective potency of an oligo is not always present, and other means have to be considered to achieve more than a passive interference of the oligo with ribosomal processing of the target messenger.

The differences between the normal and the phosphorothioate oligos in terms of their antisense efficacy in the oocyte system were instructive. While the normal oligo was found to provide antisense inhibition, the phosphorothiate was more potent (Figure 6). However, the phosphorothioate analog caused sequence non-specific inhibition of expression at concentrations higher than 3 μM (Cazenave et al. 1989). This illustrates the need to always carry out control studies with sense, random, or homo-oligomers in such studies. One notable difference between the normal phosphodiester and the phosphorothioate oligos was the effect of varying the relative order of injection of the oligo and the target mRNA into the oocyte. When the normal oligo was injected 6 hr prior to the mRNA it lost significant activity, indicative of degradation, while the phosphorothioate oligo remained active (Figure 6). A similar change occurred when the order of injection was reversed, although the differences were not as great. Clearly the best strategy with a normal oligo is to add it at the same time as the messenger, while for the phosphorothioate this requirement is not as stringent. In *in vitro* studies with normal oligos it is customary to use heat-treated medium to counter the effects of nucleases (Zamecnik et al. 1986).

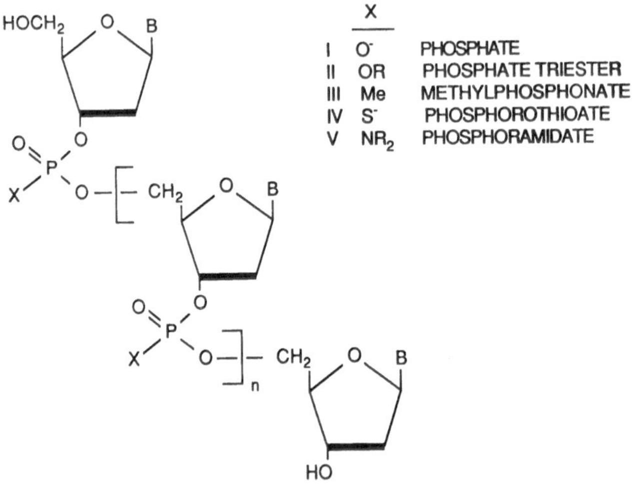

Figure 3. The structure of an oligodeoxynucleotide and the most common modifications at the phosphodiester group (B=base).

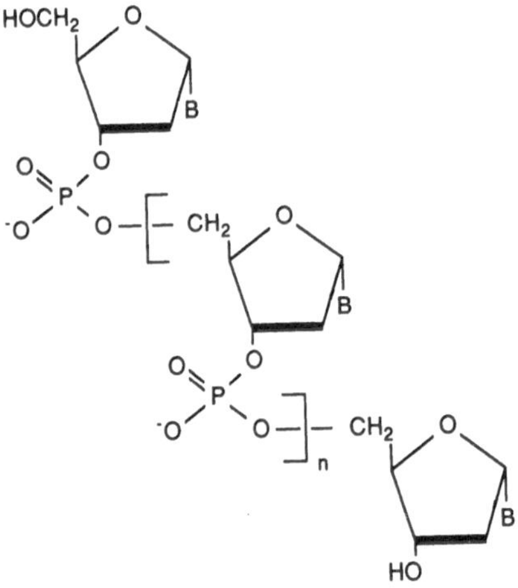

Figure 4. Structure of the non-natural glycosidic α-configuration in an oligonucleotide.

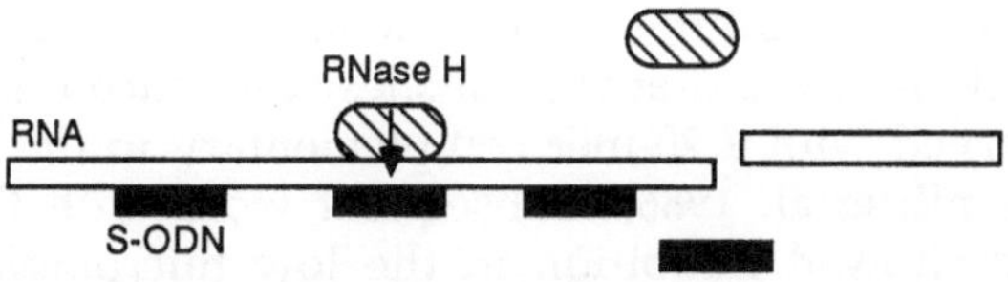

Figure 5. Function of ribonuclease-H that cleaves (*arrow*) the mRNA molecule where a DNA-RNA hybrid duplex is formed.

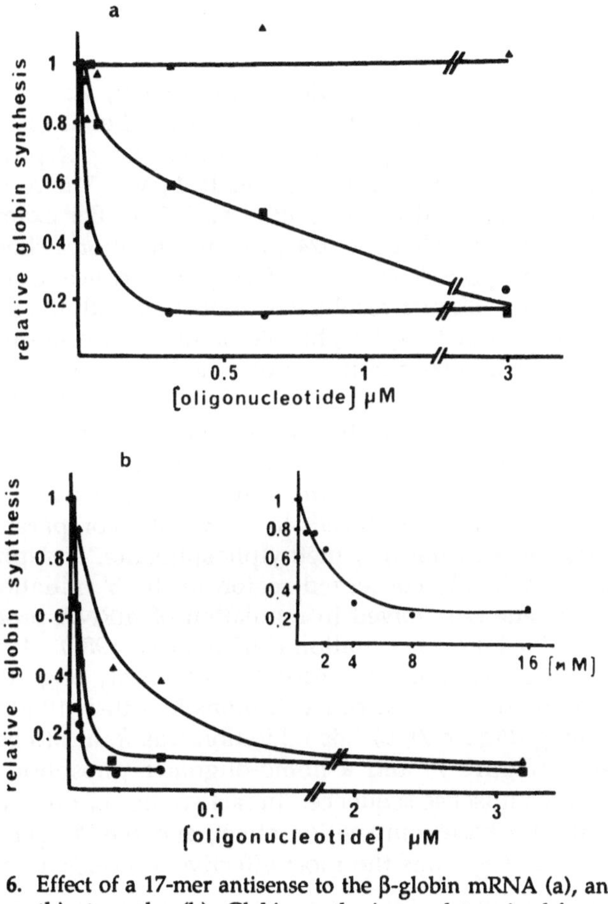

Figure 6. Effect of a 17-mer antisense to the β-globin mRNA (a), and its phosphorothioate analog (b). Globin synthesis was determined from densitometer tracings of the autoradiographs, relatively to the synthesis observed in the absence of added oligodeoxynucleotide. Triangles, oligos injected 6 hr prior to mRNA; squares, oligos injected 6 hr after mRNA; circles, oligos coinjected with mRNA. The inset in panel (b) is an enlargement of the lowest curve in (b) (from Cazenave et al. 1989).

OLIGONUCLEOTIDES AS ANTI-VIRAL AGENTS

The advent of AIDS and the identification of the retrovirus HIV-1 as its cause led to attempts to inhibit this virus with antisense oligos. The initial results with normal oligos targeted at several sites were encouraging, the greatest inhibition was observed with a 20-mer complementary to the the splice site of the *tat* gene (Zamecnik et al. 1986). Subsequent work with the phosphorothioate oligo analogs showed inhibition in the low micromolar concentration range (Matsukura et al. 1987), and this caused a great deal of attention to be focused on this area. In Table 1 are listed some applications of oligos to viral systems. Methylphosphonate oligos were earlier shown to have anti-viral activity against VSV, SV40, and HSV (Miller, 1989), although the concentrations required (30-300 μM) tend to be beyond the range of therapeutic doses.

INHIBITION OF HIV-1

Generally in the anti-HIV-1 studies carried out by Zamecnik et al. (1986); Goodchild et al. (1988), Sarin et al. (1988) and Agrawal et al. (1988), comparative inhibition was determined using a syncytia assay and p24 protein production for H9 cells at three concentrations (0.6, 3 and 15.4 μM). The results for p24 protein were generally consistent (Goodchild et al. 1988). For example, a maximal inhibition of 67% was recorded for p24 protein inhibition using a 20-mer antisense to a site in the poly-A region of HIV-1 at a concentration of 15.5 μM (Goodchild et al. 1988). Similar results were obtained with the same assay using a 15-mer (containing 13 methylphosphonate bonds) at a concentration of 22 μM (Sarin et al. 1988). A similar result was obtained using an 8-mer methylphosphonate targeted against the *tat* gene (Zaia et al. 1988); 100% inhibition in a syncytia assay was observed for both the antisense and sense sequences at concentration of 10 μM, although not for a control sequence.

Matsukura et al. (1987) used a cytopathic assay (Mitsuya and Broder, 1986) for acute HIV-1 infection in ATH8 cells to carry out a comparative study of the same 14-mer sequence as normal, methylphosphonate, and phosphorothioate oligos antisense to a highly conserved region in the 5' initiation codon of the *rev* gene. The *rev* gene is involved in regulation of mRNA processing in HIV-1, and is required for HIV-1 maturation (Malim et al. 1988). The results of this comparison were surprising in two respects: a) the only oligo that gave a dose dependent inhibition of HIV-1 at concentrations less than 10 μM was the phosphorothioate analog (Figure 7); b) this inhibition was found to be essentially the same for a sense (Figure 7) and a homo-oligomer phosphorothioate control (Figure 8) as for the antisense sequence. In fact studies of other oligos showed a significant length and base composition effect, and the 28-mer homo-oligomer of deoxycytidine, S-dC_{28}, was the most effective anti-HIV-1 compound found (Figure 8).

Matsukura et al. (1988) and Agrawal et al. (1988) also confirmed the observation of a sequence non-specific inhibition of HIV-1 by S-oligos, and Agrawal et al. (1988) also found a similar phenomenon with the phosphomorpholidates they tested. Subsequent studies with an extensive series of phosphorothioate oligos with repeating sequences (Stein et al, 1989), such as S-$d(CCG)_7$, rather than the homo-oligomers, showed that once the length is above 21 the effect of

TABLE 1. Inhibition of virus expression by antisense oligodeoxynucleotides[a]

Virus	Oligo Analog	References
Rous sarcoma	natural	Zamecnik et al. 1978
VSV	P-Me	Agris et al. 1986
	natural	Lemaitre et al. 1987
	natural	Wickstrom et al. 1986
HSV	P-Me	Smith et al. 1986
SV40	P-Me	Miller et al. 1985; Westermann et al. 1989
Influenza	natural	Zerial et al. 1987
Sendai	natural	Gupta et al. 1987
TMV	natural	Crum et al. 1988
HIV-1	natural	Zamecnik et al. 1986
	P-S	Agrawal et al. 1988; Matsukura et al. 1987, 1988, 1989a; Shibahara et al. 1989
	P-N	Agrawal et al. 1988
	P-Me	Sarin et al. 1988; Zaia et al. 1988

[a] Abbreviations used: VSV, vesicular stomatitis virus; HSV, herpes simplex virus; SV40, simian virus 40; HIV-1, human immunodeficiency virus type 1; TMV, tobacco mosaic virus; P-Me, methylphosphonate; P-S, phosphorothioate; P-N, phosphoramidate

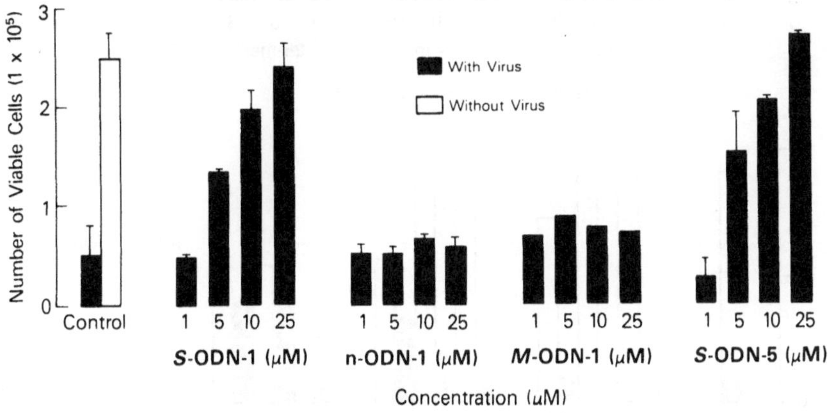

Figure 7. Comparison of inhibition of HIV-1 in the cytopathic ATH8 cell assay using normal (O), phosphorothioate (S) and methylphosphonate (M) analogs of a 14-mer oligo sequence (ODN-1) antisense to the 5'-initiation codon region of the *rev* gene. The filled bars represent viral activity, the open bars represent toxicity control (from Matsukura et al. 1987).

GC content is less pronounced (Figure 9). Experiments with basic analogs (Iyer et al. 1990) showed anti-HIV-1 action, but not as great as the original S-dC_{28}.

The absence of expression of a viral DNA fragment in Northern gels following incubation with S-dC_{28} indicated that the inhibition was occurring prior to reverse transcription. This sequence non-specific effect, while it is quite distinct from an antisense mechanism, nevertheless could be a potentially useful anti-AIDS strategy. The S-oligos have been found to be linear competitive inhibitors for the oligo primer binding site of reverse transcriptase (RT) with an inhibition constant a factor of 200 fold greater than that of the normal oligo dC_{28} (Majumdar et al. 1989). The results of inhibition of the RT polymerase activity (Figure 10), using either S-dC_n in a poly(rA)-oligo(dT) assay or S-dT_n in a poly(dI)-oligo(dC) assay (Cohen JS, Molling K, Schulze T, Subasinghe C, in preparation), also correlated with the length and composition dependence found for HIV-1 inhibition (Matsukura et al. 1987). Thus, the mechanism of the sequence non-specific inhibition of HIV-1 may arise from relatively selective inhibition of HIV-1 RT. Inhibition of surface binding of the virus, perhaps by interference with CD4 recognition as found for other negatively charged compounds could also be involved. However, similar inhibition has been found with several lentiviruses (Dahlberg JE, Archambault D, Cohen JS, Stein

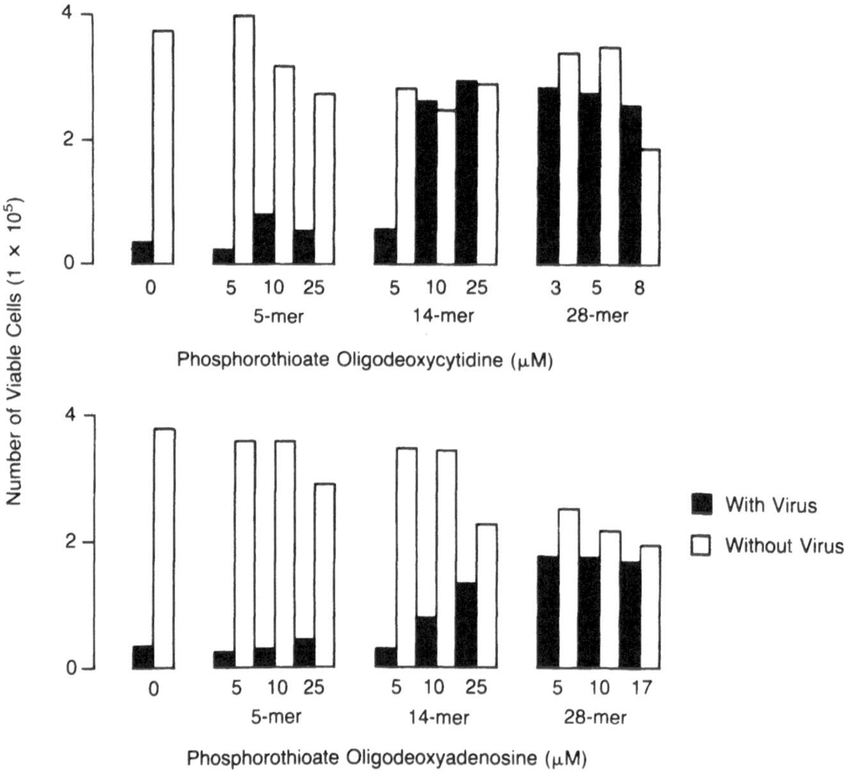

Figure 8. Comparison of inhibition of HIV-1 in the cytopathic ATH8 cell assay by S-homo-oligomers of dA and T of three lengths (from Matsukura et al. 1987).

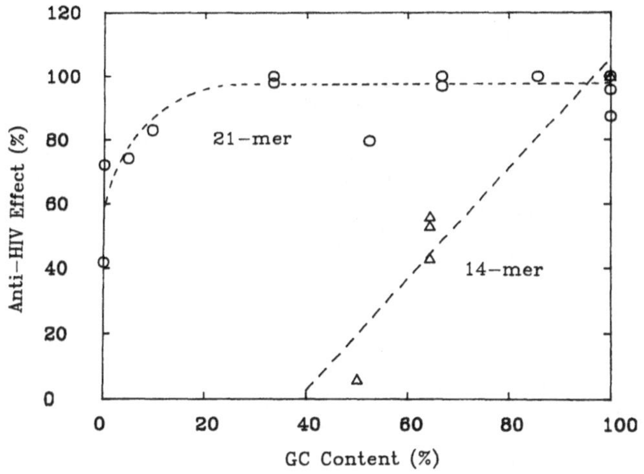

Figure 9. The number of viable ATH8 cells after >7 days incubation is plotted against the S-oligo GC content for a series of 21-mers (open circles) and 14-mers (triangles; data from Matsukura et al. 1987). Data points were obtained at the S-oligo concentration of 1 μM for the 21-mers and 5 μM for the 14-mers (Stein et al. 1989).

CA, Matsukura M, Broder S, in preparation), animal viruses of the same family as HIV-1 that bind to different cell surface targets on other cells. It is also noteworthy that S-dC_{28} is a potent selective inhibitor of HSV-2 DNA polymerase, but not that of HSV-1 (Gao et al. 1989), and this correlates with its biological activity (Cheng T-Y, Cohen JS, Gao W, Hanes RN, Stein CA, Vasquez-Padua MA, in preparation). The effectiveness of the S-oligos against HIV-1 may result from several simultaneous mechanisms, including inhibition of RT and CD4 binding.

Antisense inhibition against HIV-1 using a viral expression assay with chronically infected cells (Matsukura et al. 1989a), was observed with a 28-mer anti-*rev* sequence showed dose dependent inhibition of the expression of p24 *gag* protein (Figure 11), that was not seen with any other control, including an *N*-methyl thymidine analog, a normal analog, and sense or homo-oligomer phosphorothioates. Most anti-HIV-1 agents that have been developed, such as the di-deoxynucleosides, for example azido-thymidine (AZT) (Yarchoan et al. 1986), are not effective against HIV-1 in chronically infected cells. It was found that the sequence-specific antisense oligos remained effective longer (more than 24 hr) than the sequence non-specific oligos (Agrawal et al. 1988), and it was also found that the antisense effect persisted for up to 28 days (Matsukura et al. 1989a).

While two mechanisms are clearly operating in the case of the S-oligos with HIV-1 (Figure 12), neither that of the sequence non-specific effect nor of the antisense effect are understood in detail. The antisense inhibition could be presumed to be a classical translation arrest since a change in the mRNA profile was observed (Matsukura et al. 1989a), comparable to that of a mutant expressing low levels of *rev* (Sadaie et al. 1988). Since the S-oligos are RNase-H active (see above), there is the possibility that either the intrinsic RNase-H activity of the RT (Hansen et al. 1987) or the endogenous activity of the cell could be activated to cause translation arrest (Goodchild, 1989). In order to test this

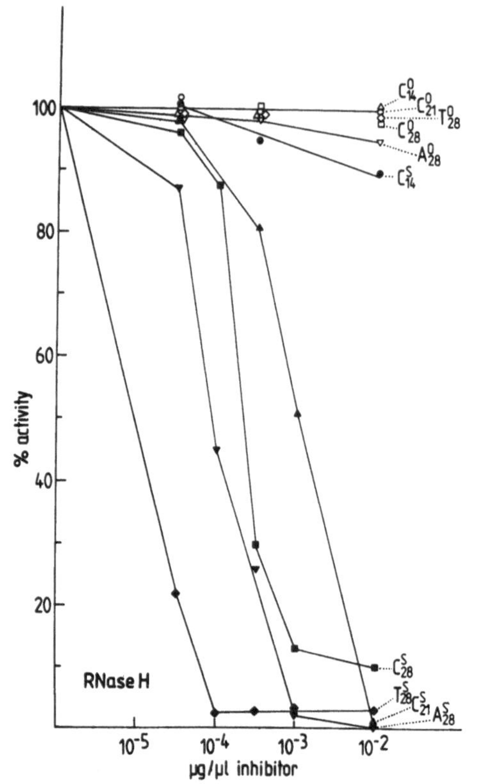

Figure 10. Inhibition of RT of HIV-1 by normal and S-oligos of various lengths using poly(rA) oligo(dT) as template and primer (Molling et al. in preparation)

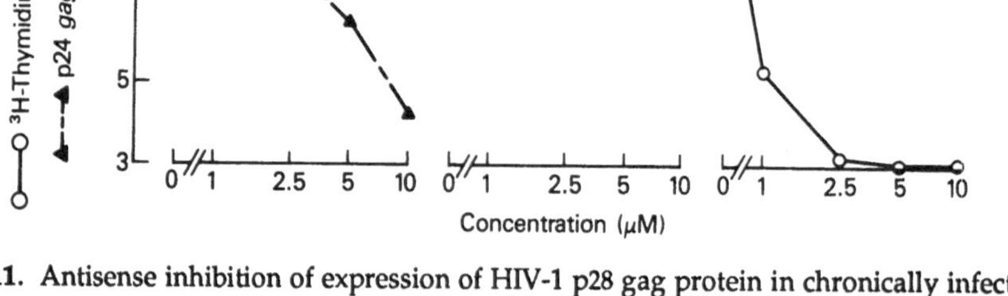

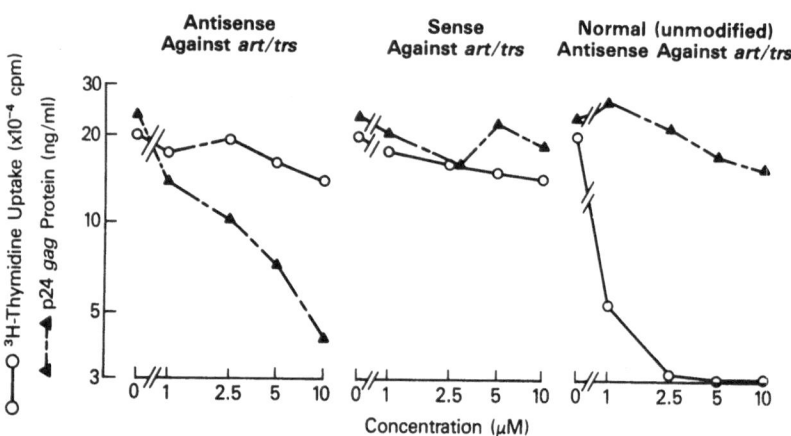

Figure 11. Antisense inhibition of expression of HIV-1 p28 gag protein in chronically infected H9 cells using the phosphorothioate 28-mer anti-*rev* sequence (left), and the absence of inhibition with the sense S-oligo and the normal antisense 28-mer; the reduction in ^{3}H-thymidine uptake in this case can be attributed to the production of cold thymidine due to hydrolysis of the normal oligo (from Matsukura et al. 1989a).

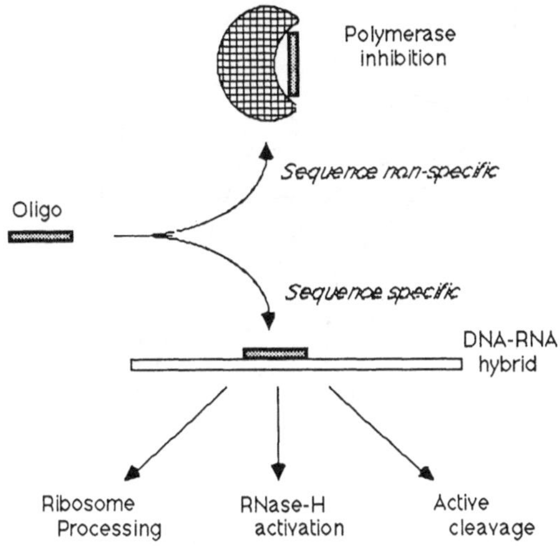

Figure 12. Representation of the two mechanisms thought to be operating in the case of inhibition of HIV-1 by S-oligos. Inhibition of RT is sequence non-specific, while inhibition of mRNA expression is an antisense sequence specific mechanism.

possibility, the RNase-H activity of RT against M13 mRNA was measured in the presence of the S-oligos of several lengths and with several normal oligo controls (Molling K, Schulze T, Subasinghe C, Cohen JS, in preparation). The results showed that RT RNase-H activity is quite different in its specificity from that of the RT polymerase activity described above (Figure 13). Thus, it is unlikely that the intrinsic RT RNase-H plays any role in the mechanism of the sequence non-specific inhibition of HIV-1 by S-oligos (Figure 12). If cellular RNase-H activity is present, it could play a role in the antisense inhibition found for normal and S-oligos, but not that of methylphosphonate oligos (Shibahara et al. 1989).

ALTERNATIVE APPROACHES

Covalent Attachment of Reactive Groups

In order to increase the potency of antisense oligos against HIV-1 several strategies can be adopted. Helene and Toulme (1989) have pioneered applications in which intercalative and/or reactive groups are attached to the termini of oligos (Thuong et al. 1989). It is possible either to covalently attach a reactive group to the oligo or to further modify the oligo, for example in the base as well as by substitution in the phosphodiester backbone. Acridine (Mori et al. 1989a) and anthraquinone (Mori et al. 1989b) 5'-linked compounds, were found to be more active than the parent anti-*rev* S-oligo analog (Matsukura et al. 1989b). It should be noted that there is also a small increase in the toxicity of these modified compounds. The mechanism of an increase in potency is not yet known, but the small increase in melting temperature on duplex formation with the anthraquinone group attached (Mori et al. 1989b), that indicates some stabilization of the duplex by intercalation, is not thought to be sufficient to give rise to

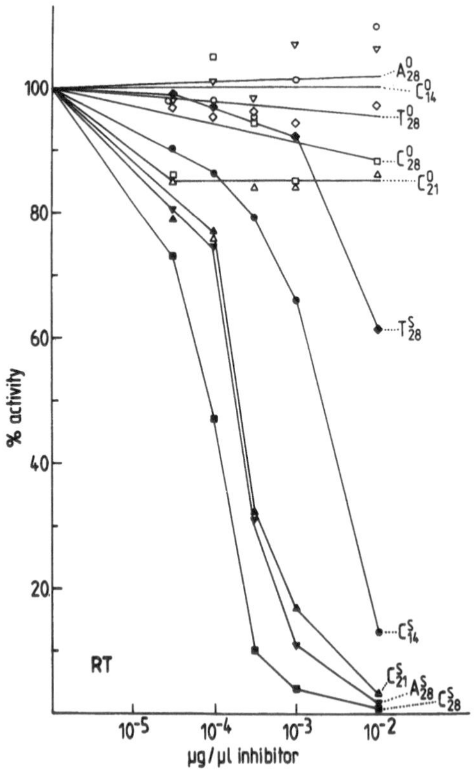

Figure 13. Inhibition of RNase-H activity of RT of HIV-1, free of cellular RNase-H. The most effective substance in this assay is S-dT28 (from Molling et al., in preparation).

the ten-fold increase in potency observed for the anthraquinone-S-oligo (Matsukura et al. 1989b). The increase may arise from destruction of *rev* mRNA by the reactive anthraquinone brought into proximity by hybridization with the oligo.

Transcription Arrest by Triplex Formation

In addition to using oligodeoxynucleotides for translation arrest by duplex formation with mRNA (Figure 1) as described above, it is possible to conceive of transcription arrest by duplex formation with DNA. However, this is considered less likely because of the stability of the DNA duplex. However the strategy of preventing transcription might be more effective than translation inhibition because one molecule of an oligo could prevent the expression of a greater number of protein products than by inhibiting the mRNA. It has been known for some time that polynucleotides form triple stranded helices (Broitman et al. 1988; Felsenfeld et al. 1957), called triplexes (Figure 14), and this could form the basis of an effective strategy for transcription arrest (Figure 2). The binding and subsequent selective cleavage of duplex DNA by pyrimidine-containing oligomers with a reactive group attached has been demonstrated by two groups (Dervan et al. 1989; Helene et al. 1989). The transcription of c-*myc* DNA into mRNA was inhibited in a cell free system by TG-containing oligomers (Cooney

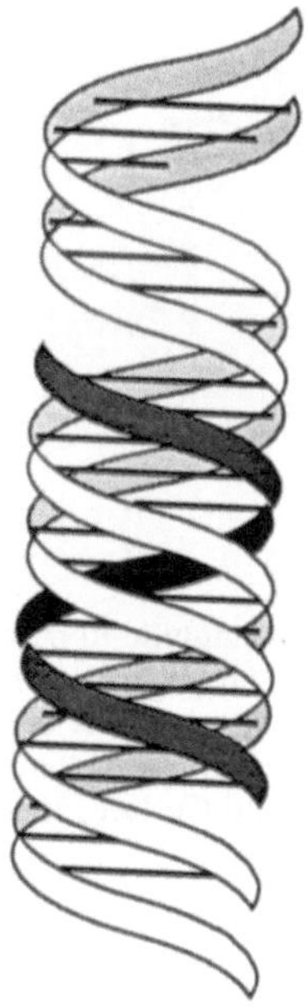

Figure 14. Formation of a triple-stranded structure, or triplex with a third oligo strand binding in the major groove of the DNA duplex.

et al. 1988). Thus, the potential for the development of an efficient means of transcription arrest exists, and it could be useful against the integrated double stranded DNA stage of HIV-1 (Orson et al. 1990). A major factor in determining the relative effectiveness of the strategy of using an oligo for duplex formation with mRNA or for triplex formation with DNA will be the relative degree of uptake into the cytoplasm and the nucleus of the cell. Up to this point the results for such distribution studies seem to depend largely upon the nature of the reporter groups used (Helene et al. 1989; Loke et al. 1989; Marti et al. 1990).

Ribozymes

Another form of antisense oligonucleotide that has been described is the self-splicing element of mRNAs (Zaug and Cech, 1986), that has recently been characterized (Haseloff and Gerlach, 1988), and termed a ribozyme (Figure 15). It has been shown that this requires the 2'-hydroxyl group of the RNase-active G residue for activity (McSwigen et al. 1989; Rajagopal et al. 1989). Recently a short ribozyme has been described (Jeffries and Symons, 1989). The effect of chemical modification on the activity of such an oligo*ribo*nucleotide is only just being investigated. It should be emphasized that the problem of nuclease-susceptibility of oligo*ribo*nucleotides is even more stringent than that of oligo*deoxy*nucleotides because of the presence of RNases *in vivo*, although these can be inhibited by substitution by a 2'-O-methyl group (Shibahara et al. 1987). But the same criteria of DNase-susceptibility, cellular uptake, and intra-

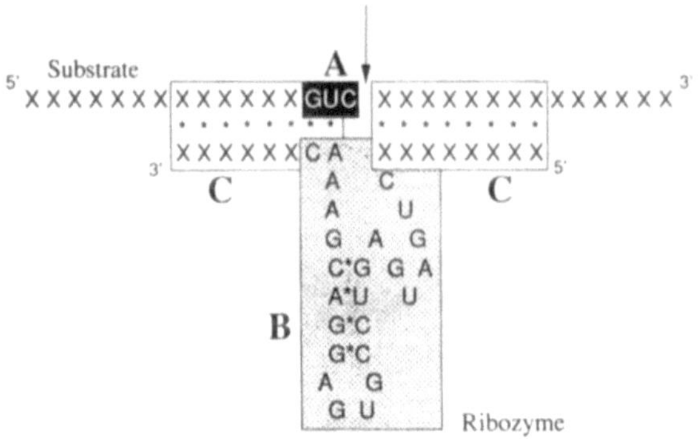

Figure 15. Structure of a catalytic RNA or ribozyme. The three structural domains are: A, target sequence in mRNA substrate adjacent to site of cleavage; B, the highly conserved sequence in the ribozyme; C, the flanking sequences that comprise the antisense segments (from Haseloff and Gerlach, 1988).

cellular distribution applies to these novel catalytic oligos as to those discussed above. Nevertheless, an effective strategy using ribozymes against HIV-1 has been reported (Sarver et al. 1990).

CONCLUSION

Antisense oligodeoxynucleotide analogs are effective as anti-viral agents *in vitro*. However, therapeutic application must be regarded as a promising but distant prospect. Much research needs to be done before this goal could be achieved. Areas in which results for oligos are very limited include all the biological areas such as cellular uptake, toxicology, and pharmacokinetics (Zon, 1989). There is also the important question of cost relative to effective dose (Zon, 1989), so that improved and cheaper methods of large scale synthesis are vital. Even if the modified S-oligos, because of their potency *in vitro* (Agrawal, 1988; Matsukura et al. 1987, 1989a,b), are found to be the first useful oligonucleotides for anti-AIDS therapy, it is highly unlikely that they will be the best or the last analog developed (Cohen, 1989b). The area of antisense oligos is currently one of great interest (Rothenberg et al. 1989), and the potential development of antisense oligo analogs for anti-viral therapy is an active research goal.

ACKNOWLEDGEMENT

The author wishes to thank Don Crisostomo for his assistance in the preparation of this review.

REFERENCES

Agrawal S, Goodchild J, Civeira MP, Thornton AH, Sarin PS, Zamecnik PC (1988) Oligodeoxynucleoside phosphoramidates and phosphorothioates as inhibitors of human immunodeficiency virus. Proc Natl Acad Sci USA 85:7079-7083.

Agris CH, Blake K, Miller P, Reddy M, Ts'o P (1986) Inhibition of vesicular stomatitis virus protein synthesis and infection by sequence-specific oligodeoxyribonucleoside methylphosphonates. Biochemistry 25:6268-6275.

Broitman SL, Im DD, Fresco JR (1988) Formation of the triple-stranded polynucleotide helix, poly (A.A.U). Proc Natl Acad Sci USA 84:5120-5124.

Buck HM, Koole LH, van Genderen MHP, Smit L, Geelen JLMC, Jurriaans S, Goudsmit J (1990) Phosphate-methylated DNA aimed at HIV-1 RNA loops and integrated DNA inhibits viral infectivity. Science 248:208-211.

Caruthers, MH (1985) Gene synthesis machines: DNA chemistry and its uses. Science 230:281-285.

Cazenave C, Stein CA, Loreau N, Thuong, NT, Neckers LM, Subasinghe C, Helene C, Cohen JS, Toulme JJ (1989) Comparative inhibition of rabbit globin mRNA translation by modified antisense oligodeoxynucleotides. Nucl Acids Res 17:4255-4273.

Cohen JS (1989b) Designing antisense oligonucleotides as pharmaceutical agents. Trends Pharmacol Sci 10:435-437.

Cohen JS (1989a) Oligodeoxynucleotides: Antisense Inhibitors of Gene Expression London/Miami: Macmillan/CRC Press.

Cooney M, Czernuszewicz G, Postel EH, Flint, SJ, Hogan ME (1988) Site specific oligonucleootide binding represses transcription of the human c-myc gene *in vitro*. Science 241:456-459.

Crum, C, Johnson, JD, Nelson, A, Roth, D (1988) Complementary oligodeoxynucleotide mediated inhibition of tobacco mosaic virus RNA translation *in vitro*. Nucl. Acids Res. 4569-4581.

Dervan PB (1989) Oligonucleotide recognition of double-helical DNA by triple-helix formation. In Cohen JS (ed.) Oligodeoxynucleoides: Antisense inhibitors of gene expression London/Miami: Macmillan/CRC Press, pp. 197-210.

Eckstein F, Gish G (1989) Phosphorothioates in molecular biology. Trends Biochem Sci 14:97-100.

Felsenfeld G, Davies DR, Rich A (1957) Formation of a three-stranded polynucleotide molecule. J Am Chem Soc 79:2023.

Gagnor C, Bertrand J-R, Thenet S, Lemaitre M, Morvan F, Rayner B, Malvy C, Lebleu B, Imbach J-L, Paoletti C (1987) a-DNA VI: comparative study of α- and β-anomeric oligodeoxyribonucleotides in hybridization to mRNA and in cell free translation inhibition. Nucl Acids Res 15:10419-10436.

Gao W, Stein CA, Cohen JS, Dutschman GE, Cheng Y-C (1989) Effect of phosphorothioate homo-oligodeoxynucleotides on herpes simplex virus type 2-induced DNA polymerase. J Biol Chem 264:11521-11526.

Goodchild J (1989) Inhibition of gene expression by oligonucleotides. In Cohen JS (ed), Oligodeoxynucleotides: Antisense inhibitors of gene expression London/Miami: Macmillan/CRC Press, pp. 53-77.

Goodchild J, Agrawal S, Civeira MP, Sarin PS, Sun D, Zamecnik PC (1988) Inhibition of human immunodeficiency virus replication by antisense oligodeoxynucleotides. Proc Natl Acad Sci 85:5507-5511.

Gupta KC (1987) Antisense oligonucleotides provide insight into mechanism of translation initiation of two Sendai virus mRNAs. J Biol Chem 262:7492-7496.

Hansen J, Schulze T, Molling K (1987) RNase-H activity associated with bacterially expressed reverse transcriptase of human T-cell lymphotropic virus-III. J Biol Chem 262:12393-12396.

Haseloff J, Gerlach WL (1988) Simple RNA enzymes with new and highly specific endoribonuclease activities. Nature 585-591.

Helene C, Toulme JJ (1989) Control of gene expression by oligodeoxynucleotides covalently linked to intercalating agents and nucleic acid-cleaving reagents. *In* Cohen JS (ed.), Oligodeoxynucleotides: Antisense inhibitors of gene expression London/Miami: Macmillan/CRC Press, pp. 137-172.

Inouye M (1988) Antisense RNA: its functions and applications in gene regulation--a review. Gene 72:25-34.

Iyer RP, Uznanski B, Boal J, Stoem C, Egan W, Matsukura M, Broder S, Zon G, Wilk A, Koziolkiewicz M, Stec WJ (1990) Abasic oligodeoxyribonucleoside phosphorothioates: synthesis and evaluation as anti-HIV agents. Nucl Acids Res (in press).

Izant JG, Weintraub H (1985) Constitutive and conditional suppression of exogenous and endogenous genes by antisense RNA. Science 229:345-352.

Jeffries AC, Symons RH (1989) A catalytic 13-mer ribozyme. Nucl Acids Res 17:1371-1377.

Kawasaki E (1985) Quantitative hybridization-arrest of mRNA in Xenopus oocytes using single-stranded complementary DNA or oligonucleotide probes. Nucl Acids Res 13:4991-5004.

Kerr MM, Stark GR, Kerr IM (1988) Excess antisense RNA from infectious recombinant SV40 fails to exhibit expression of a transfected, interferon-inducible gene. Eur J Biochem 175:65-73.

Lemaitre M, Bayard B, Lebleu B (1987) Specific antiviral activity of a poly (L lysine)-conjugated oligodeoxyribonucleotide sequence complementary to vesicular stomatitis virus N protein mRNA initiation site. Biochemistry 84:648-652.

Loke SL, Stein CA, Zhang XH, Mori K, Nakanishi M, Subasinghe C, Cohen JS, Neckers LM (1989) Characterization of oligonucleotide transport into living cells. Proc Natl Acad Sci USA 86:3474-3478.

Majumdar C, Stein CA, Cohen JS, Broder S, Wilson SH (1989) Stepwise mechanism of HIV reverse transcriptase: primer function of phosphorothioate oligodeoxynucleotide. Biochemistry 1340-1346.

Malim MH, Hauber JH, Fenrick R, Cullen BR (1988) Immunodeficiency virus rev trans-activator modulates the expression of the viral regulatory genes. Nature 335:181-183.

Marti GE, Egan W, Noguchi P, Zon G, Iversen P, Meyer A, Matsukura M, Broder S (1990) Oligodeoxyribonucleotide phosphorothioate fluxes and localization in living cells. Nucl Acids Res (in press).

Matsukura M, Mori K, Zon G, Cohen JS, Broder S (1989b) Inhibition of HIV expression: enhancement of activity of phosphorothioate oligonucleotides by chemical modification. V Intl Conf on AIDS, Montreal, Canada.

Matsukura M, Shinozuka K, Zon G, Mitsuya H, Reitz M, Cohen JS, Broder S (1987) Phosphorothioate analogs of oligodeoxynucleotides: Inhibitors of replication and cytopathic effects of human immunodeficiency virus. Proc Natl Acad Sci USA 84:7706-7710.

Matsukura M, Zon G, Shinozuka K, Robert-Guroff M, Shimada, Stein CA, Mitsuya H, Wong-Staal F, Cohen JS, Broder S (1989a) Regulation of viral expression of human immunodeficiency virus *in vitro* by an antisense phosphorothioate oligodeoxynucleotide against rev (art/trs) in chronically infected cells. Proc Natl Acad Sci USA 86:4244-4248.

Matsukura M, Zon G, Shinozuka K, Stein CA, Mitsuya H, Cohen JS, Broder S (1988) Synthesis of phosphorothioate analogues of oligodeoxyrubonucleotides and their antiviral activity against human immunodeficiency virus (HIV). Gene 72:343-347.

McSwigen JA, Cech TR (1989) Stereochemistry of RNA cleavage by the tetrahymena ribozyme and evidence that the chemical step is not rate-limiting. Science 244:679-683.

Melton DA (1985) Injected anti-sense RNAs specifically block messenger RNA translation *in vivo*. Proc Natl Acad Sci USA 82:144-148.

Miller PS (1989) Non-ionic antisense oligonucleotides. In Cohen JS (ed.), Oligonucleotides: Antisense inhibitors of gene expression. London/ Miami: Macmillan/CRC Press, pp. 79-95.

Miller PS, Agris C, Aurelian L, Blake K, Murakami A, Reddy M, Spitz S, and Ts'o POP (1985) Control of ribonucleic acid function by oligonucleoside methylphosphonates. Biochimie 67:769-776.

Miller PS, Braiterman L, Ts'o POP (1977) Effects of a trinucleotide ethyl phosphotriester Gmp(Et)Gmp(Et), on mammalian cells in culture. Biochemistry 16:1988-1996.

Miller PS, Fang KN, Kondo NS, Ts'o POP (1971) Syntheses and properties of adenine and thymine alkyl phosphotriesters, the neutral analogs of dinucleoside monophosphates. J Am Chem Soc 93:6657-6665.

Mitsuya H, Broder S (1986) Inhibition of *in vitro* infectivity and cytopathic effect of HTLV-III/LAV by 2',3'-dideoxynucleosides. Proc Natl Acad Sci USA 83:191-1915.

Moody HM, Quaedflieg PJLM, Koole LM, van Genderen MHP, Buck HM, Smit L, Jurriaans S, Geelen JLMC, Goudsmit J (1990) Inhibition of HIV-1 infectivity by phosphate-methylated DNA. Retraction. Science 250:125-126.

Mori, K, Subasinghe, C, Stein, CA, Cohen, JS (1989a) Synthesis of properties of novel 5'-linked oligos. Nucleosides & Nucleotides 8:649-657.

Mori K, Subasinghe C, Cohen JS (1989b) Oligodeoxynucleotides analogs with 5'-linked anthraquinone. FEBS Lett 249:213-218.

Orson FM, Thomas DW, McShan WM, Kessler DM, Hogan ME (1990) Inhibition of IL-2Ra mRNA transcription by colinear triplex forming oligonucleotides in intact lymphocytes. FASEB J Abstr. 971.

Rajagopal J, Doudna JA, Szostak JW (1989) Stereochemical course of catalysis by the tetrahymena ribozome. Science 244:692-694.

Rothenberg M, Johnson G, Laughlin C, Green I, Cradock J, Cohen JS (1989) Oligonucleotides as antisense inhibitors of gene expression: therapeutic implications. J Natl Cancer Inst 81: 1539-1544.

Sadaie MR, Rappaport J, Benter T, Josephs SF, Willis R, Wong-Staal F (1988) Proc Natl Acad Sci. USA 85:9224-9228.

Sarin PS, Agrawal S, Civeira MP, Goodchild J, Ikeuchi T, Zamecnik PC (1988) Inhibition of acquired immunodeficiency syndrome virus by oligonucleoside methylphosphonates. Proc Natl Acad Sci USA 85:7448-7451.

Sarver N, Cantin EM, Chang PS, Zaia JA, Ladne PA, Stephens DA, Rossi JJ (1990) Ribozymes as potential anti-HIV therapeutic agents. Science 247:1222-1225.

Shibahara S, Mukai S, Morisawa H, Nakashima H, Kobayashi S, Yamamoto N (1989) Inhibition of human immunodeficiency virus (HIV-1) replication by synthetic oligo-RNA derivatives. Nucl Acids Res 17:239-252.

Shibahara S, Mukai S, Nishihara T, Inoue H, Ohtsuka E, Morisawa H (1987) Site-directed cleavage of RNA. Nucl Acids Res 15:4403-4415.

Smith C, Aurelian L, Reddy M, Miller P, and Ts'o P (1986) Antiviral effect of an oligo(nucleoside methylphosphonate) complementary to the splice junc-

tion of herpes simplex virus type 1 immediate early pre-mRNAs 4 and 5. Proc Natl Acad Sci USA 83:2787-2791.
Stein CA, Matsukura M, Subasinghe C, Broder S, Cohen JS (1989) Phosphorothioate oligodeoxynucleotides are potent sequence non-specific inhibitors of de novo infection by HIV. AIDS Res 5:639-646.
Stein CA, Shinozuka K, Subasinghe C, Cohen JS (1988) Physicochemical properties of phosphorothioate oligonucleotides. Nucl Acids Res 16:3209-3221.
Thuong NT, Asseline U, Monteney-Garestier T (1989) Oligodeoxynucleotides covalently linked to intercalating and reactive substances: synthesis characterization and physicochemical studies. In Cohen JS (ed.), Oligodeoxynucleotides: Antisense inhibitors of gene expression London/Miami: Macmillan/CRC Press, pp. 25-51.
Walder J (1988) Antisense DNA and RNA: progress and prospects. Genes Dev 2:502-504.
Walder RY, Walder JA (1988) Role of RNase H in hybrid-arrested translation by antisense oligonucleotides. Proc Natl Acad Sci USA 85:5011-5015.
Westermann P, Gross B, Hoinkis G (1989) Inhibition of expression of SV40 virus large T-antigen by antisense oligodeoxyribonucleotides. Biomed Biochim Acta 48:85-93.
Wickstrom E, Simonet W, Medlock K, Ruiz-Robles I (1986) Complementary oligonucleotide probe of vesicular stomatitis virus matrix protein mRNA translation. Biophys J 49:15-17.
Yarchoan R, Klecker RW, Weinhold KJ et al. (1986) Administration of 3' azido-3'-deoxythymidine, an inhibitor of HTLV-III replication, to patients with AIDS or AIDS-related complex. Lancet 1: 575-580.
Zaia JA., Rossi JJ, Murakawa, GJ, Spallone PA, Stephens DA, Kaplan BE, Eritja R, Wallace RB, Cantin EM (1988) Inhibition of human immunodeficiency virus by using an oligonucleotide methylphosphonate targeted to the tat-3 gene. J Virol 62:3914-3917.
Zamecnik PC, Goodchild J, Yaguchi Y, Sarin PS (1986) Inhibition of replication and expression of human T-cells lymphotropic virus type III in cultured cells by exogenous synthetic oligonucleotides complementary to viral RNA. Proc Natl Acad Sci USA 83:4143-4146.
Zamecnik PC, Stephenson M (1978) Inhibition of Rous Sarcoma virus replication and cell transformation by a specific oligodeoxynucleotide. Proc Natl Acad Sci USA 75:280-284.
Zaug AJ, Cech (1986) The intervening sequence RNA of tetrahymena is an enzyme. Science 231:470-475.
Zerial A, Thuong NT, Helene C (1987) Selective inhibition of the cytopathic effect of type A influenza viruses by oligodeoxynucleotides covalently linked to and intercalating agent. Nucl Acids Res 15:9909-9919.
Zon G (1989) Pharmaceutical considerations. *In:* Cohen JS (ed.), Oligode-oxynucleotides: Antisense inhibitors of gene expression London/Miami: Macmillan/CRC Press, pp. 233-247

DISCUSSION

Cesario T (University of California Irvine Medical Center, Orange, CA):

Once these stable RNA's get inside the cell, and I'm thinking in terms of potential for toxicity, what is their half-life in the cell, or do they stay forever? Are they that stable?

Cohen J:

These particular phosphorothioate analogues are quite stable. We've done some studies where you look at them in the presence of pure nucleases, and show that they are broken down, but sometimes 1,000 times more slowly than the normal ones. People have worried and are worrying about antigenic effects. People are worried about other such things as incorporation, for example, if you make mononucleotides, by breakdown, are these going to be incorporated into messenger RNA and DNA in the cell. Indeed, these are problems. As far as we can tell, the incorporation is extremely slow, but not nonexistent. These are things that are going to have to be worried about in the future in terms of toxic side effects, apart from binding of the sequence to another target. I just want to make one other comment. You mentioned antisense messenger RNA. What I'm showing is oligodeoxynucleotides. The reason I'm using deoxy should be very clear, that is, those are easy to make in the automatic synthesizer and they are more stable than the RNA versions. You have to worry about RNAse activity as well as DNAse activity. For example, 2'-0- methyl substituted RNA molecules have been made, but they are still susceptible to DNAses. But, if they have those 2' hydroxyl groups, they are susceptible to RNAses and there are a lot of RNAses around, as anybody who has done any molecular biology with messenger RNA would know. So, we use the DNA molecules because they are the most stable ones we can come up with, and the easiest ones to make at the present time. The people who are working on ribozymes, for example, would be focusing on RNA molecules which can and are being synthesized in the laboratory.

de la Maza L (University of California Irvine Medical Center, Orange, CA):

Jack, can you please give us a brief description of how you are approaching this therapy *in vivo*?

Cohen J:

I should say a couple of things about this. I am, myself, not doing *in vivo* studies. There are a complex of reasons for this. I recently left NCI. About a year ago, Applied Biosystems actually gave a donation, a gift of several grams of these compounds to NCI for toxicology and pharmacology studies, principally in animals, because it's not enough to do anything in humans. In fact, it's not allowed yet until those studies are done. So, those studies are underway, they're being done by the Drug Development Program under the Division of Cancer Treatment. In addition to that, some of you may be aware that about ten companies have been set up. Small, venture capital investment pharmaceutical companies to attempt to exploit this area. Some of them, no doubt, are racing to do *in vivo* studies. The big limitation here is getting enough of the material. My attitude is that if a company is set up, it has to try and develop a particular product. I am a research scientist. What I am interested in doing is finding out which of these is the best analogue, and which works most effectively. Which is the one that is going to stay around the longest without the problem of toxic side effects. Which one will get into cells more easily. I'm not talking simply about a chemically modified backbone, I'm talking about the whole range of substitutions and adding groups at the termini and so on. My approach is more of a pharmacological approach. The whole question of the toxicity and pharmacokinetics and the potential *in vivo* application in AIDS patients is clearly something which other people are working towards very actively.

Levy J (University of California Irvine, Irvine, CA):

I am curious about the range of molar ratios that resulted in substantial inhibition of activity and if the molar ratios are very high, how do you interpret that?

Cohen J:

There are several mechanisms that these oligos can work by. As you may have noticed, in the Xenopus oocyte system, the number of messenger RNA molecules was twice as high as the number of oligonucleotides, in fact. I didn't have time to go into that. It worked at a low nanomolar range, and that's because the oocyte has a lot of RNAse H, we believe. Therefore, in effect, you're activating the effect. So, you get more activity per oligonucleotide. In some of these HIV studies, especially with the antisense effect, we're going up to 10 or 25 μM concentration, which is obviously much, much higher than the number of messenger RNA's we're attacking. I think there are very good reasons for that. One is cellular uptake. We're talking about extracellular concentration, not intracellular concentration. What is, precisely, the intracellular concentration? We're trying to answer that by using radioactively labeled compounds. We're also looking at distribution, where do the oligos go when they get in the cell? What is the distribution? As I said, people try to answer that question by putting fluorescent groups on them and we found that, in fact, the fluorescent group, that because it's a large hydrophobic group, presumably, affects the distribution. That's a whole area which people are studying. One of the problems is that if you simply have passive binding of this oligo to the mRNA, which was shown in the earlier studies, then it's always an exchange process, the oligo can come off. But it has to be there in order to prevent the mRNA being processed. That is obviously a very inefficient process. That's why we want to make activated compounds where we put a group on the end. As you saw, we have a ten-fold increase in potency. We're down now to the low micromolar range with that kind of compound. That kind of compound has not, in fact, been tested in any other system, I think, than the one we had here. Although people are making those compounds and beginning to test now, that kind of thing is only just beginning. I think that if you activate the oligo rather than just have passive, inhibitor, binding, you overcome several problems. One of the problems is the relative molar ratio. The other thing you do is you make this molecule a targeting molecule that will destroy a number of possible molecules of mRNA. There are problems, unquestionably, about making enough material to do these kinds of studies, and those kinds of concentrations may be beyond the realm of *in vivo* possibility. But there are ways, as I say, to overcome that which I think will be tested in the next few years.

Cesario T:

I'm looking again at this question of the toxicity. If you're examining synthesis of a specific compound, that is inhibited by your specific oligonucleotide. Do you ever check other proteins being synthesized in that cell to see if they are affected, to be sure that it is, in fact, a specific inhibition and not a non-specific toxic effect on cells.

Cohen J:

Yes. In fact, we didn't in that particular case, because β-globin is, as far as I remember, is not produced by those cells. So, if you see β-globin, it is selectively produced by the mRNA you're putting in. It is a really selective effect. If you see it, it is because of specific expression. If you don't see it, it's because of specific inhibition. We use controls. We use the sense and so on. But, in other cases, for example, the tritiated thymidine you saw, the question there is are you inhibiting everything? The fact is that you're not. I would go even further to say that in some cases, for example, the work that I did with John Reed on the BCL2 oncogene. I was looking for an oncogene that could be inhibited very specifically, it wasn't massively over-produced and so on. He had a very nice system. We collaborated on it, and the initial studies showed the cell count. We easily showed that cell count goes down with the oligonucleotide. The problem is what's happening intracellularly. He did those studies and showed very clearly that it wasn't. Let me mention one thing. The polymerase initiation problem, the reverse transcriptase and so on, doesn't occur in cells. That is a very useful target for an antiviral. In general, you have to be able to show whether or not if your effect is truly an antisense effect, and sequence specific.

Cesario T:

What would you guess it would cost at this point to produce this compound?

Cohen J:

The estimates vary between about 50-150 thousand dollars to produce roughly a kilogram of these compounds. That is a lot, but it's not so outrageous when you consider that it's the cost of treating one patient with genetic therapy at NIH at this point. One of the problems with genetic therapy, I don't want to go into details, but you need a lot of people to do it. You need technicians to take the cells and grow them and put them back into the patient. The price of that is never going to go down below a certain level. Whereas, with any chemotherapy, if you can find an easier way to make it, then of course, you're in a much better position. One of the things that arises in relation to these compounds is, could you make them naturally, microbiologically. The answer to that is, well, who knows. Maybe. Remember, we need a chemically modified version. But, there are tricks that one can play. One can trick organisms into doing things. We can also make compounds and then subsequently modify them. In terms of the generality of this approach, using the fact that you've got that wonderful base sequence which should be the target of modern molecular biological chemotherapeutics. That's what we should be going for, not for approaches that happen to be on the cell. It may work against HIV, it may be the one compound that really works since is is a protein inhibitor. But, in general terms, knowing what we know about molecular biology, using the specificity and the amount of drug that should work, we should be able to design something that should be able to perform that sort of tailored function.

de la Maza:

Jack, from the point of view of HIV, obviously, we are just blocking expression of the virus. The progene is still sitting in the cell. On the other hand, with these oligonucleotides, in theory, you have the ideal drug that you can go directly to the progene, with this oligonucleotide as a magic bomb, so that you can go to the progene, attach it and blow it. Do you think that you can do it?

Cohen J:

In principle, I think the answer is yes. Let me say that there are very few, if any, other compounds that work in chronically infected assays. In other words, all the dideoxyonucleoside compounds work in acute assays, they prevent reverse transcriptase working. The patient has to be given these things forever.

Now, if something works in a chronic assay, it's really preventing the expression. As you say, if you use an activated molecule, it would destroy the virus. We don't really have evidence that these compounds are working only against the mRNA, not against the viral RNA, which has the same sequence. It may be protected in some way, we don't know that. More studies need to be done to show that, but in principle, I think if you have an activated compound, it could destroy the presence of the virus and/or it's ability to express.

Levy J (University of California Irvine, Irvine, CA):

I would like to pursue that question one step further. We tend to think of oligos kind of looking at DNA sequences until they find the most stable match. In a number of contexts we've seen, they can at least transiently, bind to nucleic acids, even with very short matches. So, if you attach a bomb to a specific molecule like that, how many places will it be likely to disrupt before it disrupts the correct target?

Cohen J:

That is a very good question, and it's one that has worried people. The problem of mismatching. That is, an oligonucleotide of any particular length can bind to a complementary sequence which has one, two or three mismatches, and so on. One of the points about this is that as you increase mismatches, you lower the Tm and the Tm is used often as a measure of hybridization capability. If you have one mismatch, there is no question, you can get some binding. One thing that one should always do is check the sequence of the human genome as much as you can. In fact, what people do now is they test their sequences against all the sequences known, to see what complementary and mismatched sequences there are in all the genomes that have been determined. It's not that unusual now to be able to do a search like that fairly quickly. But, if you come up with one mismatch, you may have a problem. Two is getting a little bit more borderline, and three, usually you would say, under less stringent conditions in the cell, the likelihood goes down very significantly with three. However, there is another aspect to this, and that is that the groups that one puts on the end may be, shall we say, "made to be mild" groups. That is, they will only act if, indeed, there is a certain sort of residence time. That's going to be difficult to engineer. But, it's not beyond man's ingenuity to do that. There are many different chemical groups available. So, I would mention that this is something that we are actively pursuing.

Levy J:

Actually, my concern went a little beyond the single base mismatch. We label nucleic acids in the laboratory, often using hexamers to do random prime labeling, so we know that a six-base match is certainly sufficient to get a polymerase started, at least at certain polymerase concentrations. If you do PCR, you know that even a short 3' match can give you synthesis at the wrong site, maybe three or four bases.

Cohen J:

All I can say is that from a pragmatic point of view, inhibition, which is sequence specific, is seen. If that were a general problem, that is, parts of this oligo were binding in all sorts of different places where there were hexamers and pentamers and so on, then you'll get general toxicity. You don't see that.

Levy J:

Inhibition may depend on a fairly stable binding. A very transient binding may inhibit a polymerase very transiently. But, if you actually have a bomb attached to the molecule, and perhaps a transient binding is an option, so what I'm getting at is that the bomb would have to be very carefully designed.

Cohen J:

That's exactly right. One of the things that I like to say that this is in a way, the true magic bullet, because is really is sequence specific. It's much more specific than an antibody, for example. But, the question is getting into the cell and things like that, and it has to sort of explode at some point. We have to get over the problem of it being a passive interaction. How you design the groups that you put on, how they function, I think that is very important. You definitely have made a very perceptive point. It has to be taken very seriously.

COMPARISON OF PRIMARY AND CONTINUOUS AFRICAN GREEN MONKEY KIDNEY CELLS (AGMK) FOR A RAPID VARICELLA-ZOSTER VIRUS (VZV) CULTURE

C. Belaski, B. Brumback, M. Morris and J. Bailey

American Medical Laboratories, Inc.
Fairfax, VA, USA

Comparison of MRC-5 (human lung fibroblast), A549 (human lung carcinoma) and primary AGMK cells for rapid culture of VZV has been previously described. Since commercially available cell culture is more costly than that produced in-house, we compared the sensitivity of primary AGMK purchased from ViroMed Laboratories, Inc. to Vero (continuous AGMK) cells grown in-house. A total of 155 fresh specimens were inoculated into 1 vial each of AGMK and Vero and 1 vial of A549 for rapid VZV culture. Vials were centrifuged at 950 x g for 1 hr, refed with maintenance medium, and incubated for a minimum of 48 hr. After incubation, the coverslips were stained for VZV by indirect immunofluorescence. Of the 155 specimens cultured, 30 were positive for VZV in one or more vials. Of the 30, 25 were positive in all three vials (83%), 1 was positive in the A549 only, and 4 were positive in the Vero and A549. None were positive in the AGMK only. The sensitivities of the Vero and AGMK were 96.7% and 83.0%, respectively. Because the Vero was more sensitive for VZV and less costly than primary AGMK, our laboratory discontinued using the purchased AGMK vials. We now use a combination of Vero and A549 for rapid culture of VZV.

TRANSIENT EXPRESSION OF A CLONED NATURAL VARIANT OF HEPATITIS B VIRUS (HBV) GENOME IN A HEPATOMA CELL LINE

S.S. Bukhari and K.N. Tsiquaye

London School of Hygiene and Tropical Medicine
University of London, U.K.

Although a functional pre-C region is not required for hepadnavirus core antigen expression, it is essential for e antigen expression and secretion. Sequencing data of a number of HBV clones isolated from infected individuals have revealed an inactive pre-C region, with either an in-frame TAG stop codon or a +1 frame shift mutation. Recent sequence analysis of HBV cloned from plasma from an HBeAg-positive chimpanzee carrier has revealed an in-phase stop codon (TAG) in the pre-C region. In order to examine the replication and expression competence of this natural genomic variant, we cloned a tandem repeat of the genome in pUC19 vector and transfected the construct, a pPstbd14 in HuH7 hepatoma cell line. An expression and replication competent construct, plasmid pEcobd2 containing a tandem dimer of a wild type HBV genome (subtype adw2), was used as a control in these experiments. Data from our studies show that cells transfected with pEcobd2 and pPstbd14 replicated HBV and expressed HBsAg. In contrast to pEcobd2, no HBeAg was detected in harvested culture fluid of pPstbd14 transfected cells. Since the HBV genome in our construct was derived from virus circulating in an HBeAg-positive chimpanzee, our observations suggest that both wild type HBV and a variant with a point mutation (TGG-TAG) in the pre-C region can replicate concurrently in a host.

COMBINED USE OF DIRECT FLUORESCENT ANTIBODY AND BLOCKING ANTIBODY ASSAYS FOR CONFIRMATION AND QUALITY CONTROL OF CHLAMYDIA-ELISA RESULTS

B. Cahoon-Young, A. Chandler, B. Hill and R. Benjamin

Alameda County Health Care Services Agency
Oakland, CA, USA

Automated ELISA procedures have met the demand for widespread screening of *Chlamydia trachomatis*. However, several studies have reported unconfirmed reactive ELISA results; primarily due to poor specimen quality from female patients. It is apparent, therefore, that routine methods for confirmation and quality control must be applied to newer techniques for chlamydia detection. In this study we have evaluated the use of an in-house modified direct fluorescent antibody (DFA) assay (Ortho Diagnostic Systems, Raritan, NJ), and blocking antibody (B1-Ab) (Abbott Labs, N. Chicago, IL) for confirmation of automated Chlamydia-ELISA (Ortho Diagnostics, Raritan, NJ) assay results. The DFA procedure was modified (M-DFA) for simplification and to enable comparison of equivalent amounts of the same specimen by ELISA, B1-Ab and M-DFA. A total of 209 specimens were reactive by ELISA after screening 2,664 specimens from high and low prevalence clinics in Alameda County. All ELISA-reactive specimens were further tested by M-DFA and B1-Ab assays for agreement/confirmation. Specimen quality was determined by observation of M-DFA slides. ELISA-reactive specimens were scored as positive when one or both of the supplemental assays were reactive. Poor quality specimens (excess squamous epithelial cells or red blood cells/too few columnar epithelial cells) that were negative by both supplemental assays, were scored as unsatisfactory. The rate of agreement was 79.9% (167/209) for the two supplemental tests. Twenty-three ELISA-reactive specimens were reactive by M-DFA only and 19 specimens were reactive by B1-Ab only. Eighty-six percent (60/70) of specimens from male patients reactive by ELISA were confirmed as positive, while only 62% (86/139) of specimens from female patients were confirmed. Additionally, 32.1% (17/53) of the specimens from females which were not confirmed were unsatisfactory. These results suggested that poor specimen quality due to improper sampling led to decreased assay efficiency for females. This would account in part for the disproportionate rate of confirmation in specimens from males and females. The use of M-DFA and B1-Ab assays support the need and provide a basis for simple and convenient methods to monitor specimen quality and improve or ensure assay efficiency. Parallel studies of chlamydia culture with confirmation versus chlamydia-ELISA with confirmation are in progress.

OCULAR DISEASE IN SIMIAN IMMUNODEFICIENCY VIRUS (SIV) IMMUNOSUPPRESSED RHESUS MONKEYS

M.D. Conway, B. Davison-Fairburn, M. Murphy-Corb, L. Martin, K. Soike, P. Didier, G. Baskin and M. Insler

Tulane University Delta Regional Primate Research Center and
Louisiana State University Eye Center
Covington, LA, USA

We conducted ocular examinations in rhesus monkeys systemically infected with SIV for various periods of time as part of the ongoing studies. We found either anterior segment findings such as conjunctivitis (culture positive and culture negative for SIV), or rubeosis (iris neovascularization secondary to retinal disease). In the posterior segment, we found retinitis, optic neuritis, choroiditis and panophthalmitis both clinically and histopathologically. Investigation of the retinas by electron microscopy (EM) revealed SIV and herpes virus particles in the retina and choroid, respectively, of one animal with retinitis and SIV in the retina of the other eye. In a second animal, we documented herpesvirus particles in the choroid of both eyes. Both animals had clinical retinitis paralleling that seen in HIV-1 infected humans. Serology precluded herpes B virus or herpes simplex as the etiology of the herpes viral particles in the choroid and suggested cytomegalovirus (CMV) as the likely agent. The third animal had syncytial giant cells, histopathologic evidence of SIV infection in the entire uveal tract (iris, ciliary body, choroid) and the optic nerve. There was also retinal edema present. Approximately 50% of animals with SIV lesions in the retina or optic nerve had concomitant central nervous system lesions (meningitis or encephalitis). A fourth animal presented with yet another lesion, that of enophthalmitis secondary to pneumococcus, and a fifth had choroiditis secondary to *Histoplasma capsulatum. Mycobacterium avium-intracellulare* was found in the gut but not in the eye. The findings of ocular lesions due to SIV and opportunistic infections such as CMV and SIV retinitis, SIV conjunctivitis, pneumococcal endophthalmitis, and histoplasma choroiditis offer a useful model for ocular disease and antiviral drug testing.

THE ONSET OF LYMPHOPROLIFERATIVE DISORDERS (LPD) IN TRANSPLANT RECIPIENTS IS PRECEDED BY AN INCREASED REPLICATION OF EPSTEIN-BARR VIRUS (EBV) TYPE A

P. Diaz-Mitoma, A. Ruiz, J.K. Preiksaitis, A. Wills and D.L.J. Tyrrell

University of Alberta and
Children's Hospital of Eastern Ontario, Ottawa, Canada

Six patients developing LPD and 6 sex- and organ-matched controls were prospectively studied in order to examine the effects of the level of EBV replication in the oropharynx, and to determine whether specific EBV variants could be involved in the pathogenesis of LPD. The levels of oropharyngeal excretion of EBV were measured by a quantitative dot hybridization assay. The polymerase chain reaction was used to amplify a sequence of the EBNA-2 gene, which codes for the transforming phenotype (type A or type B) of EBV. Oligonucleotide probes were used to differentiate the EBV variants. LPDs occurred in 2 (1.4%) of 142 kidney transplants and in 4 (6.9%) of 58 heart transplants. The median times for the development of LPDs, and peak levels of EBV excretion after transplantation were 12 weeks (range: 6-26 weeks), and 6 weeks (3-16 weeks), respectively. There was a median time of 8 weeks (2-10 weeks) from the peak of EBV excretion to the diagnosis of LPD. The peak levels of EBV excretion were higher in patients with LPD than in controls (median: 1.33, range: 0.33-3.18 versus 0.088, range: 0.007-0.200 ng EBV DNA/mg hybridized DNA, $p<0.005$). All 6 patients developing LPD were infected with type A variants. Of 40 transplant recipients who did not develop LPD, 26 were infected by type A only, 6 by type A and type B, 3 by type B only, and 5 were not excreting EBV. High levels of excretion of type A variants predict the development of LPD. The identification of patients at risk of developing LPD may help in the earlier institution of antiviral therapy.

COMPARISON OF IMMUNOFLUORESCENCE (IF) AND LABELED AVIDIN D (LAD) IMMUNOPEROXIDASE (IP) ASSAYS FOR THE DETECTION OF INCLUSIONS IN SHELL VIALS

M. Espy, D. Jespersen, A.Wold and T. Smith

Mayo Clinic and Foundation
Rochester, MN, USA

IF and IP assays have previously been compared for the detection of cytomegalovirus (CMV) by the shell vial assay and found to give comparable results. A recent report indicated that use of horseradish peroxidase LAD in a IP assay was more sensitive for the detection of CMV in shell vials than IF. One hundred twelve urine specimens were inoculated into standard tube and shell vial cell cultures subsequently stained by IF and IP-LAD methods. Of 34 (30%) positive specimens, 16 (47%) were positive in all three systems. Eight (24%) specimens were detected in tube cell culture and by IF. Five (15%), four (12%), and 1 (3%) specimens were exclusively positive for CMV by IF, tube cell culture, and LAD-IP, respectively. Collectively, 29 (85%) of the CMV-positive specimens were detected by IF compared to 17 (50%) by LAP-IP. The LAD IP method is both less sensitive and more time consuming than IF and is not recommended for use in the clinical laboratory.

COMPARISON OF MRC-5 AND MINK LUNG (ML) SHELL VIALS FOR RAPID DETECTION OF CYTOMEGALOVIRUS (CMV) IN CENTRIFUGATION CULTURE

C.A. Gleaves, R.R. Bindra, D.A. Hursh and J.D. Meyers

Fred Hutchinson Cancer Research Center
Seattle, WA, USA

The sensitivity of MRC-5 and ML cells in shell vials were compared simultaneously for the detection of CMV immediate-early antigen from clinical specimens by the centrifugation culture assay. Shell vials for both cell lines were prepared in the laboratory and used from days 3 to 8 following cell seeding. Specimens were inoculated at a volume of 0.2 ml into two of each type of shell vial. Vials were then centrifuged at 700 x g for 40 min, 1.0 ml of media was added and vials were incubated at 36°C. One vial of each cell type was fixed in cold acetone and stained by indirect immunofluorescence at 20 hrs and again at 40 hrs postinoculation (pi) using a CMV specific monoclonal antibody reactive with the 72,000 MW immediate-early CMV protein (DuPont, Wilmington, DE). Of 186 specimens (126 blood, 22 tissue, 15 BAL, 12 throat and 11 urine) 32 (17.2%) were positive for CMV in centrifugation culture. At 20 hrs pi, 27 of 32 (84.4%) CMV positive specimens were detected in ML vials and 26 of 32 (81.3%) CMV positives were detected in MRC-5 vials. At 40 hrs pi, 31 of 32 (96.9%) positives were detected in ML vials and 29 of 32 (90.1%) positives were detected in MRC-5 vials. There was no significant difference in the detection of CMV by centrifugation culture using either cell line. However, in 11 of 14 CMV positive specimens that had countable foci at 20 hrs pi, there was a 2.9% increase in the number of foci in ML cells as compared to MRC-5 cells. Additionally, less toxicity was noted in ML cells as compared to MRC-5 cells, particularly in viral blood specimens. These data suggest that ML cells are comparable to MRC-5 cells for the rapid detection of CMV in centrifugation culture.

EVALUATION OF AN INDIRECT FLUORESCENT ANTIBODY (IFA) TEST KIT FOR THE DETECTION OF CYTOMEGALOVIRUS (CMV) IMMEDIATE-EARLY ANTIGEN IN CENTRIFUGATION CULTURE

C.A. Gleaves

Fred Hutchinson Cancer Research Center
Seattle, WA, USA

IFA test kit (Baxter; Bartels Diagnostic Division, Issaquah, WA) was evaluated in centrifugation culture for the rapid detection of CMV in clinical specimens. Results were compared with conventional cell cultures that were examined for CMV cytopathic effect (CPE) and subsequently confirmed as CMV using monoclonal antibodies to early and late CMV proteins. Clinical specimens were inoculated at a volume of 0.2 ml into two shell vials containing MRC-5 cells for centrifugation culture and into two tubes of HF and one tube of A549 cells for viral isolation. Shell vials were centrifuged at 700 x g for 40 min. after which 1.0 ml of MEM with 2% FBS was added to each vial and incubated at 36°C. One vial was then assayed for CMV at 20 hrs postinoculation (pi) and the second vial was assayed at 40 hrs pi using the CMV IFA test kit. Of 148 specimens (56 throat, 50 blood, 32 urine, 5 bronchoalveolar lavage, 5 tissue) 27 (18.2%) were CMV positive, 26 (17.6%) by centrifugation culture, 22 (14.9%) were positive at 20 hrs and 25 (16.9%) were positive at 40 hrs. CMV was isolated from 23 (15.5%) specimens in conventional culture. The intensity and clarity of the stain produced by the above kit was a high quality and was observed to be similar to other CMV reagents commonly used with the centrifugation culture assay. This evaluation indicates that the above CMV IFA reagent used with centrifugation culture provides for a convenient and reliable product for the rapid detection of CMV from clinical specimens.

A COMPARISON OF THE HEMAGGLUTINATION (HA) TEST AND THE HEMADSORPTION (HAd) TEST FOR INFLUENZA DETECTION

S. Johnston, K. Wellens and C. Siegel

Bellin Memorial Hospital
Green Bay, WI, USA

HA testing of the supernatant of primary rhesus monkey kidney (RMK) cell culture inoculated with clinical specimens was compared with HAd of the same cell for identifying the presence of hemagglutinin activity due to influenza. A total of 476 respiratory specimens were screened for influenza by this method. Specimens were inoculated into two RMK cell culture tubes and incubated for up to 14 days at 35°C. During this period HAd and HA testing was done 2-3 times during the first seven days of incubation and again at the end of the 14 day incubation period. A fresh suspension of 0.05% guinea pig red blood cells was used for HAd testing within the tube and for the HA testing which was performed on the supernatant in a 96-well U bottom microtiter plate. There were 127 influenza or parainfluenza positive cultures identified by the presence of cytopathic effect complemented by positive HA and HAD testing. These isolates were confirmed by staining of scraped cells with a type specific monoclonal antibody to influenza A and B and parainfluenza 1, 2 and 3. In all 127 positive cultures, there was 100% concordance between the HA and HAd tests. All tubes negative by the HA test were also negative by the HAd test and vice versa. In our laboratory, the HA method of screening for hemagglutinin activity was as sensitive as the HAd method when using fresh guinea pig red blood cells and RMK cells. In contrast to the HAd method, HA testing was performed with less tube manipulation, less hands-on time and less potential for cell culture tube contamination.

CULTURE CONFIRMATION AND TYPING OF HERPES SIMPLEX VIRUS (HSV) ISOLATES FROM CLINICAL SPECIMENS WITH A NEW MONOCLONAL ANTIBODY TYPING REAGENT

S. Johnston[1] and C. Gleaves[2]

[1]Bellin Memorial Hospital
Green Bay, WI and
[2]Fred Hutchinson Cancer Research Center
Seattle, WA, USA

FITC-labeled monoclonal antibody typing reagents for HSV (PathoDx, Diagnostic Products Corp., Los Angeles, CA) were compared to another commercially available FITC-labeled monoclonal antibody typing reagent (Syva, Palo Alto, CA) for culture confirmation and typing of HSV isolates from clinical specimens. A total of 344 clinical specimens were evaluated for culture confirmation and typing by the two systems. Two clinical trial sites were included in order to provide a diverse patient population. There were 296 HSV positive cultures identified by cytopathic effect (CPE). The PathoDx typing reagents identified 143 isolates as HSV-1, and 151 isolates as HSV-2. Two of the isolates stained with both the type 1 and type 2 reagents of both staining systems and were considered dual infections. The PathoDx typing reagents had a 100% concordance with the Syva typing reagents and were provided as a constituted liquid in an easy to use dropper bottle rather than a lyophilized reagent that needed to be reconstituted before use. The PathoDx stain typically displayed less bright fluorescence than the Syva stain, however, the interpretation of HSV-1 results was not inhibited by the HSV-2 background staining that was experienced with the Syva stain. The PathoDx HSV typing reagent performed as well as, if not better than the Syva HSV typing reagent in the clinical evaluation.

COMPARISON OF BUFFALO GREEN MONKEY (BGM) CELLS AND McCOY CELLS FOR THE ISOLATION OF *CHLAMYDIA TRACHOMATIS* IN SHELL VIAL CENTRIFUGATION CULTURE

S. Johnston and C. Siegel

Bellin Memorial Hospital
Green Bay, WI, USA

A total of 745 clinical specimens from patients attending hospitals and clinics in an area with a low prevalence of *Chlamydia trachomatis* were inoculated in parallel into shell vial cultures containing BGM and McCoy cells. The shell vial cultures were prepared inhouse and were less than five days post-seeding when inoculated. The specimens were vortexed with glass beads prior to inoculation and once inoculated, the shell vials were centrifuged at 1,500 x g for 1 hr. Following centrifugation and a 1-2 hr incubation at 35°C, the specimen was aspirated from the vials and 1 ml of medium containing 5% fetal bovine serum and cycloheximide was added. The vials were then reincubated for two days, after which they were stained with a monoclonal antibody (Syva, Palo Alto, CA). A total of 38 specimens (5%) were positive for *Chlamydia trachomatis*. In 36 of the cases, Chlamydia was detected in both the BGM and McCoy vials. In 2 cases, Chlamydia was detected in only the BGM vial. The BGM cells were more resistant to cytotoxicity and seemed to show more and larger inclusions than the McCoy cells. Furthermore, because the BGM cells displayed contact inhibition without the rounding and piling of cells that was encountered with the McCoy cells, they could be stored, ready to use, with an optimal monolayer of cells at 35°C for at least 10 days.

INFECTION OF HUMAN LYMPHOID CELLS WITH ADENO ASSOCIATED VIRUS (AAV)

F. Mileguir[1], B.J. Carter[3], Z. Grossman[1], G. Rechavi[2] and E. Mendelson[1]

[1]Central Virology Laboratory
[2]Institute of Hematology
Chaim Sheba Medical Center, Tel-Hashomer and
Sackler School of Medicine,Tel-Aviv University
Israel
[3]National Institutes of Health
Bethesda, MD, USA

AAV is a non-pathogenic human parvovirus which requires helper herpesvirus or adenovirus for efficient replication. It can also inhibit cellular transformation by other viruses. Most adults are seropositive for AAV and seroconversion occurs during the first few years of life. In the absence of helper virus AAV can integrate efficiently into the host genome. Vectors derived from AAV can stably express heterologous genes and may be useful for gene therapy. Thus, the ability of AAV to replicate in lymphoid cells is important but has not yet been demonstrated. In the present study we have examined the efficiency of AAV infection and gene expression in such cells: we used seven established T and B human lymphocyte cell lines and peripheral blood lymphocytes (PBL) from two blood donors. The infection occurred in all cell types with or without helper virus, while replication occurred only in the presence of helper virus when Rep and capsid proteins were synthesized. The yield of infectious virus was very low in all cases. Using the polymerase chain reaction (PCR) technique, we could also detect AAV sequences in the PBL of one out of forty healthy blood donors and two out of twenty-five hemophilia patients. Thus, we conclude that AAV can naturally or deliberately infect human lymphocytes which may then serve as carriers of the virus to many body tissues. The inactivity of its promoters may help stabilizing the integrated virus and prevent rescue and loss of viral sequences from the lymphocyte genome.

IMMUNE RESPONSE TO BOVINE HERPESVIRUS-1 (BHV-1) ANTIIDIOTYPIC (Anti-Id) ANTIBODIES

H.C. Minocha, D.J. Orten, O.Y. Abdelmagid and F. Blecha

Kansas State University
Manhattan, KA, USA

Anti-Id antibodies to BHV-1 are being used to study virus-cell interactions as an alternative approach to vaccine preparation. The virus causes respiratory and genital infections in cattle and establishes latency in nerve ganglia. It also produces malignant transformation of hamster embryo fibroblasts and mouse macrophages in culture. The bovine genital disease is similar to herpes simplex-2 infections in humans. Rabbit anti-Id antibodies against murine monoclonal antibodies (MAbs) specific for the major BHV-1 envelope glycoproteins were prepared, purified and characterized. Each anti-Id reagent inhibited neutralization of BHV-1 by its homologous MAb. Preincubation of tissue culture cells with any of the four anti-Id preparations inhibited BHV-1 infection. Anti-Id antibodies to a neutralizing MAb, which reacts with virus glycoprotein gIII, were used to immunize BALB/C mice. Immune responses to the purified anti-Id reagent were compared with immune responses to BHV-1. Both groups of mice produced BHV-1 neutralizing antibodies. However, cellular immune response including lymphocyte proliferation, interferon-γ, and interleukin-2 production were only specific for the immunizing antigens. The anti-idiotypes will be tested as vaccine candidates and as reagents for identification, characterization, and isolation of viral receptors.

POSTPARTUM AND POSTNATAL PRIMARY CYTOMEGALOVIRUS (CMV) INFECTION DUE TO INTRAFAMILIAL EXPOSURE DOCUMENTED BY DNA PROBE ANALYSIS

D.R. Osmon, N.K. Henry, M.P. Wilhelm, M.J. Espy and T.F. Smith

Mayo Clinic
Rochester, MN, USA

This case represents an example of primary postpartum and postnatal CMV disease and the role of streamlined DNA analysis in investigating the epidemiology of intrafamiliar and occupationally-acquired CMV infection. Using one restriction enzyme and one nucleic acid probe, we demonstrated intrafamiliar spread of CMV from a 2 year old day care attendee to her mother causing symptomatic primary CMV infection within 12 hrs postpartum. The mother, a health care worker, was seronegative intrapartum and seroconverted postpartum. Viral cervical and breast milk cultures yielded CMV. A throat culture taken at that time from the 2 year old was also positive for CMV. The term newborn developed symptomatic primary CMV infection at ten days of age. At age one month, he seroconverted, at which time a viral urine culture grew CMV. CMV strains from the three family members were identical after digestion of viral DNA with BamH1, gel electrophoresis, Southern blotting, and probing with ^{32}P-labeled 10kb probe (J. Oram). These relatively simple procedures allowed documentation of strain identity of CMV isolates from three family members within seven days, and helped differentiate intrafamilial spread from occupationally-acquired infection.

ADJUVANT THERAPY WITH ALPHA-INTERFERON (IFN) REDUCES THE RISK OF RECURRENCES OF GENITAL CONDYLOMATA AMONG PATIENTS POSITIVE FOR HPV 16 OR 18

J. Paavonen, P. Nieminen, M. Aho, M-L. Ogard, E.Vesterinen and M. Lentinen

University of Helsinki and
University of Tampere, Finland

Treatment of genital warts is a major medical problem. We designed a randomized placebo-controlled study of the treatment of genital condylomas with a combination of CO_2 laser and systemic IFN alpha-2b (Introna, Schering Co., 1.5 million units 3 times a week for 3 weeks). We correlated the treatment response with epidemiologic, cytologic, colposcopic, microbiologic, and HPV DNA findings. Overall, 50 women were treated with adjuvant IFN, and 50 women received placebo. Response to treatment was monitored 1 month, 3 months and 6 months post-treatment. Ninety-four women completed the follow-up. Although the overall rate of recurrences did not differ between the two groups, (18/47 in the IFN group, and 15/47 in the placebo group) a risk ratio analysis of selected potential predictors of recurrences showed differences between the groups.

	Relative risk for recurrences	
Risk Factor	**IFN 95% Q.I.**	**Placebo 95% Q.I.**
C. *trachomatis*	0.9 (0.4-2.3)	0.3 (0.2-0.6)
HPV DNA (any)	3.0 (1.5-6.0	2.7 (1.3-5.8)
HPV 16/18 DNA	1.8 (0.8-4.1)	7.8 (2.3-26.2)
HPV 16 E2 IgA	2.5 (1.1-5.5)	6.0 (1.7-21.4)
Smoking	0.3 (0.2-0.6)	0.9 (0.5-1.7)
Birth control pill use	1.8 (0.9-3.6)	1.7 (0.9-3.2)

In summary, adjuvant therapy with recombinant IFN alpha-2b significantly reduced the risk for recurrences in patients with genital HPV infection who were positive for HPV 16/18 DNA, or who were positive for serum IgA anti-peptide antibodies to HPV 16 open reading frame B2. Thus, specific laboratory testing for HPV might be useful in predicting the response to therapy among patients with condylomata.

EFFECT OF RIBAMIDINE THERAPY ON INFLUENZA-A VIRUS INFECTIONS OF MICE

R.W. Sidwell, J. Coombs, A. Gessaman, J. Gilbert, R. Burger, J.H. Huffman and R.P. Warren

Utah State University
Logan, UT, USA

Ribamidine (1-β-D-ribofuranosyl-1,2,4-triazole-3-carboxamidine) at dosages of 100, 320 and 1,000 mg/kg/day was evaluated for its efficacy against experimentally induced influenza A (H1N1) virus infections in mice. The animals were infected by intranasal instillation of virus, followed 4 hrs later by intraperitoneal ribamidine treatment which continued twice daily for 5 days. Lungs taken 3, 5, 7 and 10 days after virus exposure had significantly less consolidation and lower virus titers at all dosage levels. These improvements in lung disease parameters were accompanied by less virus-induced decline in blood oxygen saturation as measured by pulse oximetry using an Ohmeda Pulse Oximeter 3,740 with the whole mouse monitored daily with the instrument's finger probe. Despite these virus-inhibitory effects, the animals died of the infection. The 1,000 mg/kg/day dose significantly lowered natural killer cell activity, PHA-induced blastogenesis, and virus-specific cytotoxic T lymphocyte response, and caused marked shifts in splenic T and B cell populations. All other dosages had little effect on these key immunologic factors in the infected and uninfected mice. The 1,000 and 320 mg/kg/day dosages were eventually lethally toxic to the mice. The toxicity was accompanied by increases in serum glutamic pyruvic acid transaminase and creatinine kinase. The 100 mg/kg/day dosage was apparently well tolerated by the animals.

USE OF ACYCLOVIR FOR VARICELLA PNEUMONIA IN PREGNANCY

[1]Mark Asperilla and [2]R.A. Smego

[1]Albany Medical College
Albany, NY, USA
[2]West Virginia University
Morgantown, WV, USA

Primary varicella pneumonia occurring in pregnancy causes substantial morbidity and may carry a maternal mortality of >40%. Systemic acyclovir therapy is recommended for this condition, but data on its use during pregnancy are limited. A review of 20 such cases (5-WVU; 15 literature) was conducted to evaluate the benefit and risk of i.v. acyclovir on maternal and total outcomes. Patient ages ranged from 16 to 33 yrs. (mean 22 yrs.), and all women were in their second (11 cases) or third (9 cases) trimester. Gestational ages at onset of pneumonia and time of delivery were 27 weeks and 16 weeks, respectively. All patients had diffuse radiographic involvement and 12 required mechanical ventilation. Patients received acyclovir in dosages of 30 to 54 mg/kg/day in 3 divided doses. The mean duration of treatment was 6.8 days (range 4-12 days). Seizure activity, which developed in one women who later died, was the only possibility related maternal or fetal adverse effect. Three women (15%) died of uncontrolled infection or complications, and another died following surgery for unrelated intestinal obstruction one month after delivery. One infant was born with low-birth weight. Two infants died (whose mothers also died): one was stillborn at 34 weeks gestation, and the other died from prematurity shortly after birth at 27 weeks. No child was born with features of congenital varicella syndrome, and none developed active perinatal varicella infection, although three infants did receive zoster immune globulin after birth and two of these also received prophylactic acyclovir. Intravenous acyclovir may reduce both maternal morbidity and mortality associated with varicella pneumonia occurring during pregnancy. While fetal protective efficacy remains unproven, the drug appears to be safe for the developing fetus when given during the latter trimesters.

INDUCTION OF IMMUNE RESPONSE TO A SECRETORY TRUNCATED VARICELLA-ZOSTER VIRUS (VZV) GLYCOPROTEIN I CARRYING THE E1 EPITOPE

A. Vafai and W. Yang

University of Illinois College of Medicine
Rockford, IL, USA

VZV is the causative agent of childhood chicken pox (varicella) and shingles (zoster). VZ virions contain at least four different glycoproteins which are highly immunogenic and induce both humoral and cell-mediated immune response against VZV infection. Previously, we identified one of the major antibody-binding site (epitope) on one of the most abundant and immunogenic VZV glycoproteins designated gpI, and showed that synthetic peptides (14 amino acids) comprising this epitope (designated E1) induced antibody response which is recognized by native VZV gpI. We have now constructed a recombinant vaccinia virus which expresses a secretory truncated gpI containing the E1 epitope. This truncated gpI is of particular importance because not only does it induce antibody responses against VZV infection, but it is also secreted into the tissue culture media of the infected cells and can be produced and purified in large quantities. The potential application of such secretory truncated VZV gpI includes boosting immune response in the elderly population who are more susceptible to VZV reactivation (shingles).

VARICELLA IMMUNITY AND DISEASE IN HIV-1 POSITIVE ADULTS

J. Malone, R. Rockhill, J. Diaz and S. Pfeiffer

Naval Hospital
San Diego, CA, USA

Varicella is typically a benign illness, but can be severe in HIV-1 positive patients. We have seen four cases of adult varicella in HIV-1 positives in the past two years; the spectrum of disease ranged from mild to severe, though none were fatal. Due to continuing questions about varicella postexposure prophylaxis, we have begun to routinely measure varicella IgG antibody levels on all adult HIV positive patients. Our results are as shown:

Seropositivity by T4 Count

	T4<200	T4 200-400	T4 >400	Total
Fraction Immune	45/46	77/80	50/55	172/181
Percent Immune	98%	96%	91%	95%

Seropositivity by History

	Immune	Non-immune
Positive History	106/107 (99%)	1/107 (1%)
Negative or Unknown	48/54 (89%)	6/54 (11%)

These data suggest that serologic immunity to varicella is widespread in HIV-1 positives and persists despite dropping T4 counts. A positive varicella history in an HIV-1 positive patient correlates strongly with demonstrable antibodies; routine testing of those with negative or uncertain histories will identify a small population for who postexposure prophylaxis and/or early antiviral therapy would be appropriate.

COMPARISON OF THREE COMMERCIAL MRC-5 CELL LINES IN SHELL VIALS FOR SENSITIVITY TO CYTOMEGALOVIRUS (CMV) AND RESPONSE TO ENHANCING AGENTS

P.G. West[1,2], R.A. Hartwig[1] and W.W. Baker[2]

[1]SmithKline Beecham Clinical Laboratories
Norristown, PA, USA
[2]Villanova University
Villanova, PA, USA

MRC-5 cells grown in shell vials from three suppliers (Whittaker Bioproducts, Bartels and ViroMed) were evaluated for sensitivity to human cytomegalovirus (HCMV). Previous studies done in this laboratory showed that treating the cells with dexamethasone (DEX) and dimethyl sulfoxide (DMSO) or these agents plus calcium improved the sensitivity of cells from Whittaker Bioproducts. In this study we compare the response, treated and untreated, of the Whittaker cells to that of cells from Bartels and ViroMed when tested with HCMV AD169, a strain of wild virus and 45 clinical specimens from various sources. Progressive improvement in sensitivity with DEX-DMSO-calcium treatment was seen with AD169 and the wild virus in all cell lines, although even untreated, the Whittaker cells were much more sensitive. Most of the positive clinical specimens were enhanced by both treatments in the Whittaker cells. DEX-DMSO had little or no effect on Bartels or ViroMed cells while DEX-DMSO-calcium occasionally effected some increase in numbers of inclusions. Five specimens were recovered only on treated Whittaker cells. Protein studies were done on all groups to determine the effect, if any, of the treatments on the metabolism of the cells.

EVALUATION OF THE RECOMBIGEN HIV-1 LATEX AGGLUTINATION TEST FOR THE DETERMINATION OF ANTIBODY STATUS TO HIV-1

S.L. Aarnaes, R.L. MacDonald, L.M. de la Maza and E.M. Peterson

University of California Irvine Medical Center
Orange, CA, USA

A five minute latex agglutination (LA) test (Recombigen® HIV-1 LA, Cambridge Bioscience Corp, Worchester, MA) for HIV-1 antibody status was compared to the Abbott EIA (Abbott Laboratories, Chicago, IL) and Cambridge Bioscience Corp., Recombigen® HIV-1 EIA. The LA test uses recombigen gp41 and gp120 proteins as antigens. The LA was used according to the manufacturer's recommendations where the cards were rotated at 60 rpm for 5 min on an orbital rotator and agglutination was read with a halogen lamp with a 2x magnifier. The EIAs were also tested according to manufacturer's instructions where a positive sample needed to be repeatably positive. All positive specimens by any of the methods were further tested by Western blot (WB) (Biotech Research Laboratories, Inc., Rockville, MD and E.I. du Pont de Nemours and Co., Inc., Wilmington, DE). Of the 327 specimens tested, there were 115 positive and 203 negative by all four methods. Of the 9 discrepant samples, all were positive by LA, of which 7 were negative by Western blot and EIA, and 2 were indeterminate by the Western blot and negative by EIA. The overall sensitivity and specificity of the LA compared to the EIAs was 100% and 95.8%. There was 100% agreement between both EIAs. In this evaluation, the latex test had an improved sensitivity compared to an earlier report from our laboratory. We attribute this increase in sensitivity to the use of the orbital rotator and halogen lamp which appears to be essential when performing this test. Due to the improved sensitivity of the latex agglutination, we could recommend it as a routine screening method for HIV-1 antibody status. The LA test would have a particular advantage in situations where a rapid result is necessary.

CONTRIBUTORS

PETER AABY: Institute of Anthropology, Copenhagen, Denmark

TOBIAS BAUKNECHT: Institut für Angewandet Tumorvirologie, Deutsches Krebsforschungszentrum, Germany

JACK S. COHEN: Georgetown University Laboratories, Rockville, Maryland, USA

RICHARD CONE: University of Washington, Children's Hospital, Virology Laboratory, Virology Division, Seattle, Washington, USA

MATTHIAS DÜRST: Institut für Angewandet Tumorvirologie, Deutsches Krebsforschungszentrum, Germany

HENRY L. FRANCIS: National Institute of Allergy and Infectious Diseases, The National Institutes of Health, Bethesda, Maryland, USA

ANNE A. GERSHON: College of Physicians & Surgeons of Columbia University, Division of Pediatric Infectious Diseases, New York, New York, USA

BERT E. JOHANSSON: Mount Sinai School of Medicine, New York, New York, USA

EDWIN D. KILBOURNE: Mount Sinai School of Medicine, New York, New York, USA

SARA E. MILLER: Duke University Medical Center, Department of Microbiology and Immunology, Durham, North Carolina, USA

CYNTHIA G. PRITCHARD: Gene-Trak Systems, Amplification Research, Framingham, Massachusetts, USA

THOMAS C. QUINN: The Johns Hopkins University, Baltimore, Maryland, USA

CLAUDIA RITTMÜLLER: Institut für Angewandet Tumorvirologie, Deutsches Krebsforschungszentrum, Germany

DARRYL SHIBATA: Los Angeles County/University of Southern California Medical Center, Department of Pathology, Los Angeles, California, USA

THOMAS F. SMITH: Mayo Clinic, Department of Laboratory Medicine, Section of Clinical Microbiology, Rochester, Minnesota, USA

JAMES E. STEFANO: GENE-TRAK Systems, Framingham, Massachusetts, USA

MAGNUS von KNEBEL DOEBERITZ: Institut für Angewandet Tumorvirologie, Deutsches Krebsforschungszentrum, Germany

GIRISH N. VYAS: University of California, San Francisco, Department of Laboratory Medicine, Transfusion Research Program, San Francisco, California, USA

ARLO D. WOLD: Mayo Clinic, Department of Laboratory Medicine, Section of Clinical Microbiology, Rochester, Minnesota, USA

HARALD zur HAUSEN: Institut für Angewandet Tumorvirologie, Deutsches Krebsforschungszentrum, Germany

AUTHOR INDEX

SUBJECT INDEX